Molecules of the Cytoskeleton

L. A. Amos, B.A.(Oxon.), Ph.D.(Cantab.)
W. B. Amos, B.A.(Oxon.), Ph.D.(Cantab.)

Medical Research Council,
Laboratory of Molecular Biology,
Hills Road,
Cambridge CB2 2QH

THE GUILFORD PRESS
New York

Published in the United States of America by
The Guilford Press
A Division of Guilford Publications, Inc.
72 Spring Street, New York, NY 10012

Printed in Hong Kong

This book is printed on acid-free paper.

Last digit is print number: 9 8 7 6 5 4 3 2 1

Library of Congress Cataloging-in-Publication Data

Amos. L. A.
    Molecules of the cytoskeleton / L. A. Amos, W. B. Amos.
        p.   cm. — (Molecular cell biology series)
    Includes bibliographical references and index.
    ISBN 0-89862-404-5 (hard). — ISBN 0-89862-527-0 (pbk.)
    1. Cytoskeletal proteins.   I. Amos, W. B. (W. Bradshaw)
II. Title.   III. Series.
    [DNLM: 1. Cytoskeleton—physiology. 2. Proteins—physiology.   QH
591 A525m]
QP552.C96A46   1991
574.87′2—dc20
DNLM/DLC
for Library of Congress   91-16542
                                                                          CIP

Molecules of the Cytoskeleton

MOLECULAR CELL BIOLOGY

A Guilford Series edited by
CHRISTOPHER J. SKIDMORE, MA, DPhil
University of Reading

Molecular Genetics of *Escherichia Coli*
PETER SMITH-KEARY

Organelles
MARK CARROLL

Molecules of the Cytoskeleton
LINDA A. AMOS and W. BRADSHAW AMOS

Forthcoming

Protein Biosynthesis
ROSEMARY JAGUS

CONTENTS

	Series Editor's Preface	xiii
	Preface	xv
1	**The Organisation of Cytoplasm**	**1**
1.1	Introduction	1
1.2	What is cytoplasm?	2
	1.2.1 At light-microscope level	2
	1.2.2 'Cytosol' and 'cytoskeleton'	3
	1.2.3 Structure at the electron-microscope level	4
1.3	Cell arrangements	8
1.4	The major filament types	13
	1.4.1 Filament substructure	13
	1.4.2 Functions of filament systems	13
	1.4.3 Differences between filament types	14
	1.4.4 Functions of associated proteins	14
	1.4.5 Duplication of functions	15
1.5	Motility	15
	1.5.1 Sliding-filament systems	15
	1.5.2 Contraction at subunit level	17
	1.5.3 Calcium and the control of cytoskeletal changes	18
1.6	Evolutionary aspects	19
	1.6.1 Duplication of genes	19
	1.6.2 Origins	19
1.7	Summary	19
1.8	Questions	20
1.9	Further reading	20
2	**Intermediate Filaments and Other Alpha-Helical Proteins**	**23**
2.1	General structure of coiled coils	23
2.2	Intermediate filaments	25
	2.2.1 Various types of IF	25
	2.2.2 Sequences of IF proteins	26

v

	2.2.3	Rod-domain substructure of IFs	26
	2.2.4	IF head and tail domains	27
	2.2.5	IF molecular units	27
	2.2.6	Assembly of 10 nm filaments	27
	2.2.7	Do IFs have a polarity?	29
	2.2.8	Comparison with myosin filament structure	29
2.3		Classes of IFs expressed by different cell types	31
	2.3.1	Keratins (heteropolymers of IF polypeptide types I & II)	31
	2.3.2	Homopolymeric type III IFs	32
	2.3.3	Neurofilament proteins (IF type IV)	33
	2.3.4	Lamins (IF class V)	34
	2.3.5	Tektins and 2 nm filaments	35
2.4		IF-associated proteins (IFAPs)	35
2.5		Interaction between IFs and other cytoskeletal components	35
	2.5.1	Association of IFs with membranes	35
	2.5.2	Association of IFs with microtubules	37
	2.5.3	Association of IFs with actin filaments	37
2.6		Functions of IFs	38
2.7		Summary	38
2.8		Questions	39
2.9		Further reading	39
3		**Actin Filaments**	**42**
3.1		Structure	42
	3.1.1	Filament symmetry	42
	3.1.2	Domains of the monomer	43
	3.1.3	Subunit interactions in filaments	45
3.2		Actin filament polymerisation	47
	3.2.1	Assembly *in vitro*	48
	3.2.2	Purification of actin	48
	3.2.3	Methods of assaying assembly	48
	3.2.4	Assembly kinetics	48
	3.2.5	Initiation of assembly	50
	3.2.6	Treadmilling	50
	3.2.7	ATP hydrolysis	50
3.3		Drugs affecting actin assembly	52
3.4		Summary	53
3.5		Questions	53
3.6		Further reading	53
4		**Actin-Associated Proteins**	**56**
4.1		Calcium-binding proteins	56
	4.1.1	Calmodulin	56
4.2		Assembly-controlling proteins	57
	4.2.1	Capping activity	60
	4.2.2	Barbed-end capping proteins	60
	4.2.3	Severing activity	61
	4.2.4	Gelsolin	61
	4.2.5	Proteins related to gelsolin	61
	4.2.6	Nucleating activity	62

		4.2.7	Actin-monomer sequestration	62
		4.2.8	Profilin	63
		4.2.9	Actin-depolymerising activity	64
4.3	Force-producing proteins			64
		4.3.1	Two-headed myosin (myosin II)	64
		4.3.2	Myosin monomers, dimers and thick filaments	66
		4.3.3	Myosin heads and motive force	66
		4.3.4	Single-headed myosin (myosin I)	67
		4.3.5	Interaction of myosin I with membranous organelles	68
		4.3.6	Myosin I-like proteins from non-amoeboid cells	68
4.4	Other crosslinking and bundling proteins			68
		4.4.1	Spectrin and related proteins	69
		4.4.2	α-Actinin	70
		4.4.3	Dystrophin	71
		4.4.4	Gelation factors	72
		4.4.5	Tropomyosin	72
		4.4.6	Caldesmon	72
		4.4.7	Polar bundling	73
		4.4.8	*Limulus* acrosomal process	73
		4.4.9	Aldolase	73
4.5	Membrane-attachment proteins			74
		4.5.1	Indirect links between actin and membrane	74
		4.5.2	Integral membrane actin-binding proteins	74
		4.5.3	Ponticulin	74
4.6	Disruption of protein activity in cells			75
		4.6.1	Genetic analysis of cytoskeletal proteins	75
		4.6.2	Injection of antibodies	75
4.7	Summary			75
4.8	Questions			76
4.9	Further reading			76
5	**Actin in Cells: Structural and Contractile Roles**			**80**
5.1	Widespread and abundant distribution of actin			80
		5.1.1	Main distribution of actin in cells	80
5.2	Filament polarity in cells			81
5.3	Structural actin			82
		5.3.1	Erythrocyte membrane skeleton	82
		5.3.2	Intestinal epithelial cells	84
		5.3.3	Actin bundles in stereocilia and microvilli	85
		5.3.4	*Nitella* and *Chara*	87
5.4	Striated muscle as a model for contractile systems			88
		5.4.1	Crossbridge cycles and tension production	88
		5.4.2	Enzymatic activity of myosin	89
		5.4.3	Myosin speed and structural organisation	91
		5.4.4	Ca^{2+}-control of striated muscles	91
5.5	Smooth-muscle and non-muscle contraction			92
		5.5.1	Myosin light-chain phosphorylation	92
		5.5.2	Myosin heavy-chain phosphorylation	93

5.6		Stress fibres in non-muscle cells	93
	5.6.1	Tension versus motility	94
	5.6.2	Adhesion plaques	94
5.7		Summary	95
5.8		Questions	95
5.9		Further reading	96
6		**Actin in Cells: Roles Involving Assembly**	**98**
6.1		Polymerising activity	98
	6.1.1	*Thyone* acrosomal filament	98
	6.1.2	Platelet activation	100
6.2		Fibroblast motility	100
	6.2.1	Translocation	100
	6.2.2	Cytoskeleton arrangements	102
	6.2.3	Motile behaviour in other crawling cells	102
6.3		The cell membrane and the locomotory mechanism	103
	6.3.1	Membrane recycling and lipid-flow theory	103
	6.3.2	Intracellular route for membrane	105
	6.3.3	Could membrane flow drive motility?	105
6.4		Role for the actin cytoskeleton in locomotion	106
	6.4.1	Two-headed myosin and ruffling in amoebae	106
	6.4.2	Patching and capping	107
	6.4.3	Observations on nerve growth cones	107
	6.4.4	Analysis of glycoprotein movements	108
	6.4.5	Membrane reinsertion	109
	6.4.6	Contractile and protrusive forces	110
	6.4.7	Osmotic forces	110
	6.4.8	Possible role of inositol lipids	110
6.5		Possible locomotion cycles	111
	6.5.1	Lamellar activity	111
	6.5.2	Gelation–solation and contraction of bulk cytoplasm	112
	6.5.3	Cytoplasmic streaming	112
6.6		The amoeboid sperm of nematodes	113
6.7		Summary	113
6.8		Questions	114
6.9		Further reading	115
7		**Properties of Tubulin**	**117**
7.1		Microtubule structure	117
	7.1.1	The tubulin lattice in a microtubule	117
	7.1.2	Zinc-induced tubulin sheets	121
7.2		Molecular structure	122
	7.2.1	Monomer domains	122
	7.2.2	Tubulin sequences and post-translational modifications	123
	7.2.3	Regulation of tubulin synthesis	124
	7.2.4	Structural data from proteolysis and antibody labelling	124
	7.2.5	GTP-binding sites	124
7.3		The assembly of tubulin	125
	7.3.1	Tubulin assembly: driven by entropy	125
	7.3.2	Purification of microtubules by temperature cycles	126

7.4   Assembly initiation, elongation and depolymerisation   126
       7.4.1   Quantitative assay of assembly and disassembly   127
       7.4.2   Polymorphic assembly of pure tubulin   127
       7.4.3   Properties of lateral bonding   127
       7.4.4   Initiation of assembly *in vitro*   128
       7.4.5   Tubulin rings and oligomers   129
       7.4.6   Drugs affecting tubulin assembly   130
       7.4.7   GTP hydrolysis and its function   130
7.5   Microtubule polarity   132
7.6   Dynamic instability   132
       7.6.1   Observation of individual microtubule behaviour   133
       7.6.2   Severed microtubules   135
       7.6.3   The uniform behaviour of minus ends   136
       7.6.4   A possible modification to the simple capping
                mechanism   136
       7.6.5   The modulating effect of a GTP cap on the plus end   136
       7.6.6   A comparison with F-actin   137
       7.6.7   The effect of MAPs on assembly and disassembly rates:
                treadmilling?   138
7.7   Summary   138
7.8   Questions   139
7.9   Further reading   139

8   Proteins Associated with Microtubules   142
8.1   MAPs   142
       8.1.1   The effect of MAPs on microtubule assembly   142
       8.1.2   Variation in size and thermostability   146
       8.1.3   Tau proteins   146
       8.1.4   STOPs   147
       8.1.5   MAP 2   147
       8.1.6   MAP 1A   148
       8.1.7   Other neuronal MAPs   148
       8.1.8   Non-neuronal MAPs   149
       8.1.9   MAP 2-like proteins that bind to other structures   149
8.2   MAP structure   149
       8.2.1   Arrangement of MAPs on microtubules   150
       8.2.2   Projections *in vivo*   151
       8.2.3   Control of MAP spacing   151
       8.2.4   Possible functions of different MAP projections   151
8.3   Specialised associated proteins   152
       8.3.1   Bundling proteins   152
       8.3.2   Assembly-initiating components   152
       8.3.3   Assembly-inhibiting proteins   152
8.4   Axonemal dyneins   153
       8.4.1   Polypeptide components   153
       8.4.2   Molecular structure   153
       8.4.3   Domains of the heavy chains   155
8.5   Cytoplasmic motors   156
       8.5.1   Brain dynein   157

	8.5.2	Squid dynein and vesikin	158
	8.5.3	Sea-urchin egg dynein	159
	8.5.4	Amoeba dynein	159
	8.5.5	Nematode dynein	160
8.6	Newly-identified motor molecules		160
	8.6.1	Kinesin	160
	8.6.2	The structure of kinesin	160
	8.6.3	Dynamin	162
8.7	Kinetic behaviour of dynein and kinesin		163
8.8	*In vitro* assays of microtubule-associated motility		164
	8.8.1	Kinesin steps	164
	8.8.2	Rotation caused by 14 S flagellar dynein	165
8.9	Summary		166
8.10	Questions		167
8.11	Further reading		167

9	**The Axonemes of Cilia and Flagella**		**171**
9.1	The '9 + 2' arrangement		171
	9.1.1	Doublet microtubules	172
	9.1.2	The central structure	172
	9.1.3	Numbering of the outer doublets	176
	9.1.4	Structural variations	176
	9.1.5	ATPase activity	176
9.2	Molecular arrangements in doublet microtubules		177
	9.2.1	Longitudinal spacing of accessory structures	177
9.3	Genetic analysis of flagellar components		178
9.4	The motile mechanism		179
	9.4.1	The sliding-filament model for flagellar motility	179
	9.4.2	Roles of inner and outer arms	180
	9.4.3	Roles of spokes and nexin links	180
	9.4.4	Simple models of flagellar motility	181
	9.4.5	Coiling: a possible analogy with bacterial flagella	181
9.5	Interactions between dynein and microtubules		182
9.6	Assembly of axonemes and basal bodies		184
	9.6.1	Assembly of doublet and triplet microtubules	184
	9.6.2	Control of assembly in the axial direction	184
	9.6.3	Basal-body structure formation	185
9.7	Summary		187
9.8	Questions		187
9.9	Further reading		187

10	**Microtubules in Cells**		**190**
10.1	Microtubule networks		190
	10.1.1	Typical animal cells	190
	10.1.2	Plant cells	191
	10.1.3	Protists	192
10.2	Dynamics of microtubules in cells		193
	10.2.1	Severed microtubules	194
	10.2.2	The injection of labelled tubulin	194

10.3   Functions   194
        10.3.1   Structural supports   194
        10.3.2   Sliding of microtubules   195
        10.3.3   Movement of particles along microtubules   197
10.4   Fast axonal transport   197
        10.4.1   Fast transporting motors   197
        10.4.2   Control of transport direction   198
10.5   Slow axoplasmic transport   199
        10.5.1   Slow component a (SCa)   200
        10.5.2   Axonal microtubules   200
        10.5.3   Cold-stable microtubules   201
        10.5.4   Plasticity   201
        10.5.5   SCa transport mechanisms   202
        10.5.6   Slow component b (SCb)   202
        10.5.7   Microtubule gelation and contraction *in vitro*   202
        10.5.8   Dendrites   203
10.6   Organisation of cellular membrane systems   204
        10.6.1   Sliding of the plasma membrane relative to microtubules   204
10.7   Pigment-granule movement   205
10.8   Summary   206
10.9   Questions   207
10.10   Further reading   207

11   Changes in the Cytoskeleton during Cell Division   211
11.1   Introduction   211
        11.1.1   Stages of mitosis and cytokinesis   211
        11.1.2   Methods of chromosome separation   213
11.2   Centrosomes   213
        11.2.1   Centrosome division   213
        11.2.2   Centrosome division and nuclear chromosome division   214
        11.2.3   Does the centrosomal region contain nucleic acid?   214
11.3   Prophase   214
        11.3.1   Closed mitosis   215
11.4   Prometaphase in animal and plant cells   215
        11.4.1   Microtubule capture by kinetochores   216
        11.4.2   Variation in the spindles of different species   217
11.5   Congression of chromosomes onto the metaphase plate   217
        11.5.1   The flux of tubulin from kinetochore to pole   218
        11.5.2   Polewards transport of kinetochore microtubules   218
        11.5.3   Elimination forces   219
11.6   Anaphase   220
        11.6.1   Anaphase B: pushing poles apart   220
        11.6.2   Anaphase A: polewards transport of chromosomes   221
        11.6.3   The loss of tubulin subunits at the kinetochore during
                  anaphase   222
        11.6.4   ATP and chromosome sliding *in vitro*   222
        11.6.5   Thermal-diffusion model   223
11.7   Requirement for ATP during anaphase A *in vivo*   223
        11.7.1   Promotion of disassembly at plus ends of kinetochore
                  microtubules   224

11.8	Summary of forces involved in chromosome movement	224
11.9	Telophase	225
11.10	Cytokinesis	225
	11.10.1 The contractile ring	225
	11.10.2 What determines the position of the contractile ring?	228
11.11	Summary	228
11.12	Questions	228
11.13	Further reading	229

Appendix: Some General Methods of Studying Cytoskeletal Components **232**

A.1	Modern modes of light microscopy	232
	A.1.1 Immunofluorescence	232
	A.1.2 Microinjection of fluorescent proteins	233
	A.1.3 Resolving filaments and tubules by light microscopy	233
	A.1.4 Video-enhanced microscopy	234
	A.1.5 Confocal laser-scanning microscopy	235
A.2	Electron microscopy	235
	A.2.1 Negative staining	235
	A.2.2 Low-angle shadowing	236
	A.2.3 Visualisation of structures within cells	236
	A.2.4 Immunoelectron microscopy	237
	A.2.5 Scanning transmission electron microscopy (STEM)	237
A.3	Molecular substructure	237
A.4	The use of antibodies	238
A.5	Optical diffraction and image analysis	238
A.6	Genetic approaches to function	238
A.7	Further reading	239

Brief answers to questions	241
Index and glossary	248

SERIES EDITOR'S PREFACE

The aim of this Series is to provide authoritative texts of a manageable size suitable for advanced undergraduate and postgraduate courses. Each volume will interpret a defined area of biology, as might be dealt with as a course unit, in the light of molecular research.

The growth of molecular biology in the years since the advent of recombinant DNA techniques has left few areas of biology unaffected. The information explosion that this has caused has made it difficult for large texts to keep up with the latest advances while retaining a proper treatment of the basics.

These books are thus a timely contribution to the resources of the student of biology. The molecular details are presented, clearly and concisely, in the context of the biological system. For, while there is no biology without molecules, there is more to biology than molecular biology. We do not intend to reduce biological phenomena to no more than molecular phenomena, but to point towards the synthesis of the biological and the molecular which marks the way forward in the life sciences.

C. J. Skidmore

PREFACE

Research into the cytoskeleton has produced a long list of polypeptides, whose properties *in vitro* have been studied in varying degrees of detail. The aim of this book is to introduce students to this array of proteins and to prevailing views about the function of specific molecules in cytoplasmic processes. The topics are arranged in what is intended to be a logical manner: this is not necessarily the best order in which to *read* about them. It may be better to read the chapters on function first, in order to understand why certain proteins have been studied in great depth. Chapters 4 and 8 (as well as parts of Chapters 1 and 2) should perhaps be regarded as mini-encyclopaedias, for reference when needed. We hope the extensive cross-referencing between different sections will allow readers to bypass detail until it seems relevant.

Where many theories compete to explain the observations, we have attempted to identify the preferred ones, rather than enumerate all possibilities. Inevitably, many ideas will need to be modified in the light of new evidence, but perhaps the overall picture is now sufficiently well defined to remain recognisable in future. Readers in search of more experimental detail or alternative explanations of the observations should be able to track them down either in the reviews listed at the end of each chapter or in the selected primary references. Undergraduate students are unlikely to have time to read many of the original papers but may find a list associating the names of many of the scientists in this field with the titles of their papers a useful summary of current research activity.

We acknowledge our great debt to colleagues at the Laboratory of Molecular Biology, Cambridge, and elsewhere throughout the world for informative discussions. We especially thank those who have provided us with illustrations and those who suggested improvements to the text.

L. A. Amos and W. B. Amos

THE ORGANISATION OF *1*
CYTOPLASM

'It is ... the *organization of protoplasm* which is its predominant characteristic and which places biology in a category quite apart from physics and chemistry.... Nor is it barren vitalism to say that there is something remaining in the behavior of protoplasm which our physico-chemical studies leave unexplained.'   E. E. Just (1933)

1.1   Introduction

At the present time our knowledge of the molecular components of cell cytoplasm is growing rapidly. The number of known species of protein molecule increases with each round of biological journals and the properties of these molecules are rapidly being catalogued, including details of their amino-acid sequences and the genes that code for these. It remains true, however, that the number of different components is so great, and their interactions so complex, that we have only the vaguest impression of how cells actually grow, move and multiply.

Eukaryotic cells are organised such that different chemical processes can occur at different sites, often inside distinct **compartments** (see Figure 1.1).[12-16] The most obvious of these compartments is the **nucleus**, holding most of the cell's DNA at a high concentration. The thick mixture of proteins occupying the space around the nucleus is generally referred to as the **cytoplasm**. Other membrane-bounded compartments move slowly around in this soup. The double-layered nuclear membrane is unique in having relatively large channels (**nuclear pores**), which allow the passage of quite large macromolecules (up to about 20 kilodaltons, or 20 kD). The lipid membranes surrounding other compartments prevent the free passage of most kinds of molecule and allow different environments from that in the rest of the cell to be maintained inside.

Compartments surrounded by a single membrane include the **endoplasmic reticulum (ER)**, a complex network of membranous tubules where substances that the cell will eventually secrete are assembled; the **Golgi apparatus**, where

Figure 1.1

An epithelial layer of cells on the inner surface of the trachea (windpipe) in a mammal. Every tissue has different cells with specialised functions. Here, goblet cells synthesise mucus, which is secreted by exocytosis (fusion of vesicles with plasma membrane). Ciliated cells have many mitochondria to provide fuel (ATP) for ciliary activity, which transports the mucus along the surface. The positions and movements of these organelles are determined by the cytoskeleton.

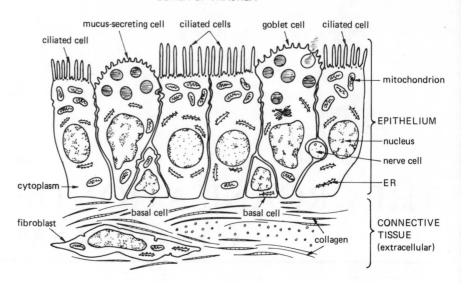

further chemical modifications take place; and **endosomes** and **lysosomes**, where endocytosed material brought into the cell is degraded. Their interiors are topologically equivalent to the outside of the cell, or to the space between the two nuclear membranes. A distinct class of organelles with double membranes includes **mitochondria**, which produce the ATP that serves as a source of energy for many processes within a cell, and the **chloroplasts** found in plant cells, which convert light into chemical energy. Organelles in this group develop from less specialised entities known as **plastids**, which are thought to have originated as symbiotic prokaryotes taken up by early eukaryotic cells, though many of their components are now coded for by nuclear genes.

This book is principally concerned with what is known at present about the components of the eukaryotic cytoplasm, but their interactions with some of the membranous organelles mentioned above will be referred to. The aim of researchers in this area of cell biology is to discover the exact mechanisms by which components of the cytoplasm control cellular processes such as cell motility, cell division, communication between neighbouring cells, and cellular morphogenesis. It is an area which is still shrouded in mystery and many of the current hypotheses are hotly disputed. What is becoming quite clear, however, as we gradually learn more about the components of the cytoplasm, is that their structures are quite beautiful and that their interactions with one another are very specific despite the fact that they occur in highly dynamic configurations.

1.2   What is cytoplasm?

1.2.1   At light-microscope level

Looking at individual cells through the light microscope tends to give an initial impression of cytoplasm as a colourless fluid, possibly with the consistency of a thin homogeneous gel. Such an impression is actually quite misleading. A cell's transparency is simply a property of its minute thickness. If a cell is barely visible by absorption in a microscope, it transmits 98 per cent of the light that falls on it, but the fraction of light that could pass through 1000 such cells would be 0.98^{1000}, which is $0.000\,000\,002$.

A better impression of the consistency of cytoplasm is provided by observing the stiffness of a gel containing 10 per cent protein. Cytoplasm is not even a homogeneous gel. It includes some more solid components which are responsible for determining the overall shape of the cell. The mechanically heterogeneous nature of cytoplasm was first shown by Harvey and Marsland in 1932, in centrifuge experiments with sea-urchin eggs. It was found that the cytoplasm could be partitioned into fractions containing components of different densities at speeds obtainable with a hand-driven centrifuge. Measurements of viscosity using tiny magnetic particles introduced into cells also showed wide variations.

It has been clear for a long time, from light-microscopic studies using phase and polarising optics, that cytoplasm contains fibrous elements, often displaying birefringent properties. But only with the development of immunofluorescent labelling methods did the long-range distributions of the fibrous components become apparent. As demonstrated by the examples in Figure 1.2, cells contain extensive arrays of long filaments. Arrays often spread out from the nucleus to reach all regions of the cell periphery or cross the cell from one side to another. Such arrangements imply that filaments may exert overall control of a cell's behaviour.

1.2.2 'Cytosol' and 'cytoskeleton'[1-11]

A large fraction of the cytoplasm is easily removed when cell membranes are disrupted by a non-ionic detergent. This fraction has been called the **cytosol**. The residue of undissolved components that is left often has the same outline as the cell and, because it has been presumed to define the cell's shape in life, has been named the **cytoskeleton**. The term is inapt since even those components that are most resistant to experimental reagents seem to undergo constant rearrangement in living cells of most types. Even more unfortunately, the term has been extended in common usage to include *all* the filamentous components of the cytoplasm, even those that readily depolymerise on cell lysis.

It may not be strictly correct, either, to refer to the apparently non-filamentous fraction of the cytoplasm as the cytosol, since evidence suggests that very few proteins are in free solution.[20-21] Small fluorescent particles, such as dextrans in a range of different sizes, have been injected into cells and their diffusion followed by photobleaching a small region and recording the rate at which the bleached area becomes blended with surrounding unbleached

Figure 1.2
Three major fibrous systems coexist in most cells. They are revealed here by specific fluorescent labelling. (a) shows actin filaments labelled with fluorescent phalloidin; (b) and (c) show the networks of microtubules and IFs labelled with specific antibodies to tubulin and vimentin respectively. The presence of the primary antibodies is revealed by fluorescently-labelled second antibodies. × 350 [Reproduced from Small *et al.*, (1988). *J. Cell Sci.* **89**, 21–24.]

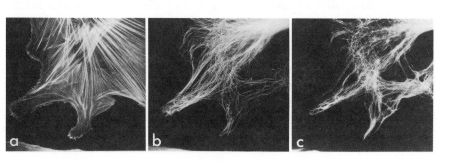

Figure 1.3

Diffusion of dextran particles injected into cytoplasm. The diffusion constant of particles of a given size that were injected into cytoplasm, relative to their diffusion constant in water, is plotted here. Very small particles (point on vertical axis) behave as if in a fluid 4× more viscous than water. Larger particles (up to 30 nm diameter) apparently find the cytoplasm increasingly viscous, as if they were in a mesh with a pore size of about 40 nm, i.e. twice the radius of gyration at which movement would cease if the initial slope continued (dashed line). However, particles of 30–100 nm diameter diffuse almost equally well, probably because the mesh is dynamic. [Redrawn from Luby-Phelps, K., Lanni, F. & Taylor, D. L. (1988). *Annu. Rev. Biophys. Biophys. Chem.* **17**, 369–396.]

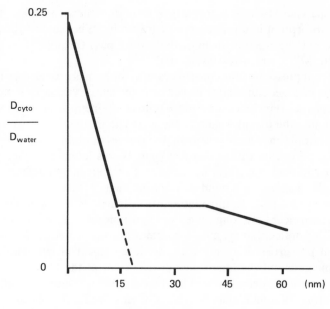

Radius of gyration of particle

fluorescence (a procedure known as **FRAP—fluorescence recovery after photobleaching**). The diffusion of even very small particles suggests that even the most fluid regions of cytoplasm are four times more viscous than water (water itself appears quite viscous at a microscopic level). The behaviour of larger particles (see Figure 1.3) suggests that there are structural barriers with an average pore size of 30–40 nm.[20]

It has been suggested that the biochemical activity inside a cell is more like 'solid phase' than solution chemistry, with enzymes, ribosomes and other small particles weakly bound to filaments and other structures. Certainly, such an arrangement would help provide the control that cells clearly exert over the spatial distribution of their components.

1.2.3 Structure at the electron-microscopic level

In the electron microscope (EM), filamentous structures seen in the cytoplasm of most cells are distinguishable into three main classes: **actin filaments** (6–10 nm in diameter), **intermediate filaments** (7–11 nm) and **microtubules** (25 nm). The protein from which they are made is itself almost transparent to electrons but can be shown up by preparing the specimens in various ways. Gentle disruption of the cytoplasm allows **negative staining** (as in Figure 1.4), which is really the formation of a microscopic amorphous cast of a heavy-metal salt. The cast follows the outline of the protein with a fidelity of about 2 nm and can be imaged. An alternative way of viewing cytoplasmic components in this much detail is to remove the liquid surrounding them and evaporate metal onto their surfaces. If the spray of metal comes from a low angle, parts of the substrate are shaded by adjacent protein structures whose shadows reveal their shapes.[10-11] Some isolated polymers are shown contrasted either by negative staining or by platinum shadowing in Figure 1.5. EM preparation techniques are discussed further in the Appendix (section A.2).

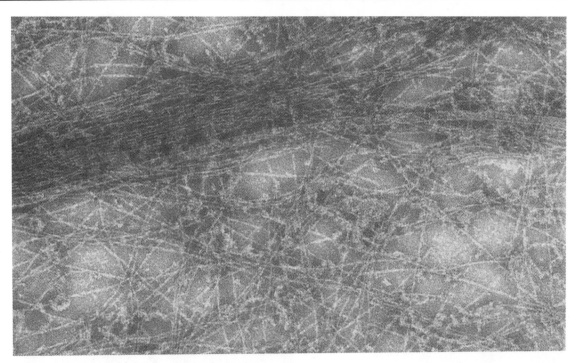

Figure 1.4

Part of the cytoskeletal network in a fibroblast, as seen by EM × 100 000. The cell membrane was dissolved away with mild detergent, with fixative present to preserve protein structures. The latter appear as white features against the negative stain (see Appendix A.2.1) which has been used to provide contrast. In the top half of the picture, a microtubule runs alongside a bundle of actin filaments. Elsewhere, a more disordered actin network is seen.
[Micrograph kindly provided by Dr Victor Small.]

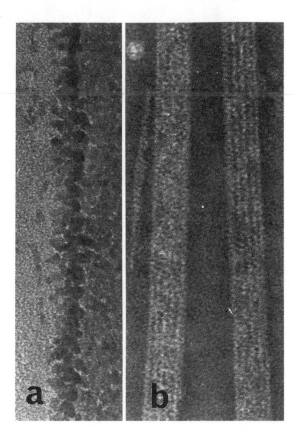

Figure 1.5

(a) An actin filament shadowed from the right-hand side with platinum (which appears dark in this positive image). The striations reveal the periodicity of actin monomers in the filament.
× 1 000 000 (b) Microtubules in negative stain, which shows protein subunits as white occlusions; the longitudinal protofilaments of subunits are clearly visible. × 350 000.

 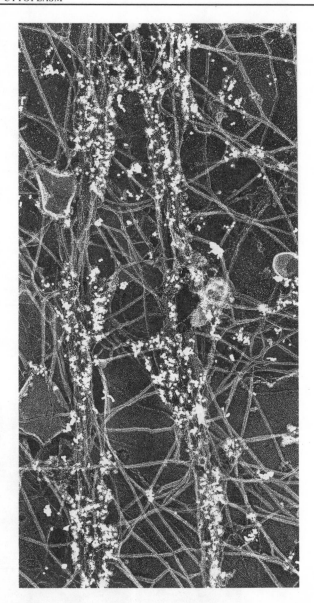

Figure 1.6

A bundle of actin filaments in a cell, as revealed by the rapid-freeze, deep-etch, shadowed-replica technique. The cell was rapidly frozen immediately after being broken open; water around the filaments was sublimed away under a vacuum, and the resulting contours were rotary-shadowed with platinum. After completion of the replica with a layer of carbon, the original specimen was dissolved away. The EM image is of the platinum-carbon replica. × 200 000. [Micrograph kindly provided by Dr Durward Lawson.]

Figure 1.7

Small actin-containing bundles in a cell, decorated with clusters of small bright dots which represent gold particles coated with anti-myosin antibodies. The filaments have been imaged by the same technique as those in Figure 1.6. × 50 000. [Micrograph kindly provided by Dr Durward Lawson; see Lawson, D. (1987). *Cell Motil. & Cytoskel.* **7**, 368–380.]

All the main cytoskeletal structures were originally identified from electron micrographs of thin sections cut from fixed intact cells embedded in plastic; such specimens are stained by techniques that leave heavy-metal salts bound within protein and carbohydrate structures, rather than coating their surfaces— that is, the specimen is **positively stained**. The three-dimensional arrangement of cytoplasmic components is less disturbed by such treatment, though individually they are less well resolved.

Microtubules appear as circular profiles in cross-section and as parallel pairs of lines in longitudinal sections. Intermediate filaments (IFs) were named on the basis of their apparently uniform diameters, intermediate between those of microtubules and actin filaments; they are also commonly referred to as **10 nm filaments**. In the absence of direct proof that they consist of actin, filaments

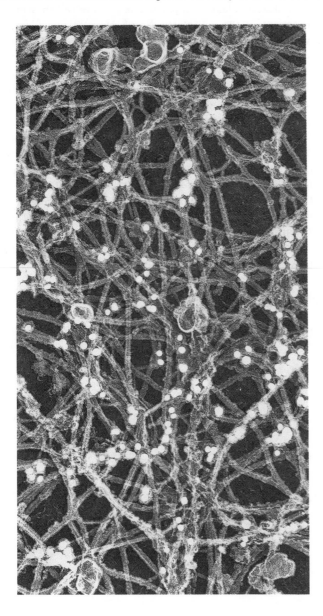

Figure 1.8

IFs in a cell, decorated with gold-labelled antibodies to an intermediate filament-associated protein. × 50 000. [Micrograph kindly provided by Dr Durward Lawson; see Lawson, D. (1983). *J. Cell Biol.* **97**, 1891–1905.]

that are seen in thin section with apparent diameters of around 6 nm may be referred to as **microfilaments**. The overall diameter of an actin filament is now thought to be greater (possibly up to 10 nm), but it differs from an intermediate filament in that its protein is most densely concentrated near the centre of the filament, so the outermost radii tend to fade out of the image.

Microtubules are fairly easy to recognise *in situ* (as in Figure 1.4) but filaments, when closely intertwined (as in the case of the actin filaments in Figure 1.6), may be less readily identified. It is useful to have some sort of specific label. Figures 1.7 and 1.8 show filaments labelled with specific antibodies attached to gold particles.

Sometimes, finer (2–5 nm) straight filaments are seen quite clearly in sectioned cells but their composition is usually unknown and it is not yet clear that they form a distinct class.[29] In particular types of cell they may even form a major component of the cytoplasm. Some of these fine filaments may be closely equivalent to the subfibrils of intermediate filaments (see Section 2.3.5). However, two cases where the constituent polypeptides are known to be much smaller than any IF polypeptides are the filaments found in the contractile organelles of the protists known as vorticellids (mentioned in Section 1.5.2) and the 2 nm filaments of nematode sperm (discussed in Section 6.6).

In thick sections of most types of cell visualised by **high-voltage transmission electron microscopy (HVEM)**, one also sees an *irregular* meshwork of material, apparently connecting together all the recognisable types of filament described above. This has been called the **microtrabecular lattice**.[23] The name provides a useful reminder of the small pore size in cytoplasm.[24] But there is no evidence that 'microtrabeculae' consist of a major cytoplasmic component that has not yet been isolated. They are more likely to represent the large array of different accessory proteins that copurify with the known major components, together with numerous other minor components of the cytoplasm that associate weakly with the cytoskeleton.

1.3   Cell arrangements[12–16]

It is impossible to describe a 'typical' cell, since there is an enormous variety of forms, even within a single layer of tissue (see Figure 1.1). Figure 1.9 shows

Figure 1.9

The structure of a fibroblast in close association with a flat surface. In this view, the cell would be moving from right to left, spreading its leading lamella out in front and drawing in its rear.

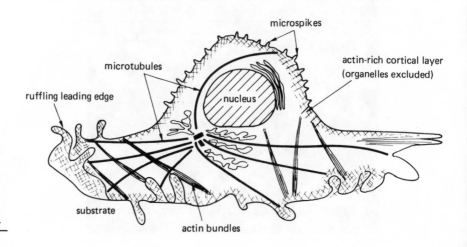

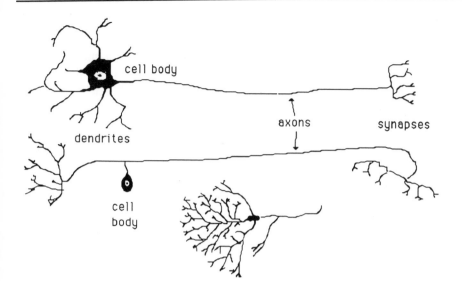

Figure 1.10

Three neural cells with varying morphologies. Unlike those of fibroblasts, each ruffling lamella (known in this case as the growth cone) is at the end of a long extension, or neurite; the cell body is left behind as the neurite grows out, rather than being pulled along behind. Usually, one neurite grows longer and becomes an axon; its growth cone eventually forms a major synapse. The other neurites branch to form complex dendritic networks.

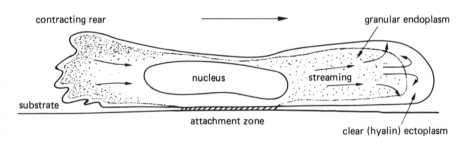

Figure 1.11

An amoeboid cell crawling in the direction indicated by the arrow; the movement can be fast enough for streaming of the cytoplasm to be directly observable in the light microscope. The actin-rich cortical layer is thick enough to be seen by light microscopy in such cells. Amoebae appear to be less firmly attached to the substrate than fibroblasts.

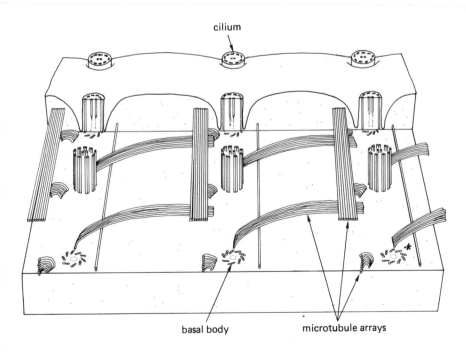

Figure 1.12

A small section of the cortical layer of the cytoskeleton of *Tetrahymena*, a ciliated protist. The rear strip shows the outside surface of the cell membrane, with the protruding cilia cut to reveal their internal '9 + 2' bundles of microtubules, or 'axonemes'. The foreground shows the surface cut away successively, to expose an intricate arrangement of microtubules. × 30 000.

a general-purpose sort of animal cell, known as a fibroblast. Such cells are motile, being able to spread out over a surface at one end (the 'leading edge') while contracting elsewhere. Figure 1.10 illustrates the extended nature of typical neural cells. In many ways a nerve cell can be thought of as an extreme form of a fibroblast; there are similarities between the growing neural process (which eventually becomes an axon or dendrite) and the leading edge of the fibroblast. The forwards movement of cytoplasm is probably due to similar processes in both cases. The motility of animal cells such as these is usually studied in culture dishes but obviously such cells also have the ability to move about within

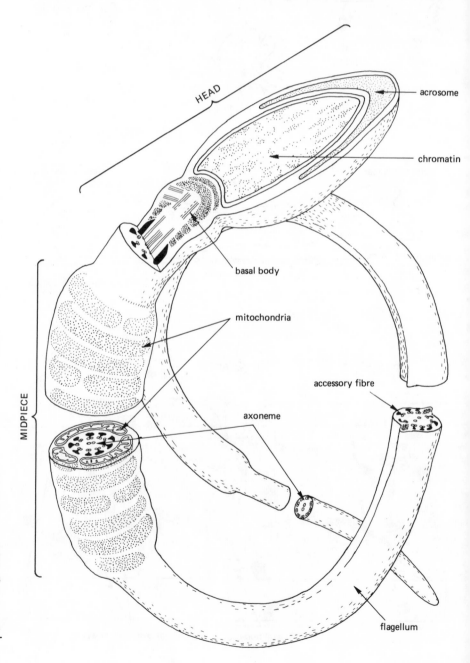

Figure 1.13

A mammalian sperm showing the structural transitions along its length from the acrosome at the tip of the head to the '9 + 2' axoneme in its tail or flagellum. The acrosome is involved in gaining entry to the egg during fertilisation. The rest of the head contains the sperm's haploid nucleus, tightly packed with chromatin (DNA plus accessory proteins). The motile axoneme is a bundle of microtubules growing out from a basal body that is closely associated with the nucleus. Below the neck, in the so-called midpiece, mitochondria (which produce the ATP required for swimming) are wrapped around the axoneme. In mammals and many other species, accessory fibres help support the ring of microtubules throughout most of the length of the axoneme. × 50 000.

tissues. The movement of amoebae, involving forwards-flowing pseudopodia (Figure 1.11), also comes into this general class of motility, though it is more diverse. Muscle cells (discussed in Section 1.6.1) are highly specialised to exhibit the contractile behaviour that is observed in other cells but seem to be incapable of protrusive activity; re-extension of muscle cells has to be produced by external forces (e.g. the contraction of other muscle cells).

Other cells produce movement by beating flagella or cilia that extend out into the surrounding medium. Many free-swimming algae and protozoa (Figures 10.2, 10.3, 1.12) use these organelles for propulsion, as do most animal spermatozoa (Figure 1.13). Layers of epithelial cells (Figure 1.1) often have cilia which they beat in a concerted way to cause fluids (e.g. mucus) to flow over the surface of the epithelium. Polar cells like these (see Figure 1.14), growing in a layer of tissue, do not move much as a whole, although there must be some rearrangement with respect to neighbours when a cell divides. Plant cells (Figure 1.15), within their rigid cellulose walls, probably move around and change their shapes least; the pattern of successive cell divisions fixes the relative position of each cell and determines the final form of the plant. However, there is often very rapid movement within the interior of plant cells (see Section 5.3.4).

Yeast cells are perhaps the simplest eukaryotes; their shapes, supported by rigid cell walls, are similar to those of bacteria. Yeast cytoplasm does, however, contain microtubules (see, for example, Figure 11.10) and actin filaments, and both proteins appear to be essential for the cells' survival. Other unicellular organisms (see Figures 1.12 & 1.16) may be exceedingly complex in structure.[19]

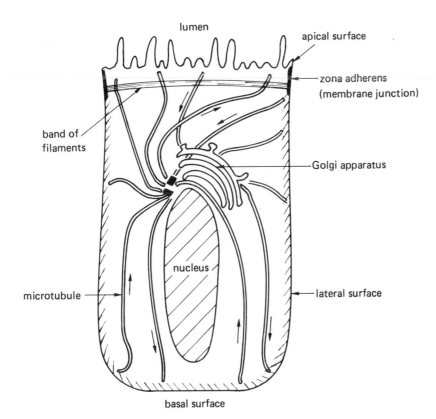

Figure 1.14

A typical polarised epithelial cell, whose apical and baso-lateral surfaces have quite distinct properties. Transporting epithelia maintain concentration gradients of ions and soluble molecules between the apical and basal compartments they separate. Arrows indicate two-way traffic of vesicles along microtubules, between each zone of the plasma membrane and intracellular compartments such as the Golgi apparatus.

Figure 1.15

A typical plant cell. Cytoskeletal structures are usually confined, by the central vacuole, to a layer under the cell membrane. The chloroplasts often move around with the aid of the cytoskeleton.

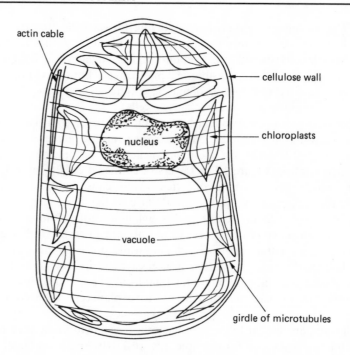

In many cases, even very elaborate organelles (such as those included in Section 10.2) are constructed from the familiar components described in detail in this book. However, the organelle shown in Figure 1.16 is a reminder that large cellular structures can also arise from novel sources.

Figure 1.16

The most complex cytoplasmic structures are found in single-celled organisms. The flagellate *Erythropsidium* (diagram (a)) possesses a contractile tentacle and an organelle, the ocelloid, which is encased in dark pigment (shown black) and equipped with a lens. Electron micrographs show a curved crystalline plate (diagram (b)), which is probably sensitive to light, and bands of contractile material, which may serve to focus the lens. The lens and photoreceptor arise during development from a single plastid resembling the precursor of a chloroplast. Nothing is known of either the mechanism or the function of vision in this protist. (b) × 20 000. [After Greuet (1977). *Protistologica* **13**, 127–143.]

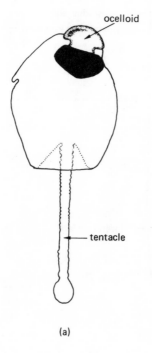

(a)

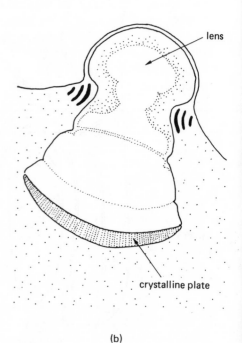

(b)

1.4 The major filament types

1.4.1 Filament substructure

Each of the three major classes of filaments can be dissociated into much smaller subunits. The major protein subunits are usually identical or very similar to each other. Actin filaments and microtubules are made of globular protein subunits (actin monomers and tubulin dimers), which can be dissociated fairly easily, especially in the case of microtubules. Individual protein molecules in these polymers can be resolved in favourable electron micrographs, such as Figure 1.5a & b. Intermediate filaments consist of elongated α-helical subunits (IF proteins), twisted together, and have a rather smooth featureless appearance (Figure 1.5c). The subunits are dissociated *in vitro* only by harsh treatments (such as in 8 M urea or very high salt concentrations), though cells seem able to disassemble and reassemble them when necessary. All three types of filaments have many kinds of protein components that copurify with them and are known as **associated proteins**. Some are quite visible, others are undetectable by EM without suitable labels (as, for example, in Figure 1.8).

Actin is normally an abundant protein in plant and animal cells, often the *major* cytoplasmic protein. It was, of course, first discovered in muscle, where it constitutes 40 per cent of the cells' protein. Other cell types have less than this. The cytoplasm of *Acanthamoeba*, for example, contains 7 mg/ml: total protein is of the order of 100 mg/ml (10 per cent). More typically the amount of actin in a cell is 1–5 mg/ml. Tubulin is often found in similar concentrations.

The core proteins such as actin and tubulin vary little throughout the eukaryotes. But evidence is accumulating to suggest that their associated proteins are much more varied and may be tailored to fit a particular location in a particular cell type. In the case of intermediate filaments, parts of the basic subunit vary according to the cell type, whilst other parts are highly conserved.

The control of assembly in the cell is beginning to be understood for actin and tubulin, somewhat less so in the case of the IF proteins. There are thought to be assembly-controlling associated proteins in each case. Microtubules seem to be designed to disassemble completely; if a microtubule is required in a new situation, assembly is usually freshly initiated using dimers or small oligomers. Actin filaments have a choice of disassembling completely or being severed into short lengths, to be reannealled elsewhere. *Some* intermediate filaments may be disassembled and their subunits reused, but others may have to be proteolytically digested for major changes to be effected.

1.4.2 Functions of filament systems

We can make a rough list of some possible functions, which will become clearer from the greater detail given in later chapters. But it is worth listing them here before describing the particular characteristics of any given filament system, since all have much in common:

(a) a framework defining cell shape;
(b) strengthening elements;
(c) contractile fibres (sliding of antiparallel filaments)—muscle (see section 1.6 and Chapter 5) provides a good example (Figure 1.17);

(d) extensile rods (sliding of parallel filaments)—cilia and flagella (Chapter 9) work in this way (Figure 1.18b);

(e) tracks for guiding transport (e.g. of particles)—transport of substances along microtubules in nerves (see Chapter 10) is a clear example;

(f) signal transmission directly along the filament—no example of this has yet been demonstrated but it is a possibility.

1.4.3 Differences between filament types

By analogy with macroscopic structures, microtubules are presumed to resist compressive forces, the other filaments to function in tension. But the functions of the different filament systems are not highly exclusive. Though intermediate filaments may appear to be the main strengthening elements in some situations, in others actin filaments may seem to carry out this role alone. Similarly, particle transport may occur mainly along microtubule tracks in some cells but along actin filament bundles in others. This must be the result of natural selection working independently in a range of cell types that started with the same basic elements.

1.4.4 Functions of associated proteins

As mentioned above, associated proteins may control assembly:

(a) by binding to the monomer subunits, preventing their polymerisation;

(b) by initiating assembly (starting off a polymer usually requires more critical conditions than the subsequent addition of subunits);

(c) by sealing off the ends of filaments (**capping**) when they reach a specified length;

(d) by stoichiometric binding to promote elongation;

(e) by causing disassembly, including severing;

(f) by enzymatically modifying the subunits of the polymers.

Other roles include:

(g) strengthening filaments by binding alongside them;

(h) cross-linking filaments into bundles and networks;

(i) formation of projections for structural interactions with other cellular components, including other filament types, cell and nuclear membranes, and so on;

(j) production of motility by tension-producing crossbridges between structures.

Associated proteins with many of these functions are found to copurify with all three types of filaments. In all three cases, proteins for crosslinking filaments into bundles and others apparently involved in controlling assembly have been found. Motility, involving sliding of interdigitating filaments or of transport of particles along filaments, occurs both for actin filaments and microtubules. There is as yet no evidence for IF-associated motility but transport along IFs cannot be excluded. Certainly, transport of IF proteins themselves occurs, as described in Chapter 2, but this may be via passive attachment to some other structure.

1.4.5 Duplication of functions

Eukaryotic cytoplasm appears to contain an unreasonably large variety of proteins, with confusingly overlapping properties. Could not natural selection have evolved a more efficient design, using, for example, a single filament system with only two or three different associated proteins to modify its properties according to whether it was required to act as a railway or a tension-producing network? In later chapters, experiments are described in which cells have been deprived of particular proteins and apparently continued to function normally.

Why, then, have cells not discarded these redundant gene products? Perhaps the answer is that the eukaryotic cell can afford to have many back-up systems so that if one system fails under a particular set of conditions, another will allow life to carry on. Eukaryotic cells are usually much larger than bacteria and can support a much larger genome in a large nucleus. Their size is not significantly limited by diffusion rates because they have evolved the means of moving large molecular complexes around in their cytoplasm at relatively high speeds. The secret of the immense diversity of eukaryotic life compared with the limited range of bacterial forms may well lie here.

1.5 Motility[1-7]

One of the most striking functions of the cytoskeleton is its participation in cell motility. As will be discussed in later chapters, movement at microscopic levels may be achieved simply through the assembly and disassembly of polymeric filaments, but the most rapid macroscopic movement appears to be produced by sliding cytoplasmic structures relative to one another. Membrane-bounded organelles are transported along filament rails and filaments slide relative to one another. Concerted changes in molecular conformation may also produce relatively large-scale movements but examples of this seem to be rare.

1.5.1 Sliding-filament systems

Striated muscle, the first system to be identified as such, is the acme of sliding-filament systems and illustrates the essential features.[18] The basic filaments consist of a series of identical elements, so that similar interactions can take place in many different positions. Crossbridges fixed to one filament (permanently, to myosin filaments, in this case), bind briefly to a neighbouring filament (F-actin) and undergo a conformational change that moves the two filaments relative to each other. Striated muscle is organised into a series of **sarcomeres**, so that simultaneous shortening of all these microscopic elements produces large-scale movements.

The **Z-line** in Figure 1.17 actually represents an extended, densely crosslinked protein layer which connects two sets of actin filaments in a bipolar arrangement. The layer is disc-shaped in the cells of some muscles and therefore called a **Z-disc**; in other cases its topology may be very complex. The actin filaments have several other proteins associated along their lengths and the resulting complexes are usually referred to as **thin filaments**. A set of bipolar myosin filaments, the **thick filaments**, which are connected together in some way on the so-called **M-line** (another *area*, in fact), interdigitate with the thin filaments. Contraction involves the Z-lines moving closer to one another.

Table 1.1
Speeds of various motile systems

System	Type of movement	Velocity ($\mu m/s$)	Comments	Chapters for reference
Muscle	Contraction due to sliding of actin and myosin filaments	up to 200	Summed movement of many sarcomeres	1, 5
Flagellar axonemes	Bend propagation due to relative sliding of doublet microtubules	10–15		1, 9
Axostyle of flagellates	Bend propagation due to sliding rows of singlet microtubules	up to 100	Summed movement of several rows	10
Spasmoneme/myoneme of vorticellids	Contraction due to Ca^{2+}-induced molecular rearrangement	23 000	Summed movement of very many molecules	1
Axopods of heliozoa	Contraction due to disassembly of microtubule bundle	>100	200 × faster than endwise loss of subunits *in vitro*	7, 10
Nitella, Chara	Cytoplasmic streaming in giant algae, along F-actin bundles	60	Table 5.1	4, 5
Thyone acrosome	Extension due to streaming and F-actin polymerisation	10	} These may all be similar phenomena due to mechanisms associated with filament polymerisation	6
Fibroblasts, nerve growth cones	Crawling by membrane protrusion, F-actin assembly/disassembly, actomyosin contraction—c.f. particles on outer surface, moving backwards or forwards	~0.02		6
	As above?	0.1–2.0		6
Amoebae (incl. *Limax*)	Crawling due to membrane protrusion, assembly/disassembly of MSP (major sperm protein)?	up to 1.6		6
Nematode gametes		1.2		6
Intracellular particles/vesicles	Fast transport along microtubules	up to 5	(See also Table 10.1)	8, 10
Slow axonal transport	Movement of axoplasmic cytoskeleton	0.002–0.1		10
Chromosomes	Separation by mitotic spindle	0.003–0.2		11

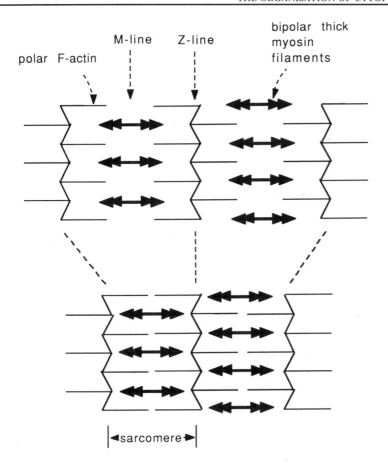

polar F-actin   M-line   Z-line   bipolar thick myosin filaments

|◄sarcomere►|

Figure 1.17

Sliding filaments as organised in striated muscle. Striated muscle is made up of many sarcomeres in series. Thick and thin filaments overlap only slightly in the relaxed state (upper half of figure) but overlap greatly after contraction (lower half).

1.5.2 Contraction at subunit level

Although the muscle cell as a whole contracts in length, possibly only a minority of protein molecules—the crossbridges in particular—alter their shapes significantly. It is possible, however, to produce reversible large movements from many shape-changing molecules arranged in series, without any sliding of filaments. This is believed to be the basis for contraction in the myonemes and spasmonemes of ciliated protozoa,[27] which contain no actin or tubulin but are instead composed of calcium-binding proteins, possibly related to calmodulin (see Section 4.1) but assembled end-to-end into fine filaments (see also Section 2.3.5).

Calcium released into the cytoplasm (to give micromolar concentrations) is thought to induce a conformational change in the individual subunits, causing the whole organelle to shorten (Figure 1.18a). Initially-straight fine filaments may coil up in segments to form either fatter filaments or helices, or a more random mass, depending on the species. It is thought that when calcium ions are pumped out of the cytoplasm back into membrane-bounded storage compartments, and the cytoplasmic level drops to $0.1-0.001$ μM, the subunits revert to their former conformation and the organelle re-extends. There is evidence for similar contractile organelles in other cell types, especially in 'rootlets' associated with the basal bodies of flagella.[28]

Figure 1.18

(a) The minute movements of individual protein molecules, when summed together, can have an effect on a macroscopic scale. In striated muscle, the contracting unit is the sarcomere (Figure 1.17), whose shortening is effected by sliding of interdigitating filaments. In other contractile systems, such as the myonemes of protozoa (see Table 1.1), the unit might be individual protein molecules, or small subassemblies of molecules, which shorten directly by changing conformation. (b) In motile systems such as cilia and flagella, relatively small amounts of inter-filament sliding are converted into bending of the bundle of filaments (microtubules). For a long bundle, even a small extent of bending can produce a large movement of the tip of the bundle.

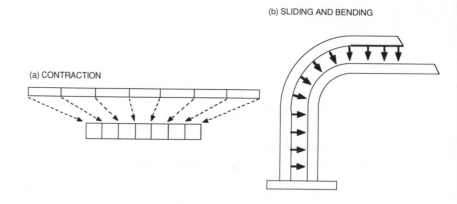

1.5.3   **Calcium and the control of cytoskeletal changes**

Intracellular signalling systems which trigger rearrangements of cytoskeletal elements very often seem to involve a change in the concentration of free calcium ions (Ca^{2+}). The normal intracellular level of free Ca^{2+} is low ($\sim 0.1\ \mu M$): this is desirable since calcium forms insoluble complexes with important ions such as phosphates. However, a transient rise in the local calcium concentration, produced by letting ions leak in from outside the cell (where calcium is usually abundant) or from internal compartments, can be tolerated. Even a modest increase in free Ca^{2+} concentration ($1-10\ \mu M$) can be detected by high-affinity Ca^{2+}-binding proteins, which change conformation after binding the ions. Thus, calcium is excellent as a messenger that can act by diffusion over short distances within the cytoplasm, in response to an external trigger of some sort. Other substances, such as cyclic nucleotides (e.g. cAMP, cGMP) or lipid derivatives (e.g. inositol trisphosphate, IP_3) are used for this purpose, often in conjunction with Ca^{2+}, forming feedback loops for controlling particular processes.[25] These control systems are complex and apparently vary from cell to cell. They remain to be worked out in full detail for any particular case.

It is not surprising, therefore, to find that numerous cytoskeletal components are high-affinity calcium-binding proteins, especially among the actin-associated polypeptides.[26] In such cases, a small increase in Ca^{2+} level can directly influence the interactions between cytoskeletal components. Higher levels of free Ca^{2+}, such as would result from cell lysis, tend to produce disassembly by various mechanisms, which include severing of actin filaments by specific associated proteins and non-specific calcium-activated proteases. This is probably important for the rapid dissolution of dead cells within a multicellular organism.

The activity of many cytoskeletal proteins depends on their state of **phosphorylation**. Enzymes known as **protein kinases** are able to transfer phosphate groups (usually from ATP) and bind them covalently to specific sites on other proteins. Another class of enzymes, the **phosphatases**, remove phosphate from proteins. Ca^{2+} may play an indirect role here by controlling the phosphorylating activity of the protein kinase. The kinase that phosphorylates the smooth-muscle myosin light chain (see Section 5.5), for example, is controlled by a Ca^{2+}-activated subunit.

1.6 Evolutionary aspects[17]

1.6.1 Duplication of genes

In the case of essential, highly conserved proteins such as actin or tubulin, most genomes carry several independently transcribable genes of slightly varying sequence. All the products readily copolymerise. The main difference is in the way their *expression*—**transcription** into messenger RNA, and **translation** from mRNA into a polypeptide chain—is controlled.[30] Thus, larger amounts of protein may be produced from some genes than from others; some genes may be active only at certain times in the cell cycle or at certain stages in the differentiation of multicellular tissues. The available evidence suggests that no individual product is rigidly reserved for a specific function within the cell. However, the slight variations in the properties of these related proteins may perhaps be important.

The associated proteins appear to have diversified much more freely and there are bewildering varieties for each filament type. Proteins that appear to have a similar function but occur in different cells may appear to have nothing in common when tested with antibodies to the protein or by attempting to hybridise bits of nucleic-acid sequence to their genes. Often, however, homologies can be detected in a set of sequences, suggesting evolution from a common ancestor.

1.6.2 Origins

Actin and tubulin have much in common: they have globular, slightly acidic protein subunits; their assembly is controlled by nucleotides; they interact with similar sorts of associated proteins. A particularly important example of the the last point is their interaction with ATPases that produce motility. It has, therefore, been suggested that they may have originated from a common ancestral protein. It also seems to be possible to classify their associated proteins into a fairly small number of families of more-or-less related polypeptides. Intermediate-filament proteins seem more closely related to some of the microtubule- or actin-associated proteins than to actin or tubulin themselves.

Whatever the original relationships between actin filaments and microtubules, the evolutionary process has produced two systems of polymers assembled from globular subunits with some important differences, as detailed in later chapters. The distinction is clearest in highly motile cell types, such as fibroblasts and amoebae. Microtubules, although individually labile, are centrally organised so that the network as a whole has a fairly even, radial distribution. The organisation of actin, on the other hand, is more subject to local conditions, which often vary in a cyclic manner. In some regions of the cell, filaments may be forming a stiff crosslinked gel, whilst in others solation may be taking place due to release of crosslinks and fragmentation of filaments. These differences are the combined effect of the properties of actin and tubulin themselves and the properties of their associated proteins.

1.7 Summary

Cells are complex and highly variable in layout. However, they tend to share the same basic elements. Their complexity hides a great deal of order and conformity at the molecular level. They frequently have duplicate systems carrying out similar functions. Actin filaments and microtubules often play similar roles to each other and may also play similar structural roles to the truly fibrous filaments (α-helical coiled-coils); nevertheless, there are important differences in their properties, and in the large range of minor components associated with them, which allow eukaryotic cells to have complex and varied behaviours.

1.8 Questions

1.1 Is the consistency of cytoplasm closer to that of a soup or to that of a stiff gel, and what types of experiment have supplied the answer to this question? What major components are responsible for the consistency?

1.2 In what main ways can the fibrous components of cytoplasm be visualised? (See also Appendices A.1–A.5.)

1.3 What similarities and differences among the three main types of fibrous component in the cytoskeleton make them suitable for various roles?

1.4 Is the vast range of the structural differences between different cell types due mainly to variations in major or minor components of their cytoskeletons?

1.5 What kinds of subcellular motility can be coordinated to move whole organisms?

1.9 Further reading

Reviews and books

1 Kirschner, M. W. & Weber, K., eds (1989 onwards). Cytoplasm and cell motility. February issues of *Curr. Opinion Cell Biol.*
   (*Annual short reviews of recent developments in a number of areas.*)

2 Bershadsky, A. D. & Vasiliev, J. M. (1988). *Cytoskeleton.* New York and London: Plenum Press.
   (*Up-to-date monograph, particularly good on actin-associated motility in a variety of cells.*)

3 Warner, F. D., Satir, P. & Gibbons, I. eds (1989). *Cell Movement*, vol. 1; Warner, F. D. & McIntosh, J. R. eds (1989). *Cell Movement*, vol. 2. New York: Alan R. Liss.
   (*These two multi-author volumes form an excellent up-to-date collection of research summaries in the field of microtubule-associated motility. Some chapters are listed individually under specific topics; the others are also worth scanning.*)

4 Schliwa, M. (1986). *The Cytoskeleton.* New York: Springer-Verlag.

5 Weber, K. & Osborne, M. (1985). The molecules of the cell matrix. *Sci. Am.* **253**(4), 92–103.

6 Lackie, J. M. (1988). *Cell Movement and Cell Behaviour.* London: Allen & Unwin.

7 Preston, T. M., King, C. A. & Hyams, J. S. (1979). *The Cytoskeleton and Cell Motility.* London: Blackie/New York: Chapman & Hall.
   (*Useful coverage of a wide range of motile systems.*)

8 Fulton, A. B. (1984). *The Cytoskeleton*: *cellular architecture and choreography*. New York: Chapman & Hall.

9 Cold Spring Harbor Symp. Quant. Biol., vol. 46: 1981 (1982). *Organization of the Cytoplasm*. New York: Cold Spring Harbor Laboratory Press.

10 Bridgman, P. (1987). Structure of cytoplasm as revealed by modern electron microscopy techniques. *Trends Neurosci.* **10**, 321–325.

11 Heuser, J. E. & Kirschner, M. W. (1980). Filament organization revealed in platinum replicas of freeze-dried cytoskeletons. *J. Cell Biol.* **86**, 212–234.

Less specialised books

The following less specialised books contain chapters on the cytoskeleton, seen from slightly varying viewpoints.

12 Alberts, B., Bray, D., Lewis, J., Raff, M., Roberts, K. & Watson, J. D. (1989). *The Molecular Biology of the Cell*, 2nd edn. New York & London: Garland.

13 Herrmann, H. (1989). *Cell Biology*: *an enquiry into the nature of the living state*. New York: Harper & Row.
   (*Specifically intended for postgraduate courses, so many primary references are included.*)

14 Darnell, J., Lodish, H. & Baltimore, D. (1990). *Molecular Cell Biology*, 2nd edn. New York: Freeman/Scientific American Press.

15 Stryer, L. (1988). *Biochemistry*, 3rd edn. New York: Freeman.

16 Wolfe, S. L. (1981). *Biology of the Cell*. Belmont, California: Wadsworth Publishing.

Background

Some background knowledge of the role of natural selection may help stimulate thoughts on how the cytoskeleton has evolved.

17 Gould, S. J. (1977). *Ever Since Darwin*: *reflections in natural history*. New York: Penguin Books.
   (*This and its sequels are recommended reading for all biologists.*)

Additional references

18 Squire, J. M., ed. (1989). *Molecular Mechanisms in Muscular Contraction.* Topics in Molecular and Structural Biology, vol. 13. New York: Academic Press.

19 Grain, J. (1986). The cytoskeleton in protists: nature, structure and functions. *Int. Rev. Cytol.* **104**, 153–249.

20 Luby-Phelps, K., Taylor, D. L. & Lanni, F. (1986). Probing the structure of cytoplasm. *J. Cell Biol.* **102**, 2015–2022.

21 Luby-Phelps, K., Lanni, F. & Taylor, D. L. (1988). The submicroscopic properties of cytoplasm as a determinant of cellular functions. *Annu. Rev. Biophys. Biophys. Chem.* **17**, 369–396.

22 Elson, E. L. (1988). Cellular mechanics as an indicator of cytoskeletal structure and function. *Annu. Rev. Biophys. Biophys. Chem.* **17**, 397–430.

23 Schliwa, M., van Blerkom, J. & Porter, K. R. (1981). Stabilization of the cytoplasmic ground substance in detergent-opened cells and a structural and biochemical analysis of its composition. *Proc. Natl. Acad. Sci., USA* **78**, 4329–4333.

(*Microtrabecular lattice.*)

24 Bridgman, R. & Reese, T. S. (1984). The structure of cytoplasm in directly frozen cultured cells. 1: Filamentous networks and the cytoplasmic ground substance. *J. Cell Biol.* **99**, 1655–1668.

25 Berridge, M. J. & Irvine, R. F. (1984). Inositol trisphosphate, a novel second messenger in cellular signal transduction. *Nature* **312**, 315–321.

26 Bennett, J. & Weeds, A. (1986). Calcium and the cytoskeleton. *Brit. Med. Bull.* **42**, 385–390.

27 Amos, W. B., Routledge, L. M. & Yew, F. F. (1975). Calcium-binding proteins in a vorticellid contractile organelle. *J. Cell Sci.* **19**, 203–213.

28 Salisbury, J. L. (1983). Contractile flagellar roots: the role of calcium. *J. Submicrosc. Cytol.* **15**, 105–110.

29 Roberts, T. M. (1987). Invited review: fine (2–5 nm) filaments: new types of cytoskeletal structures. *Cell Motil. & Cytoskel.* **8**, 130–142.

30 Breitbart, R. E., Andreadis, A. & Nadal-Ginard, B. (1987). Alternative splicing: a ubiquitous mechanism for the generation of multiple protein isoforms from single genes. *Annu. Rev. Biochem.* **56**, 467–496.

31 Simons, K. & Fuller, S. D. (1985). Cell surface polarity in epithelia. *Annu. Rev. Cell Biol.* **1**, 243–288.

32 Crick, F. H. C. & Hughes, A. F. W. (1950). Physical properties of cytoplasm. A study by means of the magnetic particle method. *Exptl. Cell Res.* **1**, 37–80.

33 Greuet, C. (1977). Evolution structurale et ultrastructurale de l'ocelloide d'*Erythropsidium pavillardi* Kofoid et Swezy (Peridinien *Warnowiidae* Lindemann) au cours de division binaire et palintomiques. *Protistologica* **13**, 127–143.

34 Just, E. E. (1933). Cortical cytoplasm and evolution. *American Naturalist* **67**, 20–29.

INTERMEDIATE *2* FILAMENTS AND OTHER ALPHA-HELICAL PROTEINS

The most truly fibrous components of cytoplasm, namely α-helical polypeptides that pair into a coiled-coil arrangement, are found in many situations. Being long rod-shaped molecules, they are ideal for contributing tensile strength to the cytoskeleton. Rod-shaped molecules can each make many more side-to-side bonds when assembled into filaments than do globular subunits like actin and tubulin.

This chapter will concentrate mainly on the intermediate or 10 nm filaments,[1-13] which form the most conspicuous group of coiled-coil assemblies and the major components of the insoluble residue for which the term 'cytoskeleton' was originally coined (see Section 1.2). Related proteins, the lamins,[14-15] appear to be universal components of the inner nuclear membrane. However, many of the proteins associated with actin filaments (Chapter 4), such as tropomyosin, paramyosin and the tail of myosin, are also similar in structure, though less closely related. Similar niches in association with microtubules are occupied by proteins called kinesin (see Section 8.7) and tektins (Section 9.3).

2.1 General structure of coiled coils

An α-helix is one of the simplest conformations taken up by polypeptide chains. The amino-acid residues jut out from the helical backbone of the chain at regular intervals. There are 3.6 residues per turn in a simple α-helix. In the case of a coiled coil, two α-helices slowly wind around each other, as illustrated in Figure 2.1. The axis of each α-helix lies on a helix with a pitch of 13–22 nm (corresponding to a length of polypeptide consisting of 90–150 residues) and each α-helix turns to contact its partner at intervals of 3.5 residues.

Figure 2.1

(a) The structure of a pair of
α-helices twisted around one
another to form a 'coiled coil'.
For simplicity, each amino acid is
represented as a sphere and the
connections between them
represent bonds between the
so-called α-carbon atoms of
successive residues. The pitch of
the right-handed α-helix
corresponds to 3.6 residues but,
when forming a left-handed coiled
coil, it interacts with its partner at
axial intervals of 3.5 residues.
(b) A view down the axis of the
filament, showing the angular
positions of successive residues a
to g in each heptad. Hydrophobic
residues in positions a and d (and
often g and e also) on each
polypeptide produce an inner core
of hydrophobic associations.
Axially, residues on one helix lie
opposite gaps between residues of
the other helix. The way these
'knobs' and 'holes' on the two
surfaces can intermesh is shown in
more detail below.

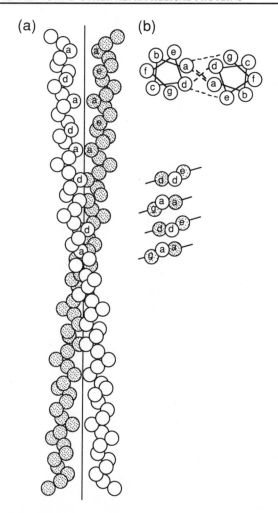

The two polypeptides are apparently held together by **hydrophobic forces**;
it has been found that hydrophobic residues occur at every seventh position
along the polypeptide chains and at positions three or four in between. This
means that there is a hydrophobic stripe which winds slowly around each
α-helix. The hydrophobic residues in one α-helix form bonds with those in the
other in such a way that the two helices are held together as a twisted molecule
(see Figure 2.1). One of the commonest residues involved in this hydrophobic
bonding is leucine; hence the name **leucine zipper** has been coined for short
segments of polypeptide of this type, found in certain DNA-binding proteins.
The zipper segments are thought to associate in pairs to produce dimer molecules
with two independent DNA-binding lobes.

Charged amino-acid residues occur mainly on the outward-facing surface
of a coiled coil. In the case of cytoskeletal coiled-coil proteins, positive and
negative charges are often arranged along the polypeptides in repeating patterns
that are thought to be important in determining their assembly into extended
filaments or, in the case of tropomyosin (Figures 4.12 & 5.11), in interaction
with actin filaments.

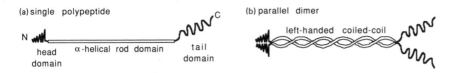

Figure 2.2
(a) A typical fibrous cytoskeletal polypeptide. A central 'rod domain', capable of forming a coiled-coil structure in parallel with a similar polypeptide (as in (b)), is often flanked by non-fibrous domains at one or both ends.

All the known fibrous proteins in this class form **polar dimers**; that is, the polypeptides line up in parallel with their N-termini together at one end, their C-termini at the other. A simple hydrophobic stripe would also allow them to associate in antiparallel. However, a set of aperiodic features along each α-helix produces an exact match only if related polypeptides line up in parallel. For example, the helical rod structure may be completely interrupted in places by short stretches of sequence that cannot conform to the coiled-coil conformation; such regions may form a protruding loop or a small 'knot' in the rod. In addition, an otherwise continuous hydrophobic stripe may occasionally be dislocated by a step of one amino-acid residue; this may produce a small extra twist in the coiled-coil molecule so that it can pack more closely into a bundle with other molecules. Specific combinations of features such as these probably make highly important contributions to the distinctive properties of the various classes of coiled-coil proteins.

Most coiled-coil proteins also have a sequence of amino acids at either end of the rod that takes up some non-helical conformation. These are known as the **head domain** and **tail domain** (see Figure 2.2) and may have enzymatic or other important properties. The head domains of myosin and kinesin are responsible for their ATPase activity (Figure 4.6). Skeletal-muscle myosin has only a very short stretch of non-helical polypeptide chain at its C-terminus, but other rod-forming myosins have a 20–30 kD domain, whose function is still uncertain. Kinesin has a distinct globular tail domain, as described in Chapter 8.

2.2 Intermediate filaments[1-13]

Intermediate filaments (IFs) are all very similar in structure, being also known as **10 nm filaments** because of their rather consistent diameters. Although cells evidently disassemble and reassemble them, they are all highly insoluble after isolation; even very high concentrations of salts such as KCl fail to dissolve them and denaturing conditions such as 8 M urea seem to be needed.

2.2.1 Various types of IF

The polypeptides are quite diverse both in molecular weight and isoelectric properties (net charge). There are several classes of filament, including **cytokeratins, desmin and vimentin, neurofilaments (NFs)** and **glial filaments.** Different classes tend to occur in different cell types but can sometimes be found together. They may form a very large proportion of the cell protein, as in epidermal cells, but seem to be completely absent in some cells. Certain IFs are attached to cell junctions, suggesting that they add to the tensile strength of tissues. It has been postulated that the main role of intermediate filaments is to form a skeletal framework. Some related proteins, the **lamins,** are found associated with the inner surfaces of nuclear membranes, where they probably perform a similar structural role. The microtubules of cilia and flagella include **tektin** filaments which seem to be distantly related to IFs and have the apparent role of structural reinforcement of microtubules (see Chapter 9).

2.2.2 Sequences of IF proteins[21,23]

A large number of IF protein chains has been sequenced; all have a central domain in which at least 93 per cent of the amino-acid residues are predicted to assume an α-helical conformation. This section is therefore referred to as the **rod domain**. There are now several lines of evidence that an IF protein molecule consists of a *pair* of polypeptide chains in axial register (instead of the three chains predicted originally). Throughout much of the sequence of an IF rod domain one finds the heptapeptide pattern, of hydrophobic and hydrophilic amino acids, expected for α-helical coiled coils. The IF proteins form a distinct family within the group because of a characteristic set of dislocations and interruptions in the coiled-coil structure that apparently cause them to assemble into very insoluble long filaments.

2.2.3 Rod-domain substructure of IFs[3]

The structure predicted for the IF rod domain consists of three or four coiled-coil segments with short linking sequences of non-specific structure (see Figure 2.3a). The links referred to as L2 consist of sequences that are very likely to form additional α-helix but the hydrophobic heptad repeat is absent. There is also a single-residue 'step' in the heptad periodicity near the middle of segment 2B in all IF types. The coiled-coil segments in IF protein chains of all types also have high contents of charged residues (30–40 per cent), with a net acidity in general. Like the rod domain of myosin (Figure 2.6a & b), they show alternating stripes of positive and negative charges, in a rough pattern with a repeat of 28 residues (eight turns of each α-helix). The combination of all these features is thought to be essential for the formation of the extremely uniform 10 nm filament structure, though the reasons are not yet understood.

The whole rod domain, including linker segments, is predicted to have a length of 46–7 nm. Rod segments 1A and 1B (of vertebrate IF proteins, at least)

Figure 2.3

(a) Although their sequences have little in common, IF polypeptides have conserved structural features. The rod domain consists of segments of coiled coil interrupted by stretches where the heptad repeat of hydrophobic residues breaks down. L12 separates the two halves of the rod. L1 and L2 subdivide each half unequally, into subdomains 1A, 1B, 2A and 2B. The heptad repeat is also slightly dislocated at a point within 2B. Lamins have an extra 42 coiled-coil residues in the middle of segment 1B, compared with vertebrate cytoplasmic IFs. (b) Two polypeptides lined up in parallel to form a polar dimer. (c) Dimers may pair to form tetramers in four possible ways: (iii) is most probable for tetramers in solution; the interaction represented by (iv) is also likely within filaments.

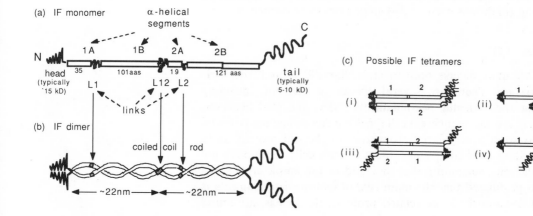

have a combined length of about 22 nm, as do segments 2A and 2B together (Figure 2.3b). Thus, the rod apparently has an approximate bilateral symmetry about its centre. There are conserved sequences at both ends. In particular, there is a highly conserved sequence of 20 amino acids in segment 2B, near the C-terminal end of the rod, which actually occurs in *all* IFs (hence there are antibodies which are able to cross-react with all IF proteins, in spite of their diversity).

2.2.4   IF head and tail domains

The N- and C-terminal non-coiled-coil (or 'head' and 'tail') domains which enclose the central rod domain vary widely in size, charge and chemical character. In some cases, the C-terminal domain is also predicted to be largely α-helical but probably does not form a direct extension to the rod. Most types of polypeptide have basic segments at both terminals.

The varying head and tail regions presumably determine the specific functions of different types of IF but very little is known about them at present. Some parts of the sequences at each end seem to be essential for assembling the protein into the characteristic filaments, perhaps by balancing the net acidity of the rod domain. One function of the tail region may be to protrude from the side of the filaments and interact with other components of the cytoplasm. There is evidence for this in the case of neurofilaments (see Section 2.3).

2.2.5   IF molecular units

As described above, an **IF dimer** is formed by two chains in axial register. Pairs of these polar molecules are thought to associate as **tetramers**, since two- and four-chain particles have been isolated from both hard and soft keratins after the filaments have been dissolved by enzymatic digestion.[26] Undigested vimentin filaments can also be disassembled into two-chain and four-chain units. There is some uncertainty as to whether tetramers are polar (dimers in parallel) or bipolar (dimers antiparallel). The approximate bilateral symmetry of the rod domains would allow all four of the possible arrangements shown in Figure 2.3c.

Various lines of evidence suggest that IF polymers are most probably bipolar. For example, structural rearrangements that can be detected by X-ray diffraction when keratin is stretched (transformation from an α-helical to a β-sheet conformation) suggest that oppositely directed polypeptide chains interact closely in filaments.[23] It seems most likely that the basic repeating unit is a **bipolar tetramer** which, in solution, probably corresponds to arrangement (iii). Within the assembled filaments there must also be associations like (iv) and some neighbouring molecules will lie parallel to one another as in (i) or (ii)—it is not possible, except when molecules are packed together in a square lattice, for *all* neighbours to be antiparallel.

2.2.6   Assembly of 10 nm filaments[3]

In current models, 10 nm filaments consist of approximately 2–3 nm diameter **protofilaments**, twisted together in some manner, like the strands of a rope (see Figure 2.4). The 2–3 nm protofilaments may be formed by end-to-end association of rod-shaped tetramers.

Figure 2.4

The most likely structural
hierarchy in IFs. Tetramers may
polymerise end-on into
'protofilaments'; protofilaments
may pair to give 'protofibrils';
four of the latter, twisted together,
may form a 10 nm filament.

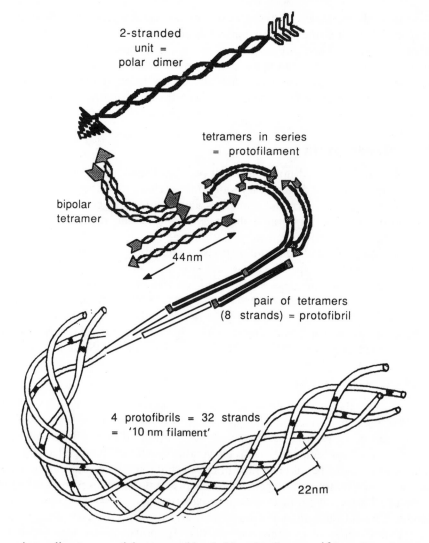

According to a model proposed by Aebi and colleagues,[16] based on electron
microscopy of fraying cytokeratin filaments, pairs of protofilaments twine
around one another to form distinct 4.5 nm **protofibrils**, and a further level of
supercoiling forms a 10 nm filament from four such protofibrils (Figure 2.4).
Electron micrographs of shadowed filaments have shown a 21–2 nm periodicity
that may indicate an arrangement of rods, each with an axial extent of 44 nm
(shortened from 47 nm by being twisted into a helix), that is *half-staggered* at
some level of assembly, perhaps between adjacent protofilaments (as in Figure
2.4). The shadowed features could arise from protruding head or tail domains.
Additional staggering, between different protofibrils for example, may lead to
a more complex arrangement than that shown in Figure 2.4 such that the
protruding domains lie on a 22 nm pitch helix. This may explain the appearance
of some types of filament when shadowed.

A rather different arrangement of protofilaments is suggested by electron
microscopy of hard (feather) keratins;[23] in cross-section, the filaments appear
to show a central core protofilament surrounded by a cylinder that could be

formed from seven or eight additional protofilaments. Hard keratin filaments may perhaps differ somewhat in structure from cytoplasmic IFs.

The patterns obtained by X-ray diffraction can potentially reveal the relative arrangement of subunits in three dimensions.[23] However, for IFs, the results are complex and difficult to interpret. Patterns obtained from hard keratin and neurofilament samples have some longitudinal spacings in common and some that differ. This supports the EM evidence suggesting that different IFs may have their molecules packed into different arrangements, despite their apparently uniform diameters.

2.2.7 Do IFs have a polarity?[5]

Whilst the arrangement of molecules within 10 nm filaments is still uncertain, the question of IF polarity remains open. If it turns out that the filaments are assembled from **polar tetramers**, then all IFs may be polar filaments. If, on the other hand, the currently favoured option of bipolar dimers is correct, the filament core must be inherently bipolar. However, most classes of IF are assembled from a mixture of different molecules, which are thought to form **heterotetramers**. Keratin filaments always contain equimolar amounts of class I and class II keratin molecules. Neurofilaments apparently assemble from three kinds of molecule of widely varying size (NF-H, NF-M and NF-L: see Table 2.1, page 32).

The distribution of dissimilar molecules within the filament is not known, but given that most other cytoskeletal structures are polar, it is likely that each polypeptide type is arranged with a single orientation, even if there is a bipolar arrangement of rod domains. This is particularly pertinent to the structure of neurofilaments,[2] whose very large C-terminal domains form long sideways projections from the filaments (Section 2.3.3). The arrangements in Figure 2.5 would cause the projecting domains to have a polar arrangement, suitable for interaction with polar microtubules, for example. Homotetramers of desmin or vimentin, on the other hand, may form totally bipolar filaments.

2.2.8 Comparison with myosin filament structure[22,30]

The rod of myosin is much longer than an IF rod and apparently not so disturbed by non-helical interruptions (Figure 2.6a). The resulting filaments are much more soluble *in vitro* and are easily disassembled and reassembled by changing the ionic conditions.

Striated muscle myosin (Section 4.3) assembles into bipolar cylindrical filaments, but each half filament is a polar assembly. A half filament may be thought of as resembling a bunch of flowers with long stems twisted together to form a 15 nm diameter shaft: the flower heads are the pairs of head domains projecting out sideways (see Figure 2.6a). The central 150 nm stretch of filament has no heads sticking out and is referred to as the **bare zone**. Only here do coiled coils of opposite polarity interact with one another, possibly forming tetramers as in IFs. The presence of the heads is thought to inhibit further antiparallel interactions, so the filaments are extended further by molecules adding in parallel. The final number of polypeptides in a cross-section is approximately double that in an IF.

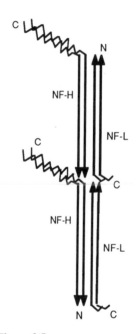

Figure 2.5

Even though IF tetramers are probably bipolar, as in Figure 2.3c (iii), heterotetramers formed from dissimilar dimers may still form polar filaments. This possibility is illustrated for neurofilaments, in which 'light' polypeptides (NF-L), similar to other IF proteins, copolymerise in an unknown manner with medium and heavy chains (NF-M, NF-H) which have unusually large C-terminal domains.

Figure 2.6

The coiled-coil α-helical rods of two-headed myosin molecules. (a) Each rod, lying between large head and small tail domains, is shown here as a row of circles. Each circle represents a 28-residue-long segment; the distribution of charged residues within each segment is similar. At intervals of 7 segments (indicated by the short vertical lines), there is a step in the hydrophobic heptad repeat (see Section 2.1). The relative stagger between different molecules in a thick filament, of 3.5 segments, produces the 14.3 nm axial spacing of the heads. (b) Schematic drawing showing the α-helices within each coiled-coil segment. Within a 28-residue, 4 nm long, segment, each α-helix turns to contact its partner 8 times. Patterns of charged residues within each segment are symbolised by shaded bars, in a complementary arrangement with adjacent coiled coils. The pitch of any coiled coil is estimated to be between 3 and 5.5 segment lengths and is shown here, for simplicity, as 4 segments. In the case of myosin rod there is evidence that the pitch is equal to the 3.5 segment stagger between rods; this would bring all the crossover points into register, while still matching up complementary charges.

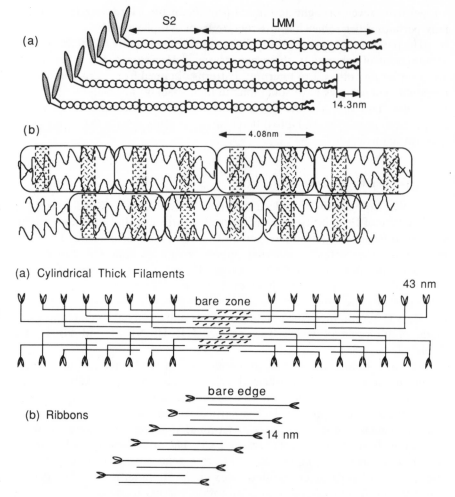

(a) Cylindrical Thick Filaments

(b) Ribbons

Figure 2.7

Assembly of two-headed myosin molecules into two kinds of polymer. (a) A schematic version of the arrangement in myosin thick filaments found in striated muscles. Only heads at 3 × the 14 nm axial spacing are shown; those in between stick out from the rod at other angles. The central 'bare zone' has antiparallel interactions between the C-termini of the coiled coils, perhaps similar to the interactions in intermediate filaments; however, a larger number of polypeptides contributes to each cross-section, giving the shaft a diameter of around 15 nm. Each half of the filament extends in a polar fashion beyond the bare zone; the large projecting heads prevent bipolar interactions between the N-terminal ends of the molecules. (b) More dynamic assemblies of myosin, such as occur in smooth muscle and non-muscle cells, effectively extend sideways rather than lengthways; the result is a simpler, ribbon-shaped, aggregate (see Figure 4.5).

Beyond the bare zone, the interactions between adjacent molecules must be different. Assembly of the long myosin filaments found *in vivo* in striated muscle appears to be complex, involving the help of various accessory proteins. These must dictate the precise arrangement and length of the filaments, which are very accurately controlled; the myosin filaments of vertebrate striated muscle are all 1.57 μm long and probably contain either 294 or 300 myosin molecules.

Smooth-muscle myosin seems to polymerise in a less complex manner (Figure 2.7), involving interactions between antiparallel filaments throughout their lengths. Since the ribbons consist of more than one layer of molecules, parallel interactions probably also occur. Small filaments grow longer by further sideways associations, apparently equivalent to those found in the nucleating structure.

2.3 Classes of IFs expressed by different cell types[4,7,8]

The IF proteins of vertebrate cells have been studied in most detail. On the basis of sequence similarities, it has been possible to group proteins found in the cytoplasm into four distinct classes of polypeptide. Lamins in the nucleus form a fifth class. The eventual list of classes may be much longer. There seem to be other types of nuclear IFs, for example, that have not yet been clearly characterised. IFs also occur in invertebrates and in plants. These have not been studied in sufficient detail to be put into classes; those sequenced so far are more similar to vertebrate lamins than to any of the vertebrate cytoplasmic IFs.

2.3.1 Keratins (heteropolymers of IF polypeptide types I & II)[23,24]

The largest subgroup of IF proteins to be described is the keratin family. Most obvious are the '**hard**' **keratins** which are used in the production of hair, hoof and horn, all of which are formed by the solidification of whole cells rather than by secretion from their surfaces. The **cytokeratins** or '**soft**' **keratins** are abundant in the cytoplasm of epithelial cells; in some cases as much as 80 per cent of the total cell protein is keratin. The number of different cytokeratins is surprisingly large; those found in human skin, for example, are coded for by at least twenty different genes.

The kinds of cytokeratin expressed in any individual epithelial cell type varies with the location of the cell and is tightly linked to the state of cellular differentiation—so much so that it is difficult to distinguish cause from effect. Investigation of the type of cytokeratin expressed, using specific antibodies, is now useful medically in determining the origin of various cancers and in following the spread of tumour cells by immunofluorescence.[12]

Keratin molecules appear always to be heterotetramers of a type I ('acidic') chain dimer and a type II (neutral/basic) chain dimer. This may, or may not, be relevant to the above discussion of filament polarity.

A characteristic of keratins not shared by the other IFs is their capacity for **bundling**, which makes them even more insoluble than other IFs. Bundling is a property thought to be determined by the end domains. Although the terminal domains are extremely variable in sequence, all keratins share marked similarities, including a high proportion of serine and glycine residues concentrated in these domains. Such domains probably interact with one another and also with similar regions in IF-associated proteins such as filaggrin. These interactions are likely to lead after cell death to an *insoluble but flexible* protective barrier, as required for an epidermis. Subtle variations in the end domains are responsible for the varying characteristics of keratins (compare, for example, the texture of the skin on your heel with that inside your mouth).

Hair and nails are composed primarily of cortical cells filled with hard keratin filaments. The generally shorter terminal domains of these IFs have

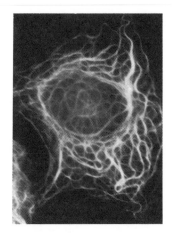

Figure 2.8

IF network in a single cell grown in tissue culture, as revealed by indirect immunofluorescence light microscopy. The network encloses the nucleus and extends out to all parts of the cell periphery. $\times$ 300.

Table 2.1
Intermediate filament and related proteins

Name	Polypeptide mass (kD)	Sources	Polypeptide class (based on sequence homologies)
IF rod proteins			
Hard keratins			
Acidic chains	40–60	Hair-, nail-,	Ia
Neutral-basic chains	55–70	hoof-producing epithelia	IIa
Soft keratins			
Acidic chains	40–60	General	Ib
Neutral-basic chains	55–70	epithelia	IIb
Vimentin (decamin)	53–4	Widespread	III
Desmin (skeletin)	53–4	Muscle cells	III
Glial fibrillary acidic protein (GFAP)	50–1	Glial cells, astrocytes	III
Peripherin	57	Peripheral neurons, neuroblastoma cells	III
Neurofilament proteins			
Light chain (NF-L)	60–8		IV
Medium chain (NF-M)	150–60	Neurons	IV
Heavy chain (NF-H)	180–210		IV
α-Internexin			IV?
Lamins A, B, C		Nuclear envelope	V
Tektins	40–60	Ciliary and flagellar microtubules, centrioles	
Some IF-associated proteins (IFAPs)			
Filaggrin		Epithelia	
Plectin	300	Fibroblasts and other cells	
Synemin	230	Muscle (assoc. with desmin)	
Paranemin	280	Muscle (with desmin and vimentin)	
BHK-IFAP	300	BHK cells	

high concentrations of cysteine residues. The latter are thought to form **disulphide bonds** not only within and between domains but also with a matrix of high-sulphur proteins. The differing mechanical properties required for nail, hair and the like depend on the composition of the matrix. The combination produces a highly crosslinked structure that is insoluble even in urea and methanol. Hydrogen peroxide is used to dissociate temporarily some of the disulphide bonds during the process of producing 'permanent waves' in hair.

2.3.2 Homopolymeric type III IFs[13]

These are a group of proteins that form IFs in various types of vertebrate non-epidermal cells. They have been defined as a subclass because of striking homologies in their sequences.

Vimentin is the major structural protein found in a wide variety of vertebrate mesenchymal cells, such as fibroblasts.[25] Certain cells have been found to be capable of expressing chains from more than one class. When this occurs, it seems that one of the chains expressed is always vimentin: it may therefore have a particularly important role. It is the most well-characterised of the non-keratin IF proteins. Sites on the C-terminal tail of the polypeptide apparently bind specifically to nuclear envelopes in a cooperative fashion. The N-terminal head interacts with a saturable number of sites on isolated plasma membranes. Vimentin filaments thus have the potential to form links between the nuclear and plasma membranes of a cell.

Recent experiments show that vimentin has distinct sites for phosphorylation by two different enzymes, protein kinase C and the cAMP-dependent kinase. After modification by the latter enzyme, vimentin filaments disassemble dramatically.[27,28] Similar results have been obtained with desmin. Phosphorylation and dephosphorylation may be one of the means by which the cell can alter the otherwise insoluble IF network, at least in the case of type III IFs.

Desmin is the major IF protein of most vertebrate muscles and is found most abundantly in smooth muscle. Here desmin filaments form an interconnected meshwork linking cytoplasmic **dense bodies** (the smooth-muscle equivalent of Z-discs) with dense plaques on the membrane. The cytoplasm of smooth-muscle cells is differentiated into two main zones, both containing actin and tropomyosin but differing in other proteins present. Myosin and proteins involved in regulating contraction occur in one, the contractile zone; while IFs, filamin (see Chapter 4) and dense bodies are found in the other, the cytoskeletal zone. This functional and spatial segregation is homologous to that found in more highly ordered types of muscle.

In striated muscle, desmin is found close to junctions between the membranes of different cells and also in Z-discs (see Section 1.5). Antibody labelling of Z-discs has shown that desmin is present with actin at the periphery, whilst the central region contains α-actinin (Section 4.4.2) and actin. It seems that Z-discs in adjacent myofibrils are linked together into a three-dimensional network by desmin connections, which thus play a very important part in maintaining the integrity of striated muscles.

Glial filaments are the specific intermediate filament type in glial cells, which are found in mammalian brain and appear to support neurons. The protein components of these filaments are known as **glial fibrillary acidic protein** (**GFAP**). GFAP, like the keratins, is useful medically as an immunological marker, in this case for detecting brain tumours.

Peripherin occurs exclusively in neurons of the peripheral nervous system and in certain central nervous system neurons. It differs from the type IV neurofilament proteins described in the next section in being more closely related to desmin and vimentin. It is not yet known whether peripherin copolymerises with type IV polypeptides, nor is it clear why it is not present in all neurons.

2.3.3 Neurofilament proteins (IF type IV)[2]

The 10 nm filaments in neurons are referred to as **neurofilaments** (**NFs**) and are usually heteropolymers of a distinctive triplet of related polypeptides. Two are unusually large for IF proteins (see Table 2.1) because they have large

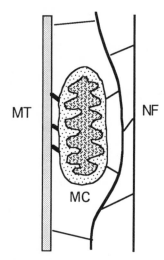

Figure 2.9

Large organelles such as mitochondria (MC) are able to move along microtubule tracks (MT), passing through apparently dense networks of filaments (the pore size in cytoplasm has been estimated to be around 40 nm—see Section 1.2). Probably the projections from neurofilaments (NF) part in front of a moving object like stems in a cornfield.

C-terminal domains (about 500 and 800 residues respectively for the medium and heavy chains, NF-M and NF-H). A recently discovered possible fourth member of the class, named **α-internexin**, is also found in the central nervous system, particularly in developing neurons.

When purified, the triplet light chain, NF-L, is able to assemble alone into filaments. Neither NF-M nor NF-H assembles alone. Their C-terminal domains appear to be highly enriched in glutamic-acid residues and may be phosphorylated on many different residues. Both of these features contribute to a significant net negative charge. The C-terminal domains may be removed by proteolytic digestion leaving apparently normal 10 nm filaments, indistinguishable from any other type of IFs.

Immunoelectron microscopy has shown that the C-terminal domains protrude from the neurofilament surface as long thin sideways projections (**NF-projections**), reaching out towards neighbouring filaments and other structures such as microtubules and vesicles. The numerous projections appear to form crosslinks, holding the filaments together. Recently it has been suggested that repulsive forces, due to the many negative charges on the projections, may be their most important feature. The charges presumably cause the domains to stretch out away from the filaments and also to repel adjacent filaments. It is likely that there are weak binding sites for microtubules and other filaments near the ends of the projections, so that the axonal cytoskeleton is loosely crosslinked together.

This idea may explain how a relatively large vesicle being pulled along a microtubule (Figure 2.9) is able to move rapidly through a thicket of surrounding filaments; the latter, being repelled by negative charges on the vesicle surface, would simply move aside temporarily, breaking a few of the weak crosslinking bonds. It is quite probable that the main role of neurofilaments is to occupy any available space within an axon, preventing the membrane from collapsing (which would ruin its electrical properties) and protecting the organelles inside it, but that they do this in a clever way which allows free passage even to very large vesicles. There could in addition be a small proportion of more permanent crosslinks, perhaps formed by IF-associated proteins, which would maintain the structural arrangement without producing significant hindrance to transport.

2.3.4   Lamins (IF class V)[14,15,29]

These are proteins localised at the inner nuclear surface in the form of an insoluble polymer system known as the **nuclear lamina**. It has been shown that the proteins are very like keratin, having a similar high content of glycine and serine residues. Antibodies against epidermal keratins cross-react with them and there are obvious similarities in their primary sequences. The most obvious difference is an insert of 42 extra α-helix-forming residues in the middle of section 1B of the rod domain (Figure 2.3a); this feature has also been found in the only invertebrate IF polypeptides that have been sequenced. In spite of these similarities, lamins do not occur *in vivo* as 10 nm filaments. Possibly this has something to do with the N-terminal and C-terminal domains; the latter also include a signal that determines the entry of lamin polypeptides into the nucleus.

2.3.5   Tektins and 2 nm filaments

Insoluble fibrous proteins are known to be associated with the highly stable microtubules of eukaryotic flagella (Section 9.6). They appear to be an integral part of these microtubules and are probably partly responsible for their stability. After tubulin has been extracted under relatively harsh conditions, the residue consists of fine, 2–5 nm filaments, probably equivalent to the protofilaments or protofibrils of IFs. The material has a high α-helical content, similar to that of IFs. Preliminary sequence data for one tektin polypeptide suggest that tektins are related to IFs but not closely enough to be classed as another real member of the family.

There is some evidence for similar proteins in association with other types of microtubule but these require further investigation. The identification of tektins with 2 nm filaments also raises the question of whether some of the other mysterious 2 nm filaments that have been seen in cells by electron microscopy might be related to IF protofilaments. The 2 nm filaments in the contractile myonemes of vorticellids (Section 1.5.2) and in the crawling sperm of nematodes (Section 6.6) are both known to be assembled from small (around 20 kD) polypeptides. Especially in the former case, one could imagine the central helices of calmodulin-like molecules (Section 4.1.1) pairing to form the equivalent of half an IF dimer.

2.4   IF-associated proteins (IFAPs)[10,11,32]

Besides the major structural subunits that make up 10 nm filaments, there also appears to be a diverse array of associated proteins. These are often specific to the cell and tissue type. **Filaggrin**, mentioned above, is a basic protein found in skin and related tissues that crossbridges keratin filaments into bundles (**tonofilaments**). A 300 kD IFAP characterised in baby hamster kidney cells (cell line BHK) appears to be responsible for a much looser network of filaments by forming intermittent crosslinks. Another large polypeptide, named **plectin**, which crosslinks vimentin filaments, is notable for having some properties in common with the high molecular weight microtubule-associated protein MAP 2 (Section 8.1.9). **Paranemin** and **synemin** are large polypeptides associated mainly with desmin filaments. Other proteins have been isolated from other tissues and classified as IFAPs because of their copurification with IF proteins and co-localisation by immunomicroscopy (see **epinemin** in Figure 1.8, for example). Little is known about their properties and functions at present.

2.5   Interactions between IFs and other cytoskeletal components

It may be helpful to return to this and Section 2.6 after reading the chapters on actin and microtubules in cells.

2.5.1   Association of IFs with membranes[5,10]

The distribution of intermediate filaments has been studied in many cell types by means of indirect immunofluorescence (e.g. Figure 2.8). Cultured cells

Figure 2.10

IFs are thought to contribute to an integrated system of filaments supporting the structure of the cell. The boundary of the nucleus is reinforced by the lamin network on the inside surface of the nuclear membrane. This network may be directly connected, through the nuclear pores, to IFs that link the nucleus to the plasma membrane, via proteins forming the so-called cortical layer lining the cytoplasmic surface.

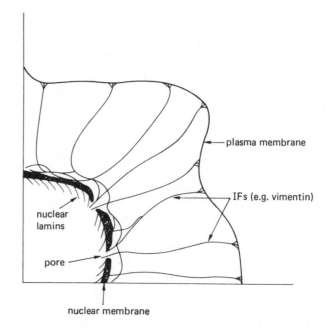

usually have an extensive network of IFs distributed throughout their cytoplasm, apparently contacting the surfaces of both the nuclear and cell membranes.

In a confluent layer of epithelial cells, the keratin-containing **tonofibrils** give the appearance of being continuous across cell boundaries wherever cells make contact with each other. This is because the filaments in each cell end on **desmosomes**, which assemble on the membranes at such contact points. By electron microscopy, desmosomes can be seen to consist of plaques of dense material on both sides of each of the two closely apposing membranes. The IFs appear to be attached to filamentous material forming the cytoplasmic layers of the plaques.

After treatment of cells with trypsin to remove them from their substrate, the IF network collapses and the filaments become concentrated in a large **juxtanuclear cap**. This is similar to what happens when a cell rounds up in order to undergo mitosis and cell division (see Chapter 11). If IF distribution is followed as the cells reattach and spread, it can be seen that the cap gives rise to fibres that radiate from the nucleus towards the cell membrane. Similar observations can be made on vimentin filaments in cultured fibroblasts (such as mouse 3T3 or baby hamster kidney cells).

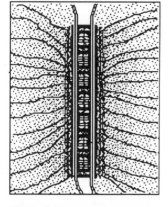

Figure 2.11

A close-up view of a small area of contact between two neighbouring cells in a tissue. A junction of this type is reinforced by a structure known as a desmosome, which appears as a set of darkly-staining layers on either side of each plasma membrane and bands connecting the external surfaces. Associations between keratin filaments and the cytoplasmic face of each plasma membrane are concentrated at desmosomes.

An interesting observation has been made by Goldman and colleagues on newborn mouse skin cells;[32] if the cells are allowed to attach, spread and divide in a medium which is deficient in calcium, the tonofibrils remain mostly round the nucleus. If the extracellular calcium is then increased to normal levels, the tonofibrils start to move out towards regions of cell surface that are in contact with neighbouring cells. Here, the re-formation of desmosomes can be followed by indirect fluorescent labelling of components such as **desmoplakin**.

IFs consisting of proteins other than keratin seem to be less stably anchored to the plasma membrane. For example, the loss of microtubules causes vimentin-containing IFs to collapse towards the nucleus, as described below, whereas keratin-containing networks are unaffected.

Arrays of both keratin- and vimentin-containing IFs appear to terminate at the nuclear surface, frequently in proximity to nuclear-pore complexes. Indeed, there is evidence that vimentin attaches to nuclear membranes[5] because of specific binding to either lamin B or polymers of lamin A and B. It seems likely that there is also mechanical continuity between the nuclear lamina and the IF network. Thus, the nucleus appears to be anchored in position within the cell by the IF network.

2.5.2 Association of IFs with microtubules

In electron micrographs of thin sections through cells, IFs often appear to run alongside microtubules as if they are associated in some manner. Crossbridges between the two structures have been seen, especially in sections through nerve axons, which tend to have a more ordered cytoplasm than most cells. There is other evidence for such an association, including the co-transportation of tubulin and NF proteins along axons, which is described in detail in Section 10.5. In tissue-culture cells, it has been observed that microtubule-disrupting agents such as colchicine or vinblastine (Section 7.4.6) cause the majority of IFs—though not keratin filaments—to collapse around the nuclear surface.

This suggests that the IF network may normally be supported in some way by the microtubule array. Is this a static interaction or does some type of interfilament sliding keep the IFs stretched out? The recovery of the keratin network in newborn mouse skin cells after calcium has been restored to the external medium, as mentioned above, suggests that IFs are able to travel along *something*. The most likely tracks for transport are microtubules. This could be due to direct sliding between IFs and microtubules; alternatively, IFs could be carried along microtubule tracks by static attachment to microtubules that slide relative to others (see Section 10.5).

2.5.3 Association of IFs with actin filaments[19]

IFs are often seen in association with microfilaments, as mentioned above in connection with the role of desmin in muscle. In non-muscle cells, IFs are found near regions of membrane interacting with the extracellular substrate via **attachment plaques** (see Chapter 4). It may be significant that trypsin cleavage of the proteins on the outer surface of the membrane in these regions releases cells from the substrate and also induces the retraction of IFs away from the plasma membrane. It has been suggested that the membrane-skeleton protein **ankyrin**, which attaches indirectly to actin filaments via spectrin (see Sections 4.4, 4.5 & 5.3.1), may also bind to the N-terminal head of vimentin or desmin molecules. There are probably other types of connection between IFs and actin polymers (**F-actin**), also.

The complete collapse of the IF network induced by microtubule-depolymerisation (Section 2.5.2) seems to be dependent on actin-associated motility; the movement towards the centre is inhibited in fibroblasts treated with both colchicine (to depolymerise microtubules) and cytochalasin (to depolymerize F-actin). Metabolic poisons inhibit this IF-network collapse during colchicine treatment, indicating that the movement also requires a supply of ATP. Hollenbeck and colleagues have suggested that, after losing their microtubule support, the IFs are carried back towards the rear of the cell by waves of contraction in the actin-rich cell cortex (see Section 6.4).

2.6   Functions of IFs[1,5,10]

The probable main function of the keratins has already been referred to, namely to contribute to the mechanical strength of an organism. The IF network holds the nucleus in position within each cell; also, in a tissue, the network of one cell is often connected to those in any neighbouring cells and to any neighbouring extracellular substrate. IFs help to produce structural continuity, especially in epithelial layers of cells (Figure 2.11).

There have been few real experiments to test the functions of IFs. In one case, the IF networks of mouse cells in tissue culture were found to collapse after microinjection of a monoclonal antibody against vimentin,[18] whereas control antibodies had no effect. Cells treated in this way moved and divided normally in culture. Similar conclusions have been made from observations on mutant cell clones that do not make IF protein. It seems, therefore, that IFs are *not essential* for the life of cells growing in the environment of a tissue-culture flask, where they are able to support themselves against a smooth plastic or glass surface. But the fact that IFs have been found in most cell types suggests that their role *is* important for cell survival in normal life. It has been suggested that they are most needed to strengthen the cytoskeleton during the development of multicellular organisms, when whole layers of associated cells undergo complex topological contortions.

There is also evidence that IFs provide a fairly specific support matrix for 'soluble' cellular organelles, such as free ribosomes. Anchorage to a filament system would reduce the likelihood of damage to growing polypeptides that might arise from shearing forces encountered in a cell under stress.

It is suspected that IFs may also have some other function, as yet unknown: their variety certainly seems excessive if they are merely strengthening the cytoplasm. Some cytokeratins are restricted to specific development lineages (e.g. the plantar epidermis of soles and palms) and are found there in embryos of all mammals, even though the ultimate fates of the epidermis are diverse (e.g. skin or hoof). Different copies of keratin genes might be expressed at different times during development in order to vary the level of protein synthesised. But genetic drift is probably not the cause of keratin diversity, since equivalent proteins in different species are more conserved than different keratins within the same species.

Neurofilaments, with their highly-charged large C-terminal domains, seem likely to have a specific function as a flexible filler in axons, as described above. Lamins and tektins are likely to act more as structural reinforcements than as space fillers. This may also be true for desmin and vimentin.

2.7   Summary

Coiled-coil proteins provide the cytoskeleton with considerable tensile strength. In muscle, for example, the thick filaments are assembled from coiled-coil myosin tails and the actin-containing thin filaments are supported by tropomyosin rods. The perimeters of such cells are supported by intermediate filaments. IFs also form a network between the nucleus and the plasma membrane in many cell types. The IF network probably connects to a shell of related proteins called lamins that encase the nucleus. At the plasma membrane, IFs often attach to intercellular junctions. The complete network probably holds and protects the cell nucleus, as well as supporting layers of cells to form tissues.

The type of IF found in the network depends on the type of cell. Keratins are found in epidermal cells, vimentin in a wide range of mesenchymal cells, desmin in muscle cells, and GFAP in non-neuronal brain cells. Neurofilaments, in nerve axons, probably support these narrow cell processes and protect the organelles within them, whilst still allowing axonal transport of the organelles. The network appears also to provide a supporting substrate for enzymic activity. Thus, it is probable that an important function is to help maintain the correct spatial order in each type of cell. All IF proteins are designed to form very stable filaments. They are further stabilised by associated proteins that form crosslinks between adjacent filaments. They can, however, be rearranged quite readily when necessary.

2.8   Questions

2.1 Compare the properties and functions of cytokeratins and neurofilaments. (It may help to have read Section 10.5 also.)
2.2 Outline current ideas of the substructure of 10 nm filaments. Why might some IFs be strictly bipolar filaments but others have at least some elements of polarity?
2.3 What obvious structural differences are there between IFs and myosin filaments?

2.9   Further reading

Some reviews

1 Klymkowsky, M. W., Bachant, J. B. & Domingo, A. (1989). Functions of intermediate filaments. *Cell Motil. & Cytoskel.* **14**, 309–331.
   (*Includes an extensive list of primary references.*)
2 Liem, R. K. H. (1990). Neuronal intermediate filaments. *Curr. Opinion Cell Biol.* **2**, 86–90.
3 Stewart, M. (1990). Intermediate filaments: structure, assembly and molecular interactions. *Curr. Opinion Cell Biol.* **2**, 91–100.
4 Osborn, M. & Weber, K. (1986). Intermediate filament proteins: a multigene family distinguishing major cell lineages. *Trends Biochem. Sci.* **11**, 469–472.
5 Geiger, B. (1987). Intermediate filaments: looking for a function. *Nature* **329**, 392–393.
6 Cohen, C. & Parry, D. A. D. (1986). α-Helical coiled-coils—a widespread motif in proteins. *Trends Biochem. Sci.* **11**, 245–248.
7 Steinert, P. M. & Roop, D. R. (1988). Molecular and cellular biology of intermediate filaments. *Annu. Rev. Biochem.* **57**, 593–626.
– Steinert, P. M. & Parry, D. A. D. (1985). Intermediate filaments: conformity and diversity of expression and structure. *Annul. Rev. Cell Biol.* **1**, 41–65.
8 Traub, P. (1985). *Intermediate Filaments: a review.* Berlin: Springer.
9 Steinert, P. M., Steven, A. C. & Roop, D. R. (1985). The molecular biology of intermediate filaments. *Cell* **42**, 411–419.
10 Goldman, R., Goldman, A., Green, K., Jones, J., Lieska, N. & Yang, H. (1985). Intermediate filaments: possible functions as cytoskeletal connecting

links between the nucleus and the cell surface. *Ann. N.Y. Acad. Sci., USA* **455**, 1–17.

11 Steinert, P. M., Jones, J. C. R. & Goldman, R. D. (1984). Intermediate filaments. *J. Cell Biol.* **99**, 22s–27s.

12 Osborn, M. & Weber, K. (1983). Tumor diagnosis by intermediate filament typing: a novel tool for surgical pathology. *Lab. Invest.* **48**, 372–394.

13 Lazarides, E. (1982). Intermediate filaments: a chemically heterogeneous developmentally-regulated class of proteins. *Annu. Rev. Biochem.* **51**, 219–250.

14 Franke, W. W. (1987). Nuclear lamins and cytoplasmic intermediate filaments: a growing multigene family. *Cell* **48**, 3–4.

15 Gerace, L. (1986). Nuclear lamina and organization of nuclear architecture. *Trends Biochem. Sci.* **11**, 443–446.

Some original papers

16 Aebi, U., Fowler, W. E., Rew, P. and Sun, T.-T. (1983). The fibrillar substructure of keratin filaments unraveled. *J. Cell Biol.* **97**, 1131–1143.

17 Aebi, U., Cohn, J., Buhle, L. & Gerace, L. (1986). The nuclear lamina is a meshwork of intermediate-type filaments. *Nature* **323**, 560–564.

18 Klymkowsky, M. W. (1981). Intermediate filaments in 3T3 cells collapse after intracellular injection of a monoclonal anti-intermediate filament antibody. *Nature* **291**, 249–251.
   (*Worth reading to see how different combinations of antibodies are used to provide controls, as well as for the main conclusion.*)

19 Hollenbeck, P. J., Bershadsky, A. D., Pletjushkina, O. Y., Tint, I. S. & Vasiliev, J. M. (1989). Intermediate filament collapse is an ATP-dependent and actin-dependent process. *J. Cell Sci.* **92**, 621–631.

Further references

20 Wang, E., Fischmann, D., Liem, R. & Sun, T.-T., eds (1985). Intermediate filaments. *Ann. N.Y. Acad. Sci., USA* **445**.

21 Parry, D. A. D. (1987). Fibrous protein structure and sequence analysis. In *Fibrous Protein Structure*, ed. Squire, J. M. & Vibert, P. J. London & San Diego: Academic Press.

22 Offer, G. (1987). Myosin filaments. In *Fibrous Protein Structure*, ed. Squire, J. M. & Vibert, P. J. London & San Diego: Academic Press.

23 Parry, D. A. D., Fraser, R. D. B., MacRae, T. P. & Suzuki, E. (1987). Intermediate filaments. In *Fibrous Protein Structure*, ed. Squire, J. M. & Vibert, P. J. London & San Diego: Academic Press.

24 Fuchs, E., Coppock, S., Green, H. & Cleveland, D. (1981). Two distinct classes of keratin genes and their evolutionary significance. *Cell* **27**, 75–84.

25 Ip, W., Hartzer, M. K., Png, S. Y.-Y. & Robson, R. M. (1985). Assembly of vimentin *in vitro* and its implications concerning the structure of intermediate filaments. *J. Mol. Biol.* **183**, 365–375.

26 Soellner, P., Quinlan, R. & Franke, W. W. (1985). Identification of a distinct soluble subunit of an intermediate filament protein: tetrameric vimentin from living cells. *Proc. Natl. Acad. Sci., USA* **82**, 7929–7933.

27 Inagaki, M., Nishi, Y., Nishizawa, K., Matsuyama, M. & Sato, C. (1987).

Site-specific phosphorylation induces disassembly of vimentin filaments *in vitro*. *Nature* **328**, 649–652.

28 Geisler, N. & Weber, K. (1988). Phosphorylation of desmin *in vitro* inhibits formation of intermediate filaments: identification of three kinase A sites in the aminoterminal head domain. *EMBO J.* **7**, 15–20.

29 McKeon, F. D. & Kirschner, M. W. (1986). Homologies in both primary and secondary structure between nuclear envelope and intermediate filament proteins. *Nature* **319**, 463–468.

30 McLachlan, A. D. & Karn, J. (1983). Periodic features in the amino acid sequence of nematode myosin rod. *J. Mol. Biol.* **164**, 605–626.

31 McLachlan, A. D. & Stewart, M. (1976). The 14-fold periodicity in α-tropomyosin and the interaction with actin. *J. Mol. Biol.* **103**, 271–298.

32 Goldman, R. D., Goldman, A. E., Green, K., Jones, J. C. R., Jones, S. M. & Yang, H.-Y. (1986). Intermediate filament networks: organization and possible functions of a diverse group of cytoskeletal elements. *J. Cell. Sci. Suppl.* **5**, 69–97.

33 Fey, E. G., Ornelles, D. A. & Penman, S. (1986). Association of RNA with the cytoskeleton and the nuclear matrix. *J. Cell Sci. Suppl.* **5**, 99–119.

3 ACTIN FILAMENTS

Actin, the most abundant protein of the cytoskeleton, is very highly conserved, with only small differences in sequence between the polypeptides produced, for example, by slime moulds and man.[1,6] This suggests that all parts of the globular actin monomer are involved in absolutely essential cellular functions. The 43 kD monomers of all types of actin assemble into filaments with a very precise structure, which has also remained constant throughout the evolution of eukaryotes.

Most organisms have more than one isotype, with differences in one or a few amino-acid residues; mammals, for example, have at least six actin genes, with subtle differences that tend to be conserved among different species. These variations may produce filaments with slightly different stabilities or strength of interaction with other proteins (such as myosin), and different isotypes are expressed in varying proportions in different cell types. In some cell types there does not appear to be any segregation of isotypes into filaments in different parts of the cytoplasm, but there is evidence for specialisation in other cells.

3.1 Structure

3.1.1 Filament symmetry

The filament is usually described as a two-stranded helix, with a repeat distance between **crossover** points of 36–40 nm. These long-pitch helices are right-handed, as shown in Figure 3.1. Two distinct single helices with shorter pitches can also be traced through successive monomers: a left-handed one with a 5.9 nm pitch and a right-handed one with a 5.1 nm pitch. The best way to see these features is to make a model (see page 44). They have been deduced by optical diffraction of electron micrographs of negatively-strained filaments; also from X-ray diffraction patterns of whole muscle and of aligned preparations of purified actin filaments.[4,26]

Long actin filaments in fixed cells appear to be very straight, which suggests that they are very resistant to bending forces. In contrast, there is evidence for significant torsional flexibility.[13] Electron microscopy of the filaments (e.g.

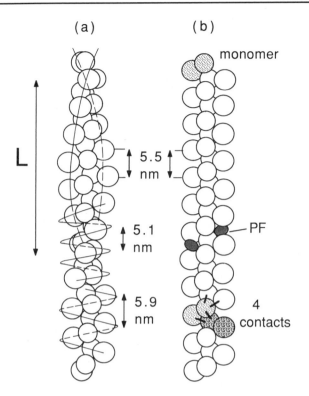

(a)   (b)

monomer

L

5.5 nm

5.1 nm

5.9 nm

PF

4 contacts

Figure 3.1

The arrangement of 43 kD monomers and their two structural domains in actin filaments. (a) Side view of a filament. Various helices can be drawn through equivalent points in the monomers; three helices of differing pitch are indicated. The shallow 5.9 nm-pitch, left-handed 'genetic' helix and the 5.1 nm-pitch, right-handed helix are fairly constant but the two long-pitch helices can vary significantly (L = 36–40 nm). (b) Arrangement of subunits in untwisted ribbons, with the two long-pitch helices completely straight—perhaps similar to actin-profilin crystals.[15] If, as the crystal structure suggests, the inner domains pack to form a single helical core, the outer domains must be mainly responsible for the obvious two-stranded appearance of F-actin (Figure 3.3). Possible bonds between domains of different monomers are indicated by the black bars for subunits near the bottom; each monomer seems to contact at least four different domains on neighbouring subunits. The position where profilin (PF) molecules are inserted in the crystal arrangement is indicated.

Figure 3.3) shows much variability in the crossover distance between the two nearly longitudinal strands of subunits. The model lattice has been drawn so that it repeats in the axial direction after exactly 28 monomers, corresponding to two 'crossovers' of the twisted strands. If the long-pitch helices are twisted a little more tightly, the crossover distance is slightly reduced and the helical lattice can be made to repeat exactly after only 13 monomers (a single crossover). Natural filaments assume a range of twists that includes these two cases. Individual filaments may have varying crossover distances along their lengths. Large changes in the crossover distance are thought to arise from only slight changes in the individual monomer configurations.

There is evidence from sections through rigor muscle, in which actin filaments may be distorted by their interactions with myosin heads, that F-actin can be untwisted, locally, to such an extent that the long-pitch helices actually run parallel to the axis (as in Figure 3.1b) over a short stretch.[19] Whether this ability to untwist has some special function in normal muscle contraction is still unknown.

3.1.2   Domains of the monomer

A variety of protein crystals containing actin monomers have been grown. Extended two-dimensional arrays of pure actin, formed in the presence of gadolinium ions, have been studied by electron microscopy and image-analysis methods.[17] There have also been X-ray crystallographic studies of actin, either complexed with DNAse I, of which actin is an inhibitor, or with profilin, an inhibitor of actin assembly (see section 4.2). The calculated structures each show

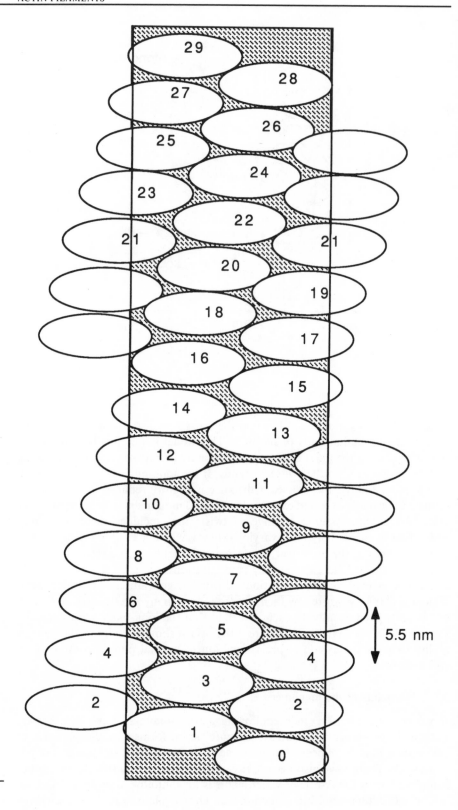

Model 3.1

A model of F-actin: Photocopy
this page (possibly with
some enlargement). Cut out the
rectangle, with an extra strip
down one side so it can be glued
to form a cylinder. To make a
longer 'filament', overlap the two
monomers of one copy with the
bottom two of another copy.

5.5 nm

a monomer consisting of two domains, one apparently slightly larger than the other.[15,16] The connection between the two domains is quite narrow and is thought to be fairly flexible.

X-ray crystallographic analyses are currently being improved to show the structure of actin monomers at atomic resolution.[28,29] Preliminary results suggest that one domain may consist of the N-terminal third of the polypeptide (see Figure 3.2) plus a short contribution from the C-terminal end, while the other domain is made up of the central section. This fits in with chemical crosslinking data, suggesting that in F-actin the two termini are near one another and both interact with other molecules such as myosin. ATP appears to bind in the cleft between the two domains. The orientation of this two-domain structure within an actin filament can be deduced from other, somewhat low-resolution information, outlined in the next section. However, the model should soon become reasonably accurate as crystal structures are improved and correlated in detail with chemical data showing which residues interact with other actin monomers or with myosin and other proteins.

3.1.3   Subunit interactions in filaments

Original attempts to analyse the subunit structure in filaments from electron micrographs produced some confusing results. The torsional flexibility of individual filaments makes them poor subjects for 3-D image reconstruction, which involves averaging data at periodic intervals along the axis. Most work was therefore carried out on actin **paracrystals** (see Figure 3.3b); the association between adjacent filaments holds them straighter and enforces a more constant crossover distance. Unfortunately, it now seems almost certain that the association involves some interdigitation of the outer domains, which were thus lost to a large extent from images of single filaments 'carved out' of the paracrystal image. There has, as a result, been some controversy over the orientation of the monomers in a filament.[4,26]

Detailed analyses of diffraction patterns derived either from electron-microscope images of individual filaments, or directly from X-ray diffraction of well-aligned filament preparations, now suggest that the long axis of the two-domain monomer structure lies roughly at right angles to the filament axis, as in Figure 3.1. The patterns show clear evidence for the protein density peaking at two distinct radial distances from the filament axis.[4]

Both types of actin-containing crystals analysed by X-ray crystallography now appear to be built up from ribbons of subunits that are essentially equivalent to untwisted filaments (as in Figure 3.1b). Detailed analysis of the actin–profilin crystals shows one domain located near the filament axis, the other towards the edge of the ribbon. Profilin, which is known to bind to the C-terminus, is associated with the domain on the outer edge of the ribbon. Additional evidence that the C-terminus lies at a high radius in F-actin is that subunits in different filaments can be crosslinked via their C-terminal residues to form dimers.[25]

Within a filament, contacts between different subunits in this orientation appear more extensive along the 'genetic' one-start helix than along the long-pitch strands, so the structure is perhaps best thought of as a zigzag. The intersubunit contacts seen in the crystal structure (Figure 3.1b) are consistent with results from a variety of chemical experiments on F-actin (see Figure 3.2b).[2]

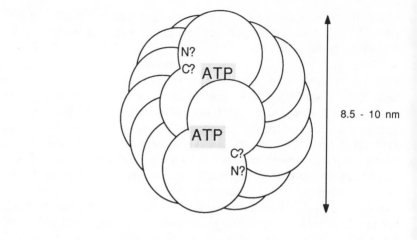

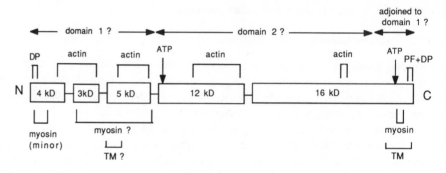

Figure 3.2

(a) View down the axis of the model filament shown in Figure 3.1a. (b) Map showing proteolytic sites (between boxes) in the actin monomer, the nucleotide-binding sites and regions that interact with other proteins, including other actin monomers.[2,23] ATP binds in a cleft between two globular domains.[16] Binding sites for myosin and tropomyosin (TM) must be exposed on the filament surface (see Figure 4.8). The binding of profilin (PF) or depactin (DP) blocks assembly of normal filaments (see Figure 4.4).

Conversion of a filament to a ribbon in an actin–profilin crystal corresponds to increasing the twist between neighbouring subunits along the helix by 13° and stretching them axially to bring the rise per subunit from 2.73 nm to 3.57 nm. The latter change presumably allows profilin to wedge into the gaps between adjacent C-terminal domains and does not seem to be a feature of normal untwisting.

The polarity of actin is not usually apparent in EM images of bare filaments but can be revealed at EM resolution by decoration with fragments of myosin, such as heavy meromyosin (HMM) or subfragment S-1, as shown in Figures 3.3 and 3.4. Myosin-fragment decoration produces the so-called arrowhead appearance, shown enhanced in Figure 3.4b; the two ends of the filaments are referred to as 'barbed' and 'pointed'. Decoration of individual filaments also improves their helical order so that 3-D image reconstruction is more reliable; recent results (summarised in Figure 4.8) confirm the overall orientation of the actin monomer shown in Figure 3.1.

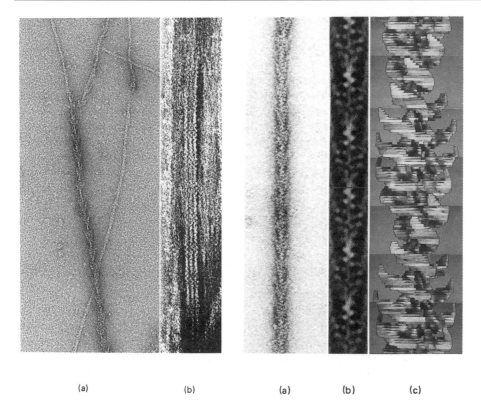

(a) (b) (a) (b) (c)

Figure 3.3

Electron micrographs of actin filaments, as seen in negative stain. (a) Filaments partially decorated with myosin subfragments (S-1 was used here). Decoration was achieved by incubating F-actin with S-1 in the absence of ATP. If the fragments are sufficiently concentrated, the filaments become decorated in a saturated fashion, with a fragment bound to each actin monomer. The decorated filaments were spun out of solution and incubated with G-actin, which polymerised onto the ends of the decorated 'seeds'. (b) An F-actin 'paracrystal', produced by leaving filaments overnight in a solution containing 20 mM $MgCl_2$. The filaments associate closely with one another, usually in antiparallel. [Micrographs taken by Dr Kenneth Taylor.]

Figure 3.4

Filaments decorated fully with myosin subfragments are as highly symmetrical as the underlying F-actin, so background noise can be reduced by averaging. The image in (a) was processed by computer to produce the 'filtered' image in (b). (c) A three-dimensional model of the structure, solved from images such as (a). Notice that individual S-1 molecules, protruding from the actin filament, are not as highly tilted as the strikingly polar arrowheads in the overall pattern they form. [Images from Taylor, K. A. and Amos, L. A. (1981). *J. Mol. Biol.* **147**, 297–324.]

3.2 Actin filament polymerisation

The assembly process is important not only for production of filaments with the correct structure for interaction with other components of the cytoplasm but also because controlled polymerisation or depolymerisation of actin filaments is required for many types of cellular activity. Some of these are discussed in Chapters 5 and 6.

3.2.1 Assembly *in vitro*

Actin filaments can be depolymerised *in vitro* by lowering the ionic strength of the medium to well below physiological levels, provided the protein concentration is not very high. The filaments (F-actin) then break down into monomers (G-actin). Spontaneous repolymerisation *in vitro* can be induced by adding suitable concentrations of salts (usually 0.1 M KCl and/or 1 mM $MgCl_2$) to a solution of G-actin in water. The presence of these ions changes the conformation of the monomer to a form that can assemble, in a process known as 'monomer activation'.

Magnesium ions bind specifically to the protein, displacing calcium ions. High concentrations of monovalent salts such as KCl may drive the protein to polymerise without binding to specific sites. The monomer also requires bound nucleotide (usually ATP) in order to polymerise. During assembly, ATP becomes hydrolysed to ADP. The significance of this is discussed in subsequent sections.

3.2.2 Purification of actin[6]

It is not practicable to purify actin by cycles of assembly and disassembly, as in the case of microtubule protein (Section 7.3.2), but a method of preparing reasonably pure actin has been known for a long time. It involves precipitating protein in acetone and drying it to a powder. Actin is relatively tough protein and is still active after such treatment, followed by being redissolved in an aqueous medium. Few of the contaminants survive the process. The actin may be purified further on suitable columns and can then be used for assembly experiments.

3.2.3 Methods of assaying assembly[3]

A solution of F-actin is very viscous because of interactions between filaments. Individual interfilament bonds may be very weak but the combined effect of many such bonds becomes significant. Paracrystals formed in the presence of divalent cations (Figure 3.3b) result from orderly series of interactions. The viscosity of actin filaments is probably a major factor in determining the consistency of cytoplasm in most cells. Also, its viscosity increases markedly when actin crosslinking proteins (Section 4.4) are added, making **viscometry** a useful technique for assaying interactions between actin and possible crosslinking factors.

Viscometry has been used to follow the assembly of G-actin into filaments but is not really reliable in experiments where the number of filaments is important, since the shearing forces involved may fragment long filaments. A spectroscopic method is now more popular, whereby fluorescence emitted by a dye bound to the protein as it polymerises can be quantitatively measured. Painstaking electron microscopy provides the most detailed information. Even if assembly is followed in bulk by spectrofluorimetry, the concentration of free ends must be estimated from electron micrographs in order to calculate subunit on- and off-rates.

3.2.4 Assembly kinetics[1,3,5,6]

The dynamics of actin filament assembly *in vitro* from actin subunits, and the control of this process, have been studied in some detail. Assembly curves such

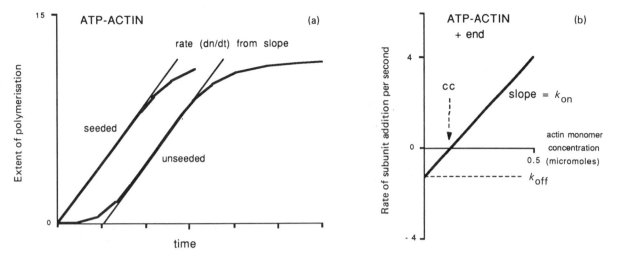

as that shown in Figure 3.5a can be explained by a simple theory, developed initially by Oosawa and Asakura: the basic assumptions are that subunits add only to filament ends and that the properties of the ends are independent of the lengths of the filaments.[12] Although the rate at which monomers add on to an end will depend on how many monomers are available for interaction, the rate at which they dissociate will be constant. The net rate of assembly at one end will be:

$$dn/dt = k_{on}c - k_{off} \qquad \text{Eqn. 3.1}$$

where c is the concentration of monomers, k_{on} the rate per mole at which monomers associate with the end of a polymer, and k_{off} the average rate at which end-monomers dissociate. The overall rates of assembly and disassembly in a bulk sample will, of course, vary according to the number of ends; a solution of short filaments will change more rapidly than one containing the same amount of protein in the form of longer filaments.

At equilibrium, there should be no net assembly and:

$$dn/dt = 0, \ c = cc = k_{off}/k_{on} \qquad \text{Eqn. 3.2}$$

This concentration of monomers is known as the **critical concentration** for elongation (cc).

Growth can be measured at a range of monomer concentrations, to give plots like those in Figure 3.5b. You can see from Eqn. 3.1 that the on-rate constant at a given end, k_{on}, is the value of the slope of the line. The off-rate constant, k_{off}, is obtained by extrapolating the line to cross the vertical axis; it corresponds to the rate at which subunits would be lost if the filaments were diluted so much that there were effectively no monomers present in the solution to add on to filament ends. Alternatively, k_{off} can be calculated from Eqn. 3.2; the critical concentration, cc, is the value where the line crosses the horizontal axis.

Figure 3.5b shows values for the barbed ends of F-actin, measured by electron microscopy as described in the next section. There are different rate constants for the pointed end, since the processes of associating or dissociating are not quite the same at the two ends (see Figure 3.8).

Figure 3.5

(a) Actin assembly from monomer solutions shows a lag time during which initiating complexes are formed. Thereafter, the polymers grow steadily until the monomer concentration begins to approach the critical concentration and the extent of assembly finally reaches a plateau. If short filaments are added to seed assembly at the start, there is no time lag before the onset of steady growth. The slope of the steadily rising part of the curve is greater for higher actin monomer concentrations, indicating a higher assembly rate. (b) If seeded growth is measured by electron microscopy, the assembly rates at a range of different monomer concentrations can be plotted for a particular type of end. The slope of the straight line obtained for each end gives a value for the rate constant k_{on} at that end. Extrapolation of the line provides values for the constant off-rate and the critical concentration (as explained in the text). At monomer concentrations below the critical concentration, points along the line represent the net loss of subunits from the end.

Figure 3.6

Actin polymerisation onto labelled 'seeds'. When short lengths of filament are decorated with fragments of myosin, such as HMM or S-1 (as in Figure 3.3a) and then incubated further in solutions containing actin monomers, fresh growth onto either end is undecorated. The ends can be distinguished from the polarity of the arrowhead decoration. Seeds incubated in very low monomer concentrations show no growth though they are stabilised against shrinkage, to some extent, by the decorating material. At high concentrations of actin monomer, growth occurs at both ends, the 'barbed' growing faster than the 'pointed' end. At intermediate concentrations, only the barbed end elongates.

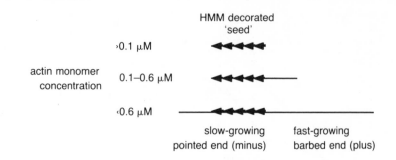

3.2.5   Initiation of assembly[3]

When monomers are first converted to a conformation that can assemble, there is a delay in the appearance of polymer (Figure 3.5a); thereafter, assembly follows a curve that can be predicted by the theory. The formation of a suitable **initiation complex**, nucleus or seed, for a filament is a more demanding process than elongation. The association between a lone pair of monomers is weaker than that in large complex; to initiate a filament they must stay associated as a dimer long enough for a third monomer to bind, and the trimer must persist long enough to form a tetramer. Thus initiation requires a series of unlikely events, which become more likely as the complex increases in size. The critical concentration for initiation is higher than for elongation at either end.

The initial delay can be abolished by 'seeding' with short fragments of pre-assembled actin filaments. The two ends of a filament can be distinguished by using seeds decorated with HMM, as in Figure 3.3a. It was established in this way that polymerisation occurs more readily at the barbed end than at the pointed end (Figure 3.6). Under optimum conditions, the critical concentration is about 0.6 μM for the pointed end, 0.1 μM for the barbed end. Other structures, such as fragments of *Limulus* acrosomal filaments (see Section 4.4.7), used as seeds, also allow the experimenter to distinguish the original seed from filaments grown off the ends. In all cases, the barbed ends can be identified by their relatively more active behaviour. Axonemes are used in a similar way to study microtubule elongation.

3.2.6   Treadmilling[8,9]

A difference in critical concentrations at its two ends allows a polymer to have a property known as **treadmilling**. At an intermediate concentration of actin monomer (about 0.2 μM in Figure 3.7a), subunits will tend to assemble onto the barbed end and disassemble from the pointed end. At *equilibrium* the net gain at the average barbed end will equal the net loss at an average pointed end. Treadmilling cannot occur in an ordinary polymer made up from identical subunits as this would be a system in perpetual motion. Ordinary polymers and crystals must have the same critical concentration at both ends, even if the individual on- and off-rates differ.

3.2.7   ATP hydrolysis[10,21]

Why does actin seem to disobey the law forbidding perpetual motion? It is possible for actin to have different critical concentrations at the two ends of a

Table 3.1
Association and dissociation rates for actin

	ATP–actin		ADP–actin	
	barbed end	pointed end	barbed end	pointed end
k_{on} ($\mu M^{-1} s^{-1}$)	11.6	1.3	3.8	0.16
k_{off} (s^{-1})	1.4	0.8	7.2	0.27
cc (μM)	0.12	0.6	1.9	1.7

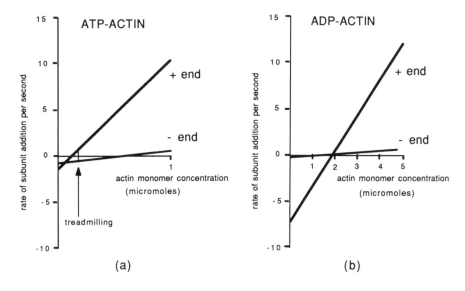

Figure 3.7

Growth rates plotted against actin monomer concentration as in Figure 3.5b, for both barbed (+) and pointed (−) ends. (a) Actin monomers started out with bound ATP. The critical concentration at the plus end is much lower than at the minus end. At a particular intermediate concentration, the rate of growth at the plus end should exactly equal the rate of loss from the minus end; the resulting flux of subunits is known as treadmilling. (b) Results obtained for actin monomers with bound ADP. The greatest difference from (a) is in the off-rate at the plus end, which leads to a large increase in critical concentration at this end. Critical concentrations are the same at each end; treadmilling is not possible in the absence of ATP hydrolysis. [Data from Pollard, T. D. (1986). *J. Cell Biol.* **103**, 2747–2754.]

filament because ATP is hydrolysed during polymerisation and this is a source of energy. Actin monomers in solution bind ATP, which is broken down to ADP and phosphate (P_i) either at the same time as the monomer joins the end of a filament or some time later. Hydrolysis of the ATP on an end monomer may be dependent upon another monomer binding to *it*. The P_i cleaved from the nucleotide is thought to remain bound for a while but to be released eventually by a kinetically slow process. ADP remains firmly bound to polymeric actin. When a monomer eventually returns into solution, the ADP is readily lost and replaced by a new molecule of ATP.

$$\text{Actin.ATP} \rightleftharpoons \text{Actin.ADP.}P_i \rightleftharpoons \text{Actin.ADP} + P_i$$

It can be seen that the subunits in a filament can be in one of at least three different states; there may be intermediate states if the energy of hydrolysis and the phosphate are lost in a number of stages. A subunit going through these stages will tend to do so with neighbouring subunits above and below it in different states, depending on which end it has added to (see Figure 3.8). Thus, it is not surprising that the net effects are unequal at the two ends.

Table 3.1 shows some values measured by Pollard for actin monomer rate constants with bound ATP or ADP, using electron microscopy as the assay method.[18] The corresponding graphs are shown in Figure 3.7. These values were obtained under buffer conditions that maximised polymerisation. If the composition of the medium is varied, all the rates can change, so none of the

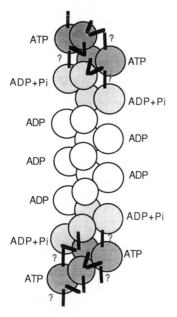

Figure 3.8

After a monomer with bound
ATP adds to the end of a
filament, the ATP is hydrolysed to
ADP and phosphate (P_i).[10] It is
not known whether hydrolysis
occurs immediately, after a certain
time lag, or as a consequence of
the addition of further monomers
on to the end. (See also Section
7.6 on tubulin assembly.) The
weakly-bound free phosphate is
thought to be lost subsequently. A
subunit adding to any end is
converted from a G to an F
conformation by interactions with
sites on its surface that differ at
the two ends, so plus-end and
minus-end rates are not the same.
But, without hydrolysis (as in the
assembly of ADP-actin),
dissociation is the reverse of
association at a particular end
and the on- and off-rates are
related, giving equal critical
concentrations at each end (see
Table 3.1).

rates shown in Table 3.1 is necessarily the same as would be found *in vivo*. Nevertheless, a set of measured rates such as this suggests that the simple theory is adequate for actin.

ATP–actin associates more rapidly than ADP–actin at either end but at the barbed end it is slower to dissociate than ADP–actin, which leaves this end quite rapidly. Thus, *in vitro*, above a free monomer concentration of 0.12 μM, addition is more probable than loss as long as the end is 'capped' by an ATP–actin subunit. But if the cap is lost, exposing an ADP–actin end, loss becomes five times more probable, until the end is rescued by the addition of an ATP–actin monomer. There may, as a result, be minor fluctuations in length at the barbed end under equilibrium conditions, but they will be negligible compared with the dynamic fluctuations that occur at the ends of microtubules (Section 7.6).

The pointed end is relatively stable, addition and loss of ATP–actin or ADP–actin being slower than at the barbed end. But the critical concentrations are significantly higher, so at steady state a net loss of subunits should occur at the pointed end. There should be a very slow flux of subunits along the filament from the barbed end towards the pointed end.

Severing and reannealing of actin filaments will have an effect on their assembly behaviour. The breakage of a filament will expose ADP-subunits and prompt a run of subunit loss. Severing and annealing also have an effect by changing the number of filament ends. This is why the method of assaying polymerisation should disturb the filaments as little as possible.

ATP-hydrolysis has a vital role in the dynamic behaviour of actin. You will see from the high on-rate (Table 3.1) that monomers to which ATP is bound are primed for rapid assembly onto barbed ends. Once incorporated in a polymer and having hydrolysed their nucleotide to ADP, their high *off-rate* now produces a potentially unstable polymer. Thus, inside a cell, filaments may be assembled rapidly when required but a small change in conditions may induce disassembly instead. *In vivo*, filament ends appear to be modified and further controlled by an array of associated proteins which may stimulate or block assembly, may promote severing or stabilise filaments.

Similar behaviour, including treadmilling, has been suggested for microtubules, which hydrolyse GTP during assembly. But, as described in Chapter 7, microtubules assembled from pure tubulin have similar critical concentrations at both ends, despite differences in individual on- and off-rates, so treadmilling is negligible in this case. However, critical concentrations are greatly affected by the presence of microtubule-associated proteins (MAPs) and treadmilling may occur in MAP-containing microtubules. Whether MAPs are present or not, the sporadic loss of large numbers of subunits from either end has a much more important effect on microtubule behaviour (see Section 7.6).

3.3   Drugs affecting actin assembly[7]

Certain fungi produce heterocyclic compounds (**cytochalasins**) which bind to actin filaments and interfere with assembly. When added to a solution of F-actin they have the effect of reducing its shear viscosity. This is because they block the elongation of barbed ends and even cause severing of filaments. In this respect, they resemble some of the actin-severing proteins, such as **gelsolin**, described in Section 4.2. Growth at the pointed end is unaffected.

In a solution of G-actin they may have the apparently opposite effect of nucleating filament assembly. This is because binding to the barbed end of a small complex stabilises it, allowing elongation from the pointed end.

Since cytochalasins are able to get into living cells from the external medium, they are very useful for research into the functions of actin. When introduced, they produce complex disruptive, but usually reversible, effects on the actin network of cells. They are also used as agents for producing multinucleate cells since they inhibit actin from producing a cleavage furrow. Inexplicably, they may also produce enucleation of cells.

A drug with the opposite effect, **phalloidin**, can be obtained from a poisonous mushroom. This binds specifically to F-actin and inhibits its disassembly. It too is useful for research into function, though microinjection is necessary to introduce it into a cell. In addition, when labelled fluorescently, it makes a superbly specific reagent for detecting actin filaments, without any background from G-actin in the surrounding cytoplasm.

3.4 Summary

Actin is a highly conserved protein that assembles into identical filaments in all known cells. ATP is hydrolysed as part of filament polymerisation, which means that the process of disassembly is not the reverse of assembly; the types of subunit involved in the two types of event tend not to be the same. The uncoupling allows the critical concentration at the ends of a filament to be different, making treadmilling possible. These properties allow actin to be a moderately dynamic polymer so that cells can rearrange their filaments without drastic changes in the medium.

3.5 Questions

3.1 What makes treadmilling possible for actin filaments?

3.2 Calculate whether monomer addition or loss occurs at each of the two ends of an average actin filament, using Eqn. 3.1 and the rates shown in Table 3.1, for ATP-actin monomer concentrations of $1.0\,\mu M$, $0.62\,\mu M$, $0.5\,\mu M$, $0.17\,\mu M$ and $0.1\,\mu M$. If filaments were *severed*, what would happen to the exposed ADP-actin ends at these concentrations of ATP-actin?

3.3 If the addition of each subunit to the end of a filament increases its length by 2.8 nm, what concentrations of free ATP-actin would account for plus-end extension rates *in vivo* of 0.1 $\mu m/s$ and 1 $\mu m/s$ (see Table 1.1, page 16; relevance discussed in Chapter 6).

3.6 Further reading

Reviews

1 Pollard, T. D. & Cooper, J. A. (1986). Actin and actin-binding proteins. A critical evaluation of mechanisms and functions. *Annu. Rev. Biochem.* **55**, 987–1035.

2 Hambly, B. D., Barden, J. A., Miki, M. & dos Remedios, C. G. (1986). Structural and functional domains on actin. *Bioessays* **4**, 124–128.

3 Frieden, C. (1985). Actin and tubulin polymerization. *Annu. Rev. Biophys. Biophys. Chem.* **14**, 189–210.

4 Egelman, E. H. (1985). The structure of the actin thin filament. *J. Muscle Res. & Cell Motil.* **6**, 129–151.

5 Oosawa, F. (1983). Macromolecular assembly of actin. In *Muscle and Non-Muscle Motility*, vol. I, pp. 151–152. New York: Academic Press.

6 Korn, E. D. (1982). Actin polymerization and its regulation by proteins from nonmuscle cells. *Physiol. Rev.* **62**, 672–737.

7 Cooper, J. A. (1987). Effects of cytochalasin and phalloidin on actin. *J. Cell Biol.* **105**, 1473–1478. (*Mini-review.*)

8 Cleveland, D. W. (1982). Treadmilling of tubulin and actin. *Cell* **28**, 689–691. (*Out of date for tubulin, but still appropriate for actin.*)

9 Neuhaus, J.-M., Wanger, M., Keiser, T. & Wegner, A. (1983). Treadmilling of actin. *J. Muscle Res. & Cell Motil.* **4**, 507–527.

10 Carlier, M.-F. (1989). Role of nucleotide hydrolysis in the dynamics of actin filaments and microtubules. *Int. Rev. Cytol.* **115**, 139–170.

11 Sheterline, P. (1983). *Mechanisms of Cell Motility.* London & New York: Academic Press.

12 Oosawa, F. & Asakura, S. (1975). *Thermodynamics of the polymerization of proteins.* New York: Academic Press.

Additional references

13 Egelman, E. H., Francis, N. & DeRosier, D. J. (1982). F-actin is a helix with a random variable twist. *Nature* **298**, 131–135.

14 Bullitt, E. S. A., DeRosier, D. J., Coluccio, L. M. & Tilney, L. G. (1988). Three-dimensional reconstruction of an actin bundle. *J. Cell. Biol.* **107**, 597–611.

15 Schutt, C. E., Lindberg, U., Myslik, J. & Strauss, N. (1989). Molecular packing in profilin:actin crystals and its implications. *J. Mol. Biol.* **209**, 735–746.

16 Suck, D., Kabsch, W. & Mannherz, G. (1981). Three-dimensional structure of the complex of skeletal muscle actin and bovine pancreatic DNase I at 6Å resolution. *Proc. Natl. Acad. Sci., USA* **78**, 4319–4323.

17 Aebi, U., Fowler, W. E., Isenberg, G., Pollard, T. D. & Smith, P. R. (1981). Crystalline actin sheets: their structure and polymorphism. *J. Cell Biol.* **91**, 340–351.

18 Pollard, T. D. (1986). Rate constants for the reactions of ATP- and ADP-actin with the ends of actin filaments. *J. Cell Biol.* **103**, 2747–2754.

19 Taylor, K. A., Reedy, M. C., Cordova, L. & Reedy, M. K. (1984). Three-dimensional reconstruction of rigor insect flight muscle from tilted thin sections. *Nature* **310**, 285–291.

20 Bonder, E. M., Fishkind, D. J. & Mooseker, M. S. (1983). Direct measurement of critical concentrations and assembly rate constants at the two ends of an actin filament. *Cell* **34**, 491–501.

21 Korn, E. D., Carlier, M.-F. & Pantaloni, D. (1987). Actin polymerization and ATP hydrolysis. *Science* **238**, 638–644.

22 Otey, C. A., Kalnoski, M. H. & Bulinski, J. C. (1987). Identification and quantification of actin isoforms in vertebrate cells and tissues. *J. Cell Biochem.* **34**, 113–124.

23 Bertrand, R., Chaussepied, P., Kassab, R., Boyer, M., Benjamin, Y. & Roustan, C. (1987). Cross-linking of the skeletal myosin subfragment-1 heavy chain to the NH_2-terminal region of actin within residues 40–113. *J. Muscle Res. & Cell Motil.* **8**, 70.

24 Gaertner, A., Ruhnau, K., Schroer, E., Selve, N., Wanger, M. & Wegner, A. (1989). Probing nucleation, cutting and capping of actin filaments. *J. Muscle Res. & Cell Motil.* **10**, 1–9.

25 Millonig, R., Salvo, H. & Aebi, U. (1988). Probing actin polymerization by intermolecular crosslinking. *J. Cell Biol.* **106**, 785–796.

26 Amos, L. A. (1985). Structure of muscle filaments studied by electron microscopy. *Annu. Rev. Biophys. Biophys. Chem.* **14**, 291–314.

27 Taylor, K. A. & Amos, L. A. (1981). A new geometry for the geometry of the binding of myosin crossbridges to muscle thin filaments. *J. Mol. Biol.* **147**, 297–324.

Late additions

28 Kabsch, W., Mannherz, H. G., Suck, D., Pai, E. F. & Holmes, K. C. (1990). Atomic structure of the actin:DNase I complex. *Nature* **347**, 37–44.

29 Holmes, K. C., Popp, D., Gebhard, W. & Kabsch, W. (1990). Atomic model of the actin filament. *Nature* **347**, 44–49.

(*The new data in these two papers should resolve some of the uncertainties in Figures 3.1 and 3.2.*)

4 ACTIN-ASSOCIATED PROTEINS

The properties of purified actin are greatly modified if certain associated proteins are added. Many such proteins have now been isolated (categorised as far as possible in Tables 4.1 & 4.2): it seems probable that almost all cells contain dozens of proteins that interact with actin, often competitively. The properties of some of the most intensively studied ones will be described. Others tend to have similar properties but may have them in different combinations.

4.1 Calcium-binding proteins[49-52,40]

Many actin-associated proteins are controlled by calcium, so it is appropriate to consider the known structure of some highly specific Ca^{2+}-binding sites. A type of site first identified by X-ray crystallographic studies of **parvalbumin** (a soluble protein that buffers calcium in fish muscle) consists of a loop between two α-helices at right angles (Figure 4.1a). The configuration is called an **EF hand**, simply because the two helices forming one of the two Ca^{2+}-binding sites in parvalbumin had been designated E and F, and are perpendicularly oriented like the thumb and forefinger of a hand. Ca^{2+} fits into the loop and binds to six coordinating positions on residues at specific intervals along the polypeptide. Homologous stretches of sequence have been found in many high-affinity Ca^{2+}-binding proteins. A site with optimal residues in all the coordinating positions is occupied by a calcium ion when the free Ca^{2+} concentration is above 10^{-6} M. Some variations produce weaker binding sites.

4.1.1 Calmodulin[40,51,52]

This ubiquitous 17 kD protein, once called CDR (calcium-dependent regulator), contains four EF hands. They are arranged in pairs so that the molecule, as crystallised, has two globular domains connected by a long α-helix (Figure 4.1b). The pair in one domain are high-affinity sites, those in the other have a lower affinity for Ca^{2+}. The way that a pair of EF hands interact with each

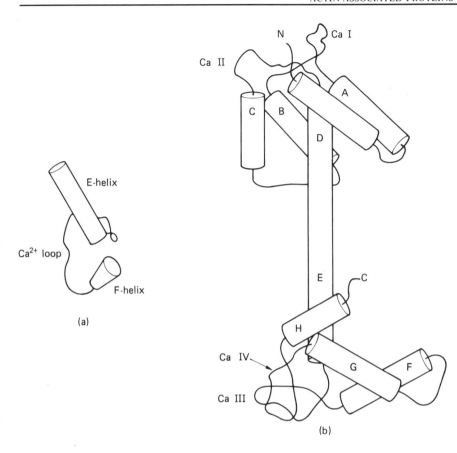

(a)

(b)

Figure 4.1

(a) An EF-hand structure. The
two α-helices are represented as
cylinders. The Ca²⁺-binding site is
in the loop between the helices.
(b) The three-dimensional
structure of troponin-C, as
crystallised. [Based on Herzberg,
O. and James, M. N. G. (1985).
Nature **313**, 653–659; and
Sundralingham *et al.* (1985).
Science **227**, 945–948.] The crystal
structure of calmodulin is very
similar. Two pairs of EF hands
are separated by a long central
helix to produce a dumbbell
shape. There is evidence that
calmodulin assumes a compact
globular structure in solution,
probably through bending of the
central helix (note how the
α-helical rod of myosin is able to
bend—Figure 4.5). All the
members of the calmodulin-like
family of proteins may adapt their
overall conformations to suit the
enzymes to which they bind.

other is altered by Ca²⁺-binding; the resulting conformational change allows
calmodulin to bind to and activate a large variety of enzymes, including the
kinase that phosphorylates smooth-muscle and non-muscle myosin regulatory
light chains (RLC) near their N-termini and thereby controls their activity, as
described in Sections 4.3 and 5.5.

Troponin C, the molecule that responds to calcium in vertebrate striated
muscle and initiates the changes that allow myosin heads to interact with F-actin
(see Figure 4.12), has very similar properties and crystal structure to
calmodulin.[49,50]

Myosin light chains are more weakly homologous to calmodulin, mostly
binding calcium to only one of their four potential EF hands, and fairly weakly
even here (Section 4.3).

4.2   Assembly-controlling proteins[1–5,7,8,61]

Proteins that control the assembly state of actin were originally classified
separately as **capping proteins**, **monomer-sequestering proteins** and **depolymerising
factors**. It may be more appropriate to regard these activities as different aspects
of a single mechanism, since most of the proteins exhibit two or all three of
these properties.

Table 4.1
Proteins that bind to actin

Name	Polypeptides (kD)	Sources	Molecules
Motor proteins			
Myosin II	175–220, LCs = 25, 22	Muscle, non-muscle	160 nm dimer
Myosin I	125–135, LC = 17–27	Acanthamoeba, Dictyostelium	30 nm monomer
Brush border 110 kD	110–119	Vertebrate gut	30 nm monomer
Proteins controlling F-actin interactions (with myosin, etc.)			
Tropomyosin	35	Muscle and non-muscle	40 nm dimer
	30	Non-muscle	35 nm dimer
Troponin I, T, C	30	Striated muscle	Complex
Caldesmon	140?	Smooth muscle	
	70–90	Non-muscle	Ca^{2+}-sensitive bundling?
Limulus 50 kD	50	Acrosomal process	5 nm rod
Actin-membrane proteins			
Ponticulin	17	Dictyostelium	Integral to membrane
Dictyostelium 24 kD	24	Dictyostelium	Monomers—also bind lipid and spectrin
Calpactin I, II (calelectrins, annexins)	30–40	Vertebrate cells	
Isotropic network-forming proteins (gelation factors)			
Spectrin family	220, 240	Red blood cell	196 nm tetramer
(Fodrin)	235, 240	Brain, avian lens	200 nm tetramer
(TW260/240)	240, 260	Intestinal epithelium (incl. brush border)	260 nm tetramer
	260	Acanthamoeba	80–230 nm rod
Filamin	250–270	Smooth muscle, avian gizzard, myocytes	160 nm dimers
Actin-binding protein		Platelets, macrophages, Xenopus oocytes, HeLa	
HMWP	230	Physarum	60 nm strands
120 kD gelation factor	120	Dictyostelium	35 nm rod (homodimer)
Bipolar filament bundling proteins			
α-Actinin family	95	Striated and smooth muscle	48 nm dimer
	105	Vertebrate non-muscle cells (incl. brush border)	48 nm dimer
	95	Sea-urchin eggs	48 nm dimer
	95	Dictyostelium	48 nm dimer
	85	Acanthamoeba	?
(actinogelin)	115	Liver, ascites cells	?
Polar filament bundling proteins			
Fascin	57	Sea-urchin eggs	6 nm crosslinks?
	53	Porcine brain	(probably monomer)
	57	Starfish sperm	(probably monomer)

Protein	kD	Location	Structure
Limulus crosslink	60	Acrosomal process	(globular dimer)
Fimbrin	68	Intestinal epithelium and many other cells	globular monomer?
Villin	95	Intestinal epithelium, toad oocytes	Ca^{2+}-variable rod (monomer)
30 kD	30	*Dictyostelium* filopodia	Monomer
Other bundling proteins			
Band 4.9	48, 52	Erythrocytes	Monomer
MAP 2	200	Neurons	100–180 nm monomer
Synapsin I	76, 78	Presynaptic junctions	Heterodimer?
Aldolase		Vertebrate muscle	
Pointed (−) end-blocking proteins			
β-Actinin	35–37	Kidney, skeletal muscle	Monomer
Simple barbed (+) end-blocking proteins			
Capping protein	28–31	*Acanthamoeba*	Heterodimer
cap Z	32, 34	Muscle	Heterodimer
HA1	?		?
ADP-ribosylated actin		Actin modified by bacterial toxins	
Barbed end blocking and severing proteins			
Gelsolin	90	Mammalian cells	Monomer
Villin	95	Mammalian and avian epithelia, amphibian eggs	Monomer rod
Fragmin (Cap 42b)/Severin	40–45	*Physarum, Dictyostelium,* sea-urchin eggs	Monomer rod
...not part of cytoskeleton—secreted form of gelsolin			
Brevin	93	Blood plasma	
Monomer-sequestering proteins			
Profilin	12–15	Most cells	Monomer
Actobindin (caps oligomers)	12	*Acanthamoeba*	Dimer
Monomer-sequestering and filament-severing proteins			
Depactin	18	Starfish oocytes	Monomer
19 kD protein/ADF/Cofilin	19–21	Vertebrate brain	Monomer
Destrin	19	Kidney	Monomer
Actophorin	14	*Acanthamoeba*	Monomer
...probably irrelevant to living cytoskeleton			
Vitamin D-binding protein (Gc-globulin)		Blood plasma	
DNase I		Bovine pancreas	

Table 4.2 **Some proteins secondarily-** **associated with F-actin**	***Protein***	***Function/location***
	Calmodulin	Ubiquitous Ca^{2+}-dependent regulator (CDR)
	Myosin light-chain kinase	Regulates smooth muscle and non-muscle myosins by phosphorylation
	Talin	Links vinculin to membrane proteins at adhesion plaques
	Vinculin	Binds to α-actinin, may also bind to actin
	Ankyrin Band 4.1 (= fibroblast 80 kD protein; related to synapsin)	} Bind to spectrin in red blood cells and elsewhere
	Titin (connectin) Nebulin	} Two very large proteins with elastic properties; hold Z-discs and myosin bands together in muscle

4.2.1 Capping activity

Control of actin filament length *in vivo* may be achieved by selective blocking of one or other of the two ends. Many proteins with appropriate *in vitro* effects have been discovered. If the cytoplasmic G-actin concentration is midway between the two critical concentrations (see Section 3.2), a protein which 'caps' or blocks the barbed end will cause shortening, while one which caps the pointed end will cause elongation. Note that with such controls one set of actin filaments could be growing whilst another set in the same environment was in the process of disassembly. This is probably an important requirement for the structural rearrangements that are constantly occurring in the cytoplasm, especially in the case of highly motile cells.

Pointed-end capping proteins seem to form a separate class from the other proteins controlling assembly and disassembly but have not, as yet, been well characterised. They should be especially useful in situations where filaments need to be maintained intact in the presence of rather low concentrations of actin monomer, as in muscle. **β-Actinin**, which can be isolated from skeletal muscle and elsewhere, has a molecular weight similar to that of actin.

4.2.2 Barbed-end capping proteins

Caps at the plus ends of actin filaments probably act to stop them from growing indefinitely, since the level of free actin found in most cells is well above the plus-end critical concentration. They can also act as nucleation sites for filament growth when the actin monomer level is greater than the minus-end critical concentration. Barbed-end capping activity has been found in actin extracts from a variety of cell types, often in the form of a heterodimer of 32–34 kD protein subunits.

Vinculin, which is included in the dense plaques at adhesion zones (see Sections 4.5 & 5.6.2), was originally classed as a barbed-end capping protein. But at least some of the capping activity has since been shown to belong to contaminants of vinculin preparations. These contaminating polypeptides have been isolated and are referred to as HA1. There is evidence that these small proteins (∼20 kD) may be proteolytic products of a much larger (150 or 200 kD) polypeptide, which has not yet been characterised. Vinculin itself binds only weakly to actin under conditions *in vitro*.

4.2.3 Severing activity

All known actin-filament severing proteins are calcium-binding proteins. Although they are activated by calcium-ion concentrations at a similar low level to that required by calmodulin (μM), there is no evidence of EF-hand-like stretches of sequence, so the calcium-binding sites may be of a different class.

4.2.4 Gelsolin[2-4]

This well-studied protein is both a **barbed-end capping protein** and a calcium-activated **filament-severing agent**. It is found in the cytoplasm of many vertebrate cells and is also secreted into blood, where it may help to destroy actin released by dead cells. After cleavage of the leader sequence that enables it to be secreted, plasma gelsolin is identical to cytoplasmic gelsolin; transcription of messenger RNA is thought to be from the same gene in each case, but initiated at two different sites.

An important role for cytoplasmic gelsolin has been identified in the crawling types of cell motility (as described in Section 6.2), in which regions of gelated actin need to be solated before the cell can move forward. It has been shown for some cells that an initially low expression of gelsolin is increased up to fifty times after the cells differentiate to a flat spreading form, while the levels of actin and other associated proteins may only double. Antibody labelling has localised gelsolin at the cell periphery, in the central cytoplasmic mass, and on thin fibres (microspike actin bundles?—Figure 6.3) radiating from the central mass. It is not found in stress fibres. When activated by calcium ions, gelsolin is thought to sever actin filaments into short lengths and then to cap the barbed end, thus preventing further growth or annealing (see Figure 4.2). Vesicles consisting of inositol lipids, such as phosphatidyl inositol bisphosphate (PIP_2), can bind to the protein and inhibit the severing activity. The possible relevance of this is covered in Section 6.4.

The sequence of gelsolin seems to have arisen by a process of gene duplication of a small segment, with modifications that result in a rather complex molecule with a varied repertoire of behaviour. Segments on either side of a short central link are closely related and within each half a conserved 50-amino-acid motif is repeated three times (see Figure 4.3). Points at the end of the central link and also between the first and second of the six segments are easily proteolysed, suggesting there are three *functional* domains. The small first domain binds by itself to G-actin and has properties resembling those of profilin (Section 4.2.8). The larger second domain binds to F-actin, apparently to the sides of filaments. The third and largest domain binds on its own only to monomers. Both the first and second domains are required for filament-severing but without the third domain they are active whether Ca^{2+} is present or not.

4.2.5 Proteins related to gelsolin[2,4]

Under normal cytoplasmic conditions, **villin** is an actin-bundling protein which assists another protein called **fimbrin** (see Section 4.4.6) in forming polarised actin bundles in brush-border microvilli (see Section 5.3.2). Under the influence of Ca^{2+} ions, it has a filament-severing property, which may be useful for dismantling filaments in worn-out cells. The sequence of villin shows that it is

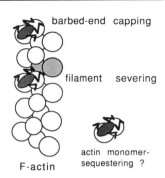

barbed-end capping

filament   severing

actin monomer-
sequestering ?

F-actin

Figure 4.2

Scheme for the activity of gelsolin.[1-4] Severing of actin filaments probably involves a molecule of gelsolin interacting with two or more adjacent actin monomers in some way that weakens the bonds between them. If gelsolin remains attached to the monomer that forms a new barbed end, it will prevent further growth here (barbed-end capping). Binding to individual monomers will prevent their participation in assembly (monomer sequestration).

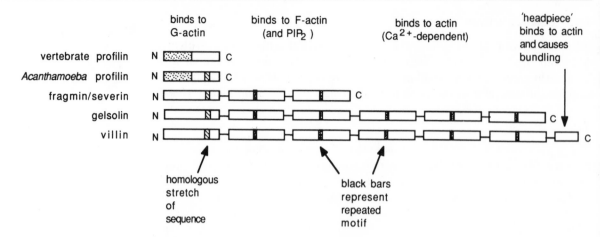

Figure 4.3

Comparison of some related actin-binding proteins.[1,2,4] The sequence of *Acanthamoeba* profilin seems to be related to N-terminal domains of many larger actin-binding proteins, all of which include a homologous region in a similar position. A further conserved motif is found repeated twice in the C-terminal sections of fragmin and severin, and five times in gelsolin and villin. The C-terminal half of vertebrate profilin seems to have diverged from the pattern conserved in the other proteins. Villin possesses an extra C-terminal domain containing a different type of actin-binding site.

closely related to gelsolin but has an extra, unrelated, C-terminal domain carrying another actin-binding site which is responsible for the bundling property.

The proteins **fragmin** and **severin** isolated from lower eukaryotes appear to be very similar to the first half of gelsolin and have both severing and capping activities.

4.2.6   Nucleating activity[61]

Gelsolin and related cutting proteins also nucleate actin polymerisation from the pointed end, in much the same way as the cytochalasins (Section 3.3). When they are added to G-actin, polymerisation occurs without the lag required for normal nucleation. Presumably the nucleating property is due to their multiple binding sites for actin monomers.

4.2.7   Actin-monomer sequestration

A high proportion of actin is in a soluble form in many types of cells. Although the actin may be at a sufficiently high concentration to support spontaneous self-assembly, it is apparently prevented from doing so by proteins that form 1:1 complexes with G-actin. The best studied is **profilin**, which is discussed in the next section. Others (Section 4.2.9) sequester monomers after severing them from F-actin. **Actobindin** does not sever but appears to inhibit the formation of actin-filament initiation complexes. It may cap small oligomers in addition to sequestering monomers.

4.2.8   Profilin[5,7,8]

A small (12–15 kD) basic protein known as **profilin** was first identified in thymus extracts. It occurs in large amounts in blood platelets (see Section 6.1.2). Proteins of a similar size and with similar properties have been found in a wide range of other cell types, including amoebae and yeast. All act as **monomer-sequestering proteins**. *In vitro* studies of the interaction with actin have shown that actin polymerisation is reduced in proportion to the amount of profilin present. Profilin is apparently an essential cytoplasmic protein: disruption of the profilin gene in yeast produces cell death.

The proteins that have been sequenced are homologous over their N-terminal halves but the C-terminal segment of vertebrate profilin is apparently unrelated to those of *Acanthamoeba* and yeast (Figure 4.3). The C-termini of the latter class show some homology with the C-terminal part of the first domain of gelsolin. Although both vertebrate and *Acanthamoeba* profilins seem to bind to the C-terminus of actin, it is not surprising to find subtle differences in their interactions with actin.

Detailed studies of the properties of *Acanthamoeba* profilin by electron microscopy have led Pollard to suggest that this protein may weakly cap the barbed end of F-actin. Both nucleation and growth at the pointed end are strongly inhibited in the presence of actin–profilin complexes, but growth at the barbed end is only weakly affected. If profilin blocks the site on the monomer that ought to bind to a pointed end, the complex might be able to bind to a barbed end, allowing profilin to dissociate after the actin monomer changes to an F-actin conformation. Alternatively, the lack of effect at the barbed end might just reflect the low critical concentration for elongation there (see Table 3.1).

Profilin is tightly bound to actin in the free complex but can be dissociated by the binding of PIP_2 to profilin. This reaction is thought to be of great importance for controlling the polymerisation of actin filaments onto sites associated with cell membranes (Figure 4.4). Splitting of the actin–profilin bond by PIP_2 in the membrane would release actin monomers for addition to nearby barbed ends (see Section 4.6). If *Acanthamoeba* profilin is indeed a weak barbed-end capping protein, the complex might add to the end of a filament already anchored to the membrane before the removal of profilin by PIP_2, thus making the whole process even more efficient.

Figure 4.4

Scheme for the release of actin monomers from complexes formed with the monomer-sequestering protein profilin.[59] The complex can be cleaved by PIP_2 in membranes,[25] leaving the actin monomer free to add to filament ends. The interaction with PIP_2 may be promoted by the binding of complexed actin to specific proteins in the membrane. Membrane-bound actin monomers may then diffuse over the membrane surface to associate with the barbed ends of bound polymers. Clustering of integral-membrane actin-binding proteins, as proposed for ponticulin in *Dictyostelium*, may assist in filament initiation.

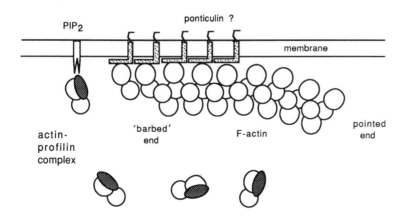

4.2.9 Actin-depolymerising activity[4]

Whereas gelsolin interacts most strongly with F-actin, there are small sh (~19 kD) actin-binding proteins that *sever* actin filaments and then play a *monomer-sequestering* role. They have been called **actin depolymerising factor** (**ADF**), a major 19 kD protein found in mammalian and avian brain (also known as **cofilin**); **destrin**, from kidney; **depactin**, identified from starfish oocytes; and **actophorin** from *Acanthamoeba*. They could combine the roles of gelsolin and profilin that are discussed in Section 6.4.

4.3 Force-producing proteins

Enzymatically, myosins are all fairly closely related, though there are functional variations (see Section 5.4). The complete molecules fall into two distinct classes, namely those that polymerise into 'thick filaments' and those that do not. We know much more about the first class since this type of protein is so abundant in muscle and because its ability to assemble into filaments and disassemble again into monomers *in vitro* makes it easy to purify.

4.3.1 Two-headed myosin (myosin II)[9,10,33,34,36]

Molecules that are very similar in their properties to the myosins of striated and smooth muscle can be isolated from non-muscle cells (see Figure 4.5). Soluble molecules occur as pairs of heavy chains, together with two pairs of light chains. The C-terminal halves of the heavy chains associate in a polar fashion to form a long α-helical coiled coil, as described in Section 2.1. The N-terminal halves fold up into two separate globular heads which possess the actin-activated ATPase activity.

A number of myosin heavy chains have been sequenced and there have been numerous studies of proteolytic subfragments, providing some idea of the functions of different domains within the heads. This is summarised in Figure 4.6. How the head domains fold up in three dimensions is still not quite certain (Figure 4.7).

Figure 4.5

Smooth-muscle myosin molecules shadowed with platinum and supported on a layer of carbon: (a) molecules in folded conformations; (b) & (c) extended molecules; (d) & (e) molecules associating to form small filaments. (a)–(c) × 250 000; (d), (e) × 90 000. [Micrographs provided by Drs Robert Cross & John Kendrick-Jones.]

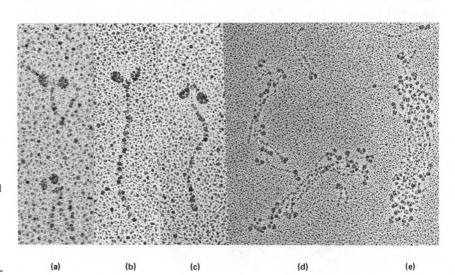

(a)   (b)   (c)   (d)   (e)

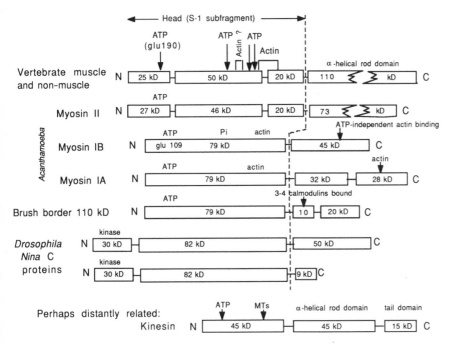

Figure 4.6

Comparison of the primary structures of different types of myosin and related proteins. Proteolytic sites lie between boxed segments; regions are indicated where actin and nucleotide may bind. Single-headed amoeboid myosin I molecules, the 110 kD ATPase found in vertebrate intestinal microvilli, and the Nina C proteins expressed in *Drosophila* retina all possess domains highly homologous to the N-terminal head portions of filament-forming myosins.[10,11,18,19,31–33,35] The microtubule-associated enzyme, kinesin (Section 8.6), is not closely related to myosin but has a somewhat similar structure, including a long rod-domain.

There are two different light chains associated with each head. One can be removed without destroying the ATPase activity, the other cannot. For this reason, the latter has been referred to as the **'essential' light chain** (ELC). The other, which is often required for on-off regulation, is referred to as the **regulatory light chain** (RLC). The light chains have been located in the neck region of the myosin head, with the ELC nearest the rod, but both may stretch up towards the tip of the head. Each appears to have some homology to calmodulin (see Section 4.1), though in most cases the high-affinity calcium-binding properties seem to have been lost. In spite of this, removal of the RLC from calcium-sensitive

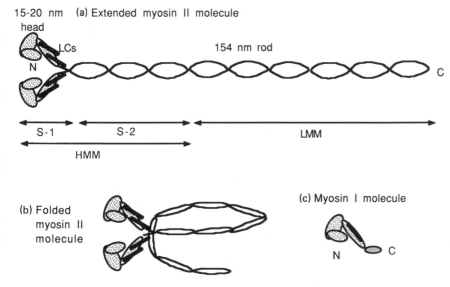

Figure 4.7

(a) Scheme for the structure of a myosin II molecule, showing the main subfragments into which it can be cleaved.[9,10] The long α-helical C-terminal stretch can be cut without disrupting the pairing of the two heavy chains. One product, light meromyosin (LMM), retains the capability of forming into filaments. The other product, heavy meromyosin (HMM), is a two-headed molecule with a short tail and retains the full ATPase activity. Cleavage of HMM in the neck region leaves each head (subfragment-1, or S-1, shown in more detail in Figure 4.8) as an entity, possessing half the ATPase activity of the original molecule, and a short rod section (S-2). The latter domain is thought to lie close to the myosin filament shaft in resting muscle but to swing out in activated muscle, allowing the heads to interact with actin filaments. (b) Myosin II molecules from tissues other than striated muscles fold up so that part of the tail contacts the junction between the two heads (see Figure 4.5a). This inhibits filament formation by the tail and also switches off the ATPase activity of the heads (see also Figure 5.12).[9] (c) A myosin I molecule consists of a single head, equivalent to S-1, and a distinct C-terminal domain which can apparently bind either to actin in an ATP-independent fashion or to the surface of a lipid membrane.[11]

Table 4.3 **Effect of RLCs on** **actin-activated ATPase activity** **of scallop myosin**[62] **(without** **Ca^{2+}/with Ca^{2+})**	*Source of RLC*	*RLC unphosphorylated*	*RLC phosphorylated*
	Non-muscle cells and vertebrate smooth muscle	$-/+$	$+/+$
	Scallop striated muscle	$-/+$	NA
	Vertebrate striated muscle	$-/+$	NA

myosins (see Section 5.4), leaves the ATPase activity switched on whatever the calcium level. Calcium activation apparently involves the binding of Ca^{2+} to a site on the *heavy* chain; if Ca^{2+} is not bound, the RLC is able to inhibit ATPase activity further up the molecule. Phosphorylation of the RLC can remove this ability.

The properties of different RLCs can be tested by adding them to scallop muscle myofibrils from which the native RLCs have been extracted (Table 4.3).

4.3.2 Myosin monomers, dimers and thick filaments

In the field of muscle research, the heavy-chain *pair* is referred to as a **myosin monomer** (whereas comparable pairs of other proteins are often referred to as *dimers*: this can lead to confusion). A bipolar myosin 'dimer' consists of two antiparallel α-helical rods (four heavy chains) with a pair of heads at each end.

The monomers usually assemble to form larger bipolar aggregates, with shafts resembling IFs, though thicker in diameter (see Section 2.2.8). *In vitro*, purified myosin assembles into bipolar filaments of varying lengths. Myosin heads stick out all round the end segments but there is a **bare zone** in the middle of each filament (Figure 2.8a).

Smooth muscle and non-muscle myosins apparently assemble into simpler ribbon-like aggregates (Figures 2.8b & 4.5). Their assembly is readily reversible and control of the equilibrium between monomers and filaments is an important part of the control of contraction (see Section 5.5).[9,10,33] In the case of vertebrate myosins, light-chain phosphorylation regulates filament assembly. If the light chains are *not* phosphorylated, the tail has an affinity for the heads and tends to fold up as shown in Figure 4.5b. Heavy-chain phosphorylation has been shown to be important in controlling the assembly of amoeboid myosins; assembly is inhibited if the heavy chain *is* phosphorylated, at a site near the C-terminus.

4.3.3 Myosin heads and motive force[28,29]

The formation of filaments is undoubtedly important for organising contractile activity, but individual heads seem to be sufficient for force production. Myosin heads proteolytically removed from their tail domains and known as S-1 subfragments show the full ATPase activity, which is enhanced in the presence of actin filaments to the same extent as that for whole molecules.

The interaction with F-actin can be assayed for production of motility by light-microscopic methods. The initial technique of Sheetz and Spudich used fluorescent beads, to which myosin fragments were covalently bound. The beads were observed moving along actin cables obtained from *Nitella* (see Section 5.3.4) in solutions containing ATP. A simpler method has been developed more

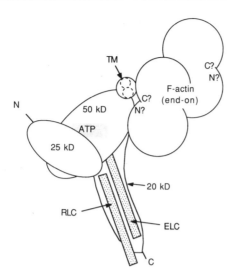

Figure 4.8

A possible arrangement of domains in a myosin head (S-1). ATP is thought to bind in a cleft between the N-terminal 25 kD and central 50 kD domains of S-1 (see Figure 4.6). Binding to actin apparently extends over the junction between the 50 kD domain and the 20 kD C-terminal, competing with tropomyosin (TM is shown in a position that does not block myosin binding—see Figure 5.11). F-actin is shown relative to S-1 in the most likely of four possible orientations compatible with the most recent 3-D image.[23] The essential and regulatory light chains (ELC and RLC) are thought to bind to the 20 kD domain, contributing to the neck region. The presence of RLC makes S-1 appear a little longer, so this light chain probably overlaps with the N-terminus of S-2.[48] [Modified from Vibert, P. J. and Cohen, C. (1988). *J. Muscle Res. & Cell Motil.* **9**, 296–305.]

recently, whereby myosin fragments, such as S-1, are stuck to a suitably-charged glass surface: fluorescently-labelled actin filaments can be observed sliding over the surface.

Estimates of the distance moved during each 'stroke' of the crossbridge cycle (see Section 5.4) still vary widely. The longitudinal spacing of equivalent binding sites is 5.5 nm, from one actin monomer to the next along on the same side (see Figure 3.1), and it seems unlikely that a myosin head could move more than this distance in a single conformational cycle. But it is not yet clear whether a molecule of ATP is hydrolysed for each such step (as is assumed in most models of the crossbridge cycle—see Section 5.4) or whether several steps can be taken using the same burst of energy.[60]

4.3.4 Single-headed myosin (myosin I)[11,18]

A different kind of molecule was first isolated from *Acanthamoeba* along with a two-headed myosin similar to that found in muscle. The heavy chains are significantly smaller (140 and 125 kD for two homologous gene products, myosins 1A and 1B) and, together with a single light chain per heavy chain (17 kD for myosin 1A, 27 kD for myosin 1B), they form molecules with single heads and very short tails (Figure 4.7c). These molecules do not polymerise into filaments. Nonetheless, they possess the most important property of myosin, namely an ATPase activity that is greatly enhanced in the presence of F-actin. The N-terminal part of each heavy chain appears to be homologous to this region of myosin II, as summarised in Figure 4.6. The short tail is quite different from a myosin II tail and has a second, ATP-insensitive, actin-binding site. This means that monomeric myosin I molecules should be able to move actin filaments relative to one another in much the same way as dynein molecules are able to produce a shearing force between microtubules (see Section 9.2.2).

A protein with similar enzymic and antigenic properties to *Acanthamoeba* myosin IA has been isolated from *Dictyostelium* amoebae, which also have a polymerisable myosin II. Myosin I molecules from the two species of amoebae appear to be more similar to one another than are myosins I and II from the same species, to judge from their antigenic cross-reactivities.

4.3.5   Interaction of myosin I with membranous organelles[11]

Adams and Pollard purified vesicles from *Acanthamoeba* and found a significant fraction of the cells' myosin I associated with the preparation. The vesicles were assayed for motile activity along actin cables from *Nitella*. In the presence of ATP, the vesicles moved along the cables at speeds of up to 0.5 μm/s. Antibodies to myosin I, though not those to myosin II, inhibited the transport. Myosin I was probably bound directly to the vesicle membrane, since membranes treated with 0.1 M sodium hydroxide to strip off attached proteins were still able to bind to myosin I, though not to myosin II which could presumably bind *only* via actin on the membrane. There is also recent direct evidence that myosin I can bind to lipid head groups.

4.3.6   Myosin I-like proteins from non-amoeboid cells[11,18,35]

The fact that myosin I-like molecules have not yet been isolated from many cell types does not mean they are absent since the standard purification procedure for myosins selects for molecules that polymerise into filaments. Similar proteins are gradually being identified in specialised systems involving relatively small varieties of proteins such as intestinal microvilli and retinal rods.

A **110 kD protein** isolated from vertebrate gut microvilli appears to be highly homologous in sequence to amoeboid myosin I, though its ATPase activity is significantly lower than the standard range of myosins (but these are quite variable, as shown in Section 5.4). Its size, its shape and the way it decorates F-actin are similar to these properties of myosin I. One fragment even cross-reacts antigenically with antibodies to the ATP-binding site of *skeletal* myosin. In place of the calmodulin-like light chains of bona fide myosins, this molecule has three or four actual calmodulin molecules bound near the tail (see Figure 4.6). It is not yet known how calmodulin regulates the activity of the 110 kD protein.

Two related proteins in the retinal microvilli of *Drosophila* have been identified from their gene sequences.[35] Mutants deficient in these proteins have shorter microvilli and reduced stores of rhodopsin. A novel feature of both proteins is an additional N-terminal domain with a sequence suggesting that it has protein kinase activity. It has been speculated that the function of the hybrid molecules could be to control some retinal activity by phosphorylation; the ability to move along filaments would help in concentrating kinase activity in the microvilli.

4.4   Other crosslinking and bundling proteins[1,2,5,7,22]

These have the property *in vitro* of causing **gelation** of actin filaments. One subclass forms **isotropic networks**. These may increase the viscosity of a solution of actin filaments so much that the test tube may be turned upside-down without loss of its contents. In the absence of ATP, myosin II can also cause gelation of a suspension of actin filaments but it does not fit into the same structural subclass. Two other subclasses form anisotropic **bundles** of filaments, either all orientated with the same polarity or alternating in polarity. What all these proteins have in common is that they possess two or—in the case of myosin filaments—more binding sites for actin. Actually, the proteins forming bipolar bundles and anisotropic networks are similar to each other; the subclasses are both dimeric rods, with equivalent actin-binding sites at either end, but differ in being either relatively rigid or more flexible. Polar bundles are formed by monomer crosslinks with distinct actin-binding sites at each end.

4.4.1 Spectrin and related proteins[5,7,12,13,26,39,57]

Spectrin was first discovered in red blood cells, but similar proteins have since been isolated from many other sources. Some have been given other names, such as brain spectrin which was named **fodrin** at the time of its discovery.

The structure of spectrin has been analysed in some detail. The protein always consists of equimolar quantities of two homologous polypeptide chains, α- and β-spectrin, each about 230 kD. From the amino-acid sequences it has been proposed that each polypeptide chain consists of a series of homologous rigid segments of 106 amino acids ($\sim 12\,000$ kD) joined by short disordered regions. In this model, the secondary structure of the rigid segments is a *triple helix* in which the polypeptide chain folds back on itself twice (Figure 4.9a). The resulting series of segments bears some resemblance to a string of sausages. This string is thought to pair in antiparallel with a homologous string, to form a **bipolar heterodimer** (Figure 4.9d). It is quite distinct, therefore, from the parallel coiled-coil structure proposed for myosin rod, tropomyosin and intermediate-filament dimers and is consistent with the thicker, yet more flexible, rods that are observed by electron microscopy.

The spectrin molecules found *in vivo* are thought to consist predominantly of two heterodimers joined end-to-end (bipolar tetramers), though hexamers and single heterodimers also occur (Figure 4.10). The shape of the four-chain molecule can vary widely from a flexible thin filament up to 194 nm long to a more compact form that is only moderately asymmetrical. The combination of a flexible structure with a highly acidic amino-acid composition means that spectrin varies its shape reversibly with ionic strength, being more extended at lower ionic strength. This property is probably important for the role of spectrin in forming the tensile skeleton that lines the red-blood-cell (erythrocyte) membrane (Section 5.3).

Spectrin-like proteins in non-erythroid cells also tend to be concentrated in the cortical layer of cytoplasm next to the membrane,[13] although none appears to be directly involved in actin-membrane linkage. The properties of spectrins vary quite significantly according to the type of cell in which they occur; it has been suggested that this may be related to the varying degrees of membrane rigidity required by cells.

Figure 4.9

(a) Model for the structure of the 104-amino-acid repeat found in spectrin and α-actinin. Three short helices (represented as cylinders) probably fold back on each other as shown, to form a structural unit. (b) The central rod of an α-actinin monomer consists of four segments. The N-terminal domain binds actin, while an EF hand structure has been predicted for the C-terminal domain (though muscle α-actinin may have lost the ability to bind Ca^{2+}). (c) Monomers are arranged in antiparallel to give a dimer an actin-binding site at each end. In the case of non-muscle α-actinin, binding to actin by the N-terminal domain of one polypeptide is probably controlled by Ca^{2+} binding to the C-terminal domain of its partner. (d) Spectrin monomers are thought to have 18 triple-helical segments.[12] α and β monomers pair with each other to form heterodimers. The N-terminal domain of β-spectrin is homologous to the actin-binding domain of α-actinin; the C-domain of α-spectrin is homologous to the C-domain of α-actinin.[27]

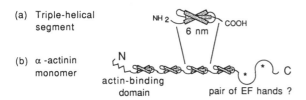

(a) Triple-helical segment

(b) α-actinin monomer

actin-binding domain   pair of EF hands ?

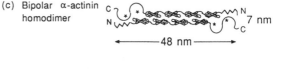

(c) Bipolar α-actinin homodimer

48 nm

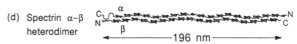

(d) Spectrin α–β heterodimer

196 nm

Figure 4.10

Spectrin heterodimers may lie end-to-end to form long bipolar tetramers (functionally equivalent to very elongated α-actinin dimers) or associate in threes so that two-dimensional networks can be formed (Figure 5.2).

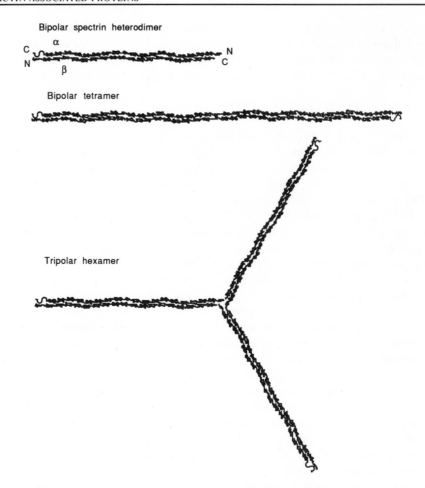

Bipolar spectrin heterodimer

Bipolar tetramer

Tripolar hexamer

4.4.2 α-Actinin[6,7,21,26,44]

α-Actinin was originally identified as a component of the Z-lines in striated muscle (see Section 1.6) and seems to be a key element in maintaining the structure of the sarcomeres. Related proteins have since been found concentrated in dense bodies in smooth-muscle cells, in adhesion plaques of motile cells (Section 5.6), and in other dense patches of protein near to cell membranes where close contacts occur between cells; it is also found in the cleavage furrows of mitotic cells. Unlike the crosslinked mesh produced by spectrin, actin in striated muscle occurs in well-ordered bundles, apparently with α-actinin as one of the main crosslinking agents. It apparently crosslinks filaments of opposite polarity so, in an appropriate rotational orientation, its bipolar ends can make identical interactions with F-actin (Figure 4.11). It seems likely that it interacts with actin in a similar way in other cell types, though it may not be essential for the filaments to lie exactly antiparallel.

It is consistent with a bipolar type of interaction that α-actinin can be isolated as a dimer of apparently *identical* polypeptides, whose molecular weights lie between 90 and 100 kD. In the electron microscope the molecule appears as a rod 48 nm long and 7 nm wide, with globular domains at each end.

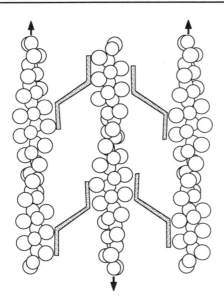

Figure 4.11

In the Z-discs of striated muscles,[58] α-actinin is thought to crosslink pairs of actin filaments of opposite polarity (from adjacent sarcomeres), a function well suited to a simple bipolar molecule. A similar arrangement probably occurs in stress fibres (Figure 5.13). Other proteins, such as tropomyosin, are also involved in making highly ordered arrangements.

Acanthamoeba α-actinin (originally given the name **gelation protein**) seems to be special in having an extra globular domain in the middle of the rod and has a stronger tendency to split longitudinally into two subunits.[5]

EM images suggest that the two polypeptides lie antiparallel to each other and consist of a globular head at one end with a tail that extends at least as far as the edge of the globular head at the other end. The central domain appears to be related to that of spectrin but with a smaller number of 'sausages' in the string (Figure 4.9b). Following an actin-binding domain at the N-terminus related to the N-terminus of β-spectrin, the sequence of α-actinin contains four repeats of around 120 residues that are homologous to the 106-residue repeats of spectrin.

Unlike the protein in muscle, which is insensitive to calcium concentration, crosslinking of actin by non-muscle α-actinin is completely inhibited by micromolar calcium. Its globular C-terminal domain appears to have strong homologies with calmodulin and to include two EF hands (see Figure 4.1). A model suggested for the dimer (see Figure 4.9c) has the calcium-binding globular domain of one polypeptide regulating the actin-binding site of the other chain in a dimer.[16] The C-terminal domain of α-spectrin appears to be related to the equivalent domain in α-actinin.

α-Actinin in non-muscle cells probably contributes to the calcium-regulated gelation–solation cycle observed in cells (see Section 6.4). Its interaction with F-actin is most easily observed in stress fibres (Section 5.6). Bundles of actin filaments may be stabilised by α-actinin crosslinks until these are loosened by calcium and the bundle allowed to contract. This would apply to disordered bundles and even meshworks of filaments as well as to stress fibres.

4.4.3 Dystrophin[26,27]

This enormous protein (425 kD) was discovered because of its association with the genetic disease called **Duchenne muscular dystrophy**. The gene sequence suggests that it belongs in the same family as spectrin and α-actinin. The homology with α-actinin is strongest, especially in the putative actin-binding domain at the N-terminus. This is followed by 25 spectrin-like repeats. After a putative Ca^{2+}-binding domain, there is finally an additional C-terminal tail domain that may be involved in binding to membrane. Muscular dystrophy is the result of genetic deletions within the sequence that presumably lead to production of a non-functional molecule. The normal protein has been localised at the muscle cell surface and may anchor actin filaments to the plasma membrane.

4.4.4   Gelation factors[5,7,27]

Actin-binding protein (so called because it was the first such protein identified) from macrophages and **filamin** from smooth muscle appear to be homodimers in the form of long, relatively flexible, bipolar rods. Their crosslinking of actin filaments is calcium-insensitive. Actin filaments often appear to make T-junctions as well as intermeshing with one another at a range of angles.

The sequence of *Dictyostelium* 120 kD gelation factor shows an N-terminal actin-binding domain homologous to that of α-actinin. The structure of the rod domain is predicted to be somewhat different from that of spectrin and α-actinin. Instead of coiling into α-helices, the sequence shows a potential to form a β-sheet structure, an extended polypeptide chain that repeatedly folds back on itself. The interaction with F-actin presumably leads to gelation rather than bundling because the rod domain does not hold the two actin-binding end-domains firmly at a relative angle of 180°.

4.4.5   Tropomyosin[6,41,43,53,56]

Not a bundling protein in itself, **tropomyosin** is mostly found in actin bundles. The α-helical coiled-coil rods, each around 40 nm long, line up end-to-end alongside each of the two long-pitch helices of F-actin (Figure 4.12). Tropomyosin's best-known role is in striated muscle, where it is thought to block the interaction between actin and the myosin crossbridges, unless moved out of the way by the calcium-dependent troponin complex (Section 5.4).

Myosin activity does not seem to be controlled in this manner in non-muscle cells, yet tropomyosins are found in most cell types. Short actin filaments supported by tropomyosin rods are an important feature of the red-blood-cell cytoskeleton (see Section 5.3.1), for example. In fibroblasts and such cells, stress fibres include tropomyosin (Section 5.5). Transformed rat kidney cells that lack tropomyosin do not form stress fibres and disruption of the only tropomyosin gene in yeast results in cells lacking actin cables. Thus, tropomyosin may be involved in organising the assembly of actin filaments into neat bundles; alternatively, it may protect the filaments to which it binds from severing activity.

4.4.6   Caldesmon[6,63]

A protein with an apparent molecular weight of 120 kD (but whose sequence corresponds to only 87 kD) isolated from smooth muscle co-localises with tropomyosin. Immunological studies of non-muscle cells and proteins extracted from them have identified 70–90 kD polypeptides with related characteristics. Non-muscle caldesmon has been localised in stress fibres with a periodic

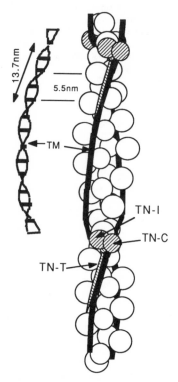

Figure 4.12

Tropomyosin molecules (TM) bind head-to-tail, with a small overlap, along each side of an actin filament. Each molecule is a polar dimer, in the form of an α-helical coiled coil (see Section 2.1). Charged amino-acid residues on the outer surface of TM are arranged in a pattern (represented as bars on the molecule shown on the left) to match the actin monomer repeat. In vertebrate striated muscles, a troponin complex binds to the C-terminal end of each TM molecule. Troponin-T (TN-T), an elongated molecule, binds directly to TM;[6,43] troponin-I (TN-I) binds to TN-T; troponin-C (TN-C) binds to both TN-T and TN-I. Calcium ions are picked up by the TN-C subunit but TN-I is essential for the inhibitory effect (see Figure 5.11).

distribution. Monomers, which are elongated molecules, associate into oligomers through disulphide bonds. Their ends can bind to actin and cause bundling. The binding to actin is enhanced in the presence of tropomyosin. The natural stoichiometry seems to be 1 caldesmon molecule to about 14 actin monomers.

Actin-activated myosin ATPase activity is inhibited in the presence of caldesmon, but calmodulin, when activated by Ca^{2+}, can bind to caldesmon and remove the inhibition. In this respect, caldesmon is similar in behaviour to the troponin complex of striated muscle and may provide another control in addition to the one acting directly on myosin filaments (Section 5.5). There is also some evidence that caldesmon may have a myosin-binding site and thereby be responsible for the so-called 'catch mechanism' which allows some types of muscle to maintain a steady tension for indefinite periods.

4.4.7 Polar bundling[6,7]

Fascin, from sea-urchin eggs, where it is involved in the formation of actin bundle-supported microvilli, and **fimbrin**, a similar protein from the brush border microvilli of intestinal epithelium, both bind actin filaments into polar bundles (Section 5.3). The molecules consist of single polypeptides with two different actin-binding sites which are complementary in that they bind to diametrically opposite sides of an actin filament (see Figure 4.13). They form 9 nm-long crosslinks between two filaments at axial intervals corresponding to the actin crossover repeat. There are probably similar molecules in the actin bundles of stereocilia and of microvilli in general.

4.4.8 *Limulus* acrosomal process

This is a highly ordered bundle of actin filaments which initially forms a neat coil in the sperm head. During acrosome activation, the coil straightens and extends out to interact with an egg. The structure of the bundle has been studied by reconstructing three-dimensional images of filaments from electron micrographs.[22]

Apart from actin, the main components are 60 kD globular molecules and 50 kD rod-shaped molecules that appear to bind to every actin monomer to form a 1:1:1 complex. The globular molecule is associated mainly with the inner domain of the actin monomer, the rod-shaped molecule entirely with the outer domain. When the complexed filaments are associated into a bundle, there appear to be contacts between some 60 kD molecules, which thus form a bipolar crosslink between filaments, and also some 50 kD molecules contact 60 kD molecules on neighbouring filaments (Figure 4.14). DeRosier and colleagues have postulated that a conformational change in the 50 kD protein may be responsible for a change in the twist of the bundle that is thought to be responsible for making the coil extend during activation.

4.4.9 Aldolase

Aldolase is an abundant enzyme that catalyses a key step in the pathway of glycolysis and also bundles actin *in vitro*.[46] Its binding to actin illustrates an important function for the cytoskeleton in providing a solid support matrix for cytoplasmic enzymes.[47] Glycolysis is, of course, very important in striated muscle since it provides a reserve method of synthesising ATP.

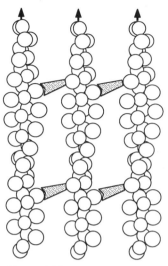

Figure 4.13

Fascin forms crosslinks in polar actin bundles. The binding at each end is such as to bring adjacent filaments exactly into parallel register. This may help them to cooperate in forming small rigid neo-hexagonal bundles (Figure 5.6) as well as large bundles with a random cross-sectional packing.

Figure 4.14

End-on view of the apparent arrangement of 60 kD and 50 kD crosslinking proteins that interact with actin filaments in the acrosomal filament bundle of a *Limulus* sperm. [Based on Bullitt *et al.* (1988). *J. Cell Biol.* **107**, 597–611.]

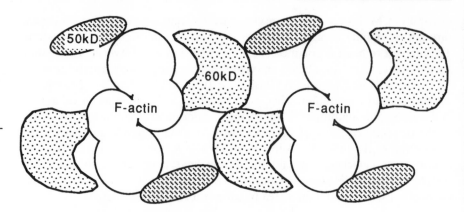

4.5   Membrane-attachment proteins

4.5.1   Indirect links between actin and membrane[14,37,38]

There is good evidence for indirect connections between membranes and actin filaments, especially at dense structures such as adhesion plaques and Z-discs, where forces tend to be concentrated. **Vinculin** is a soluble protein found fairly widely distributed in cytoplasm but is enriched near adhesion plaques (see Section 5.6.2). It binds to actin indirectly through an association with α-actinin and may also bind directly. A protein called **talin** is believed to connect vinculin to proteins embedded in the cell membrane (see Figure 5.14).

Actin appears to be closely associated with the whole cytoplasmic surface of the cell membrane, not just at the above-mentioned specialised structures. In the red blood cell (Section 5.3.1), actin is connected via spectrin and **ankyrin** to integral membrane proteins distributed over the whole cell membrane. In other cell types, there are calcium-binding proteins that bind to both actin and spectrin as well as to anionic phospholipids. These proteins, known as types I and II of a class of membrane-associated proteins called **calpactins**, **calelectrins** or **annexins**, may help connect the cortical cytoskeleton to the membrane. Myosin I and the intestinal 110 kD protein (Sections 4.3.6 & 5.3.3) also interact with membrane lipids without being embedded in the membrane as integral components. All these proteins probably interact with transmembrane proteins as well as lipids, but such interactions have not been investigated fully.

4.5.2   Integral membrane actin-binding proteins

Integral membrane proteins that connect actin filaments directly to membranes have been sought in various kinds of cells. There are several that may fit into this category, some better characterised than others.

4.5.3   Ponticulin[24]

There is good evidence that this 20 kD protein, isolated from *Dictyostelium* by an actin-binding method, is an integral membrane glycoprotein. It has domains that can be labelled on either side of the plasma membrane. The cytoplasmic domain has been shown to bind directly to actin filaments. Luna and colleagues have also shown that chemically modified actin monomers, which assemble poorly even in the presence of normal actin, show enhanced assembly onto purified *Dictyostelium* membranes. Their interpretation is that ponticulin molecules bind to the sides of actin monomers and, by clustering together in the membrane (see Figure 4.4), assist in filament nucleation and elongation.

4.6 Disruption of protein activity in cells

4.6.1 Genetic analysis of cytoskeletal proteins[15–17,34,35,55,56]

Actin, like tubulin, has been shown to be an essential protein; its absence is lethal to temperature-sensitive mutants of microorganisms such as yeast. Genetic approaches are just beginning to yield valuable information about the functions of actin-associated proteins. Mutations in non-essential proteins are less easy to detect, so techniques of targeting specific genes (Appendix A.6) offer most possibilities for future work.

As mentioned above, profilin appears to be necessary for the viability of yeast cells. Yeast cells that lack tropomyosin have been found to survive but grow more slowly than normal. Disruption of the expression of myosin II in *Dictyostelium* is described in Section 6.4.1. The cells have difficulty in dividing and move much more slowly. Clearly they would not survive in competition with normal cells.

Gerich's group has been using antibodies to assay for mutant *Dictyostelium* clones lacking actin-binding proteins and have found clones lacking α-actinin, severin and the 120 kD gelation factor. None of the three appears to be essential for cell division and the deficient amoebae are still able to crawl normally. It is not known whether this would be true for amoebae lacking both crosslinking proteins. Nor is it known whether they possess an alternative to severin. Possibly filament-severing activity is not absolutely required; F-actin may sever spontaneously with sufficient frequency to allow rearrangements of the cytoskeleton to take place. In any case, it seems likely that each of these mutants would be at some disadvantage in competition with wild-type cells, even though their disability is not obvious to us. The overlapping roles of all the competing actin-binding proteins are likely to become clearer as more of them are targeted in this way.

4.6.2 Injection of antibodies

An alternative method of disrupting the function of a particular protein is to microinject antibodies into the cell. This too has only been attempted in a few cases. Antibodies to myosin were known to inhibit cytokinesis (Section 11.10) long before the genetic targeting of myosin II in *Dictyostelium* produced a similar result. Surprisingly, injection of antibody to spectrin into a variety of living cell types caused no detectable alterations in microfilaments or microtubules, although spectrin was precipitated inside the cells and intermediate filaments collapsed.[57]

4.7 Summary

Although pure actin assembles into polymers in a very specific manner to form a relatively invariant filament structure, a large variety of actin-binding proteins can produce a seemingly endless variety of supramolecular forms, from highly ordered bundles to random gels. Other proteins control assembly by assisting or discouraging polymerisation, or by severing existing filaments. Yet others connect actin filaments to structures such as membranes or other types of filaments. In the case of the two types of myosin, these connections are able to generate sliding motility. The state of actin in a cell at any particular place and time will depend on the dynamic balance of effects due to the accessory proteins present.

4.8   Questions

4.1 Describe means by which cells might (a) prevent actin filaments from depolymerising or from growing, or (b) encourage them to depolymerise or to grow.

4.2 What variations in the properties of actin-crosslinking proteins might produce gels with differing characteristics?

4.3 Which activities of actin-binding proteins are controlled by calcium ions?

4.4 In which interactions between actin and its associated proteins do lipids or lipid bilayers seem to be involved?

4.5 What range of structural forms does the myosin family include?

4.9   Further reading

Some recent reviews

1 Vandekerckhove, J. (1989). Structural principles of actin-binding proteins. *Curr. Opinion Cell Biol.* **1**, 15–22.

2 Matsudaira, P. & Jamney, P. (1988). Pieces in the actin-severing protein puzzle. *Cell* **54**, 139–140.
      (Compact review of gelsolin and villin.)

3 Yin, H. L. (1987). Gelsolin: calcium and polyphosphoinositide-regulated actin modulating protein. *Bioessays* **7**, 176–179.

4 Way, M. & Weeds, A. (1990). Actin-binding proteins: cytoskeletal ups and downs. *Nature* **344**, 292–294.

5 Pollard, T. D. & Cooper, J. A. (1986). Actin and actin-binding proteins. A critical evaluation of mechanisms and functions. *Annu. Rev. Biochem.* **55**, 987–1035.

6 Bretscher, A. (1986). Thin filament regulatory proteins of smooth and non-muscle cells. *Nature* **321**, 726–727.

7 Stossel, T. P., Chaponnier, C., Ezzell, R. M., Hartwig, J. H., Janmey, P. A., Kwiatkowski, D. J., Lind, S. E., Smith, D. B., Southwick, F. S., Yin, H. L. & Zaner, K. S. (1985). Nonmuscle actin-binding proteins. *Annu. Rev. Cell Biol.* **1**, 353–402.
      (Encyclopaedic but growing out of date.)

8 Korn, E. D. (1982). Actin polymerization and its regulation by proteins from nonmuscle cells. *Physiol. Rev.* **62**, 672–737.

9 Citi, S. & Kendrick-Jones, J. (1987). Regulation of non-muscle myosin structure and function. *Bioessays* **7**, 155–159.

10 Warrick, H. M. & Spudich, J. A. (1987). Myosin structure and function in cell motility. *Annu. Rev. Cell Biol.* **3**, 379–421.

11 Adams, R. J. & Pollard, T. D. (1989). Membrane-bound myosin I provides new mechanisms in cell motility. *Cell Motil. & Cytoskel.* **14**, 178–182.

12 Marchesi, V. T. (1985). Stabilizing infrastructure of cell membranes. *Annu. Rev. Cell Biol.* **1**, 531–561.
   (Properties of spectrin and family.)

13 Moon, R. T. & McMahon, A. P. (1987). Composition and expression of spectrin-based membrane skeletons in non-erythroid cells. *Bioessays* **7**, 159–164.

14 Mangeat, P. H. & Burridge, K. (1984). Actin-membrane interaction in fibroblasts: what proteins are involved in this association? *J. Cell Biol.* **99**, 95s–103s.

15 De Lozanne, A. (1989). Gene targeting and cell motility. *Cell Motil. & Cytoskel.* **14**, 62–68.

16 Noegel, A. A., Leiting, B., Witke, W., Gurniak, C., Harloff, C., Hartmann, H., Weismuller, E. & Schleicher, M. (1989). Biological roles of actin-binding proteins in *Dictyostelium discoideum* examined using genetic techniques. *Cell Motil & Cytoskel.* **14**, 69–74.

17 Bray, D. & Vasiliev, J. (1989). Networks from mutants. *Nature* **338**, 203–204.

Some recent papers

18 Hoshimaru, M., Fujio, Y., Sobue, K., Sugimoto, T. & Nakanishi, S. (1989). Immuno-chemical evidence that myosin I heavy chain-like protein is identical to the 110-kilodalton brush border protein. *J. Biochem. (Japan)* **106**, 455–459.

19 Coluccio, L. M. & Bretscher, A. (1988). Mapping of the microvillar 110 kD calmodulin complex: calmodulin-associated or -free fragments of the 110 kD polypeptide bind F-actin and retain ATPase activity. *J. Cell Biol.* **106**, 367–374.

20 Atkinson, M. A. L., Lambooy, P. K. & Korn, E. D. (1989). Cooperative dependence of the actin-activated Mg^{2+}-ATPase activity of *Acanthamoeba* myosin II on the extent of filament phosphorylation. *J. Biol. Chem.* **264**, 4127–4132.

21 Wallraff, E., Schleicher, M., Modersitzki, M., Rieger, D., Isenberg, G. and Gerisch, G. (1986). Selection of *Dictyostelium* mutants defective in cytoskeletal proteins: use of an antibody that binds to the ends of alpha-actinin rods. *EMBO J.* **5**, 61–67.

22 Bullitt, E. S. A., DeRosier, D. J., Coluccio, L. M. & Tilney, L. G. (1988). Three-dimensional reconstruction of an actin bundle. *J. Cell Biol.* **107**, 597–611.

23 Milligan, R. A. & Flicker, P. F. (1987). Structural relationships of actin, myosin and tropomyosin revealed by cryo-electron microscopy. *J. Cell Biol.* **105**, 29–39.

24 Schwartz, M. A. & Luna, E. L. (1988). How actin binds and assembles onto plasma membranes from *Dictyostelium discoideum. J. Cell Biol.* **107**, 201–209.

25 Lassing, I. & Lindberg, U. (1988). Evidence that the phosphatidylinositol cycle is linked to cell motility. *Exptl. Cell Res.* **174**, 1–15.

26 Byers, T. J., Husain-Chishti, A., Dubreuil, R. R., Branton, D. & Goldstein, L. S. B. (1989). Sequence similarity of the amino-terminal domain of

Drosophila beta-spectrin to alpha-actinin and dystrophin. *J. Cell Biol.* **109**, 1633–1641.

27 Noegel, A. A., Rapp, S., Lottspeich, F., Schleicher, M. & Stewart, M. (1989). The *Dictyostelium* gelation factor shares a putative actin binding site with α-actinins and dystrophin and also has a rod domain containing six 100-residue motifs that appear to have a cross-beta conformation. *J. Cell Biol.* **109**, 607–618.

28 Toyoshima, Y. Y., Kron, S. J., McNally, E. M., Niebling, K. R., Toyoshima, C. and Spudich, J. A. (1987). Myosin subfragment-1 sufficient to move actin filaments *in vitro*. *Nature* **328**, 536–539.

29 Hynes, T. R., Block, S. M., White, B. T. & Spudich, J. A. (1987). Movement of myosin fragments *in vitro*: domains involved in force production. *Cell* **48**, 953–963.

Additional reviews and reports

30 Vibert, P. J. & Cohen, C. (1988). Domains, motions and regulation in the myosin head. *J. Muscle Res. & Cell Motil.* **9**, 296–305.

31 Mornet, D., Bonet, A., Audemard, E. & Bonicel, J. (1989). Functional sequences of the myosin head. *J. Muscle Res. & Cell Motil.* **10**, 10–24.

32 Cooke, R. (1989). Structure of the myosin head. *Cell Motil. & Cytoskel.* **14**, 183–186.

33 Korn, E. D. & Hammer, J. A. (1988). Myosins of nonmuscle cells. *Annu. Rev. Biophys. Biophys. Chem.* **17**, 23–45.

34 Emerson, C. P. & Bernstein, S. I. (1987). Molecular genetics of myosin. *Annu. Rev. Biochem.* **56**, 695–726.

35 Montell, C. (1989). Molecular genetics of *Drosophila* vision. *Bioessays* **11**, 43–48.

36 Harrington, W. F. & Rodgers, M. E. (1984). Myosin. *Annu. Rev. Biochem.* **53**, 35–76.

37 Glenney, J. R. (1987). Calpactins: calcium-regulated membrane-skeletal proteins. *Bioessays* **7**, 173–175.

38 Burgoyne, R. D. (1988). Calpactin in exocytosis? *Nature* **331**, 20.

39 Coleman, T. R., Fishkind, D. J., Mooseker, M. S. & Morrow, J. S. (1989). Functional diversity among spectrin isoforms. *Cell Motil. & Cytoskel.* **12**, 225–247.

40 Kretsinger, R. H. (1980). Structure and evolution of calcium-modulated proteins. *CRC Crit. Rev. Biochem.* **8**, 119–174.

41 Zot, A. S. & Potter, J. D. (1987). Structural aspects of troponin-tropomyosin regulation of skeletal muscle contraction. *Annu. Rev. Biophys. Biophys. Chem.* **16**, 535–559.

42 Cote, G. P. (1983). Structural and functional properties of the nonmuscle tropomyosins. *Mol. Cell. Biochem.* **57**, 127–146.

43 Hitchcock-DeGregori, S. E. (1989). Structure-function analysis of thin filament proteins expressed in *Escherichia coli*. *Cell Motil. & Cytoskel.* **14**, 12–20.

44 Blanchard, A., Ohanian, V. & Critchley, D. (1989). The structure and function of α-actinin. *J. Muscle Res. & Cell Motil.* **10**, 280–289.

45 Pagliano, L. & Taylor, D. L. (1988). Aldolase exists in both fluid and solid phases of cytoplasm. *J. Cell Biol.* **107**, 981–992.

46 Masters, C. (1984). Interactions between glycolytic enzymes and components of the cytomatrix. *J. Cell Biol.* **99**, 222s–225s.

47 Lambooy, P. K. & Korn, E. D. (1988). Inhibition of an early stage of actin polymerization by actobindin. *J. Biol. Chem.* **263**, 12 836–12 843.

48 Flicker, P. F., Walliman, T. & Vibert, P. (1983). Electron microscopy of scallop myosin: location of regulatory light chains. *J. Mol. Biol.* **169**, 723–741.

49 Herzberg, O. & James, M. N. G. (1985). Structure of the calcium regulatory muscle protein troponin-C at 2.8 Å resolution. *Nature* **313**, 653–659.

50 Sundralingham, M., Bergstrom, R., Strasburg, G., Roa, S. T., Roychowdury, P., Greaser, M. & Wang, B. C. (1985). Molecular structure of troponin C from chicken skeletal muscle at 3-Ångstrom resolution. *Science* **227**, 945–948.

51 Babu, Y. S., Sack, J. S., Greenhough, T. J., Bugg, C. E., Means, A. R. & Cook, W. J. (1985). Three-dimensional structure of calmodulin. *Nature* **315**, 37–40.

52 Bayley, P., Martin, S. & Jones, G. (1988). The conformation of calmodulin: a substantial environmentally sensitive helical transition in Ca^{2+}–calmodulin with potential mechanistic function. *FEBS Lett.* **187**, 160–166.

53 Phillips, G. N., Fillers, J. P. & Cohen, C. (1986). Tropomyosin crystal structure and muscle regulation. *J. Mol. Biol.* **192**, 111–131.

54 Horowits, R., Kempner, E. S., Bisher, M. E. & Podolsky, R. J. (1986). A physiological role for titin and nebulin in skeletal muscle. *Nature* **323**, 160–164.

55 Magdolen, V., Oechsner, U., Muller, G. & Bandlow, W. (1988). The intron-containing gene for yeast profilin (PFY) encodes a vital function. *Mol. Cell. Biol.* **8**, 5108–5115.

56 Liu, H. & Bretscher, A. (1989). Disruption of the single tropomyosin gene in yeast results in the disappearance of actin cables from the cytoskeleton. *Cell* **57**, 233–242.

57 Mangeat, P. H. & Burridge, K. (1984). Immunoprecipitation of non-erythrocyte spectrin within living cells following microinjection of specific antibodies: relation to cytoskeletal structures. *J. Cell Biol.* **98**, 1363–1377.

58 Cheng, N. & Deatherage, J. F. (1989). Three-dimensional reconstruction of the Z-disk of sectioned bee flight muscle. *J. Cell Biol.* **108**, 1761–1774.

59 Petrucci, T. C. & Morrow, J. S. (1987). Synapsin I: an actin-bundling protein under phosphorylation control. *J. Cell Biol.* **105**, 1355–1364.

60 Harada, Y. & Yanagida, T. (1988). Direct observations of molecular motility by light microscopy. *Cell Motil. & Cytoskel.* **10**, 71–76.

61 Gaertner, A., Ruhnau, K., Schroer, E., Selve, N., Wagner, M. & Wegner, A. (1989). Probing nucleation, cutting and capping of actin filaments. *J. Muscle Res. & Cell Motil.* **10**, 1–9.

62 Sellers, J. R., Chantler, P. D. & Szent-Gyorgyi, A. G. (1980). Hybrid formation between scallop myofibrils and foreign regulatory light chains. *J. Mol. Biol.* **144**, 223–245.

63 Fujii, T., Imai, M., Rosenfeld, G. C. & Bryan, J. (1987). Domain mapping of chicken gizzard caldesmon. *J. Biol. Chem.* 2757–2763.

5 ACTIN IN CELLS: STRUCTURAL AND CONTRACTILE ROLES

5.1 Widespread and abundant distribution of actin

Actin has been found in all eukaryotic cells that have been studied in detail and is often the most abundant protein in plant and animal cells. It is less abundant than in the highly specialised cells of striated muscle, which contain around 40 mg/ml actin, but is still the major protein in many cells; the cytoplasm of *Acanthamoeba*, for example, contains 7 mg/ml. Myosin molecules have also been isolated from most types of cell and in each case have been shown to have properties in common with muscle myosin (see Section 4.3). It is probable that small bundles of actin filaments do contract in non-muscle cells in a similar manner to the contraction of myofibrils. But the concentration of myosin in non-muscle cells is very low; the molar ratio of myosin to actin in *Acanthamoeba*, for example, is 1:200, as compared with 1:5 in muscle. This fact alone suggests that interfilament sliding, effected by myosin filaments, may not be the dominant function of non-muscle actin filaments.

5.1.1 Main distribution of actin in cells[5,34]

The distribution of actin in cells can be studied by light microscopy using fluorescently labelled reagents. Figure 5.1 shows a cell labelled with fluorescent phalloidin, which is highly specific for F-actin. The distribution of both filaments and monomers can be revealed by immunofluorescence, using anti-actin antibodies, or by microinjection of fluorescent actin.

A layer of actin on the cytoplasmic surface of the cell membrane, known as the **cortical layer**, seems to be a universal feature of eukaryotic cells.[22] Since organelles are excluded from it, it may show up in the light microscope as a clear layer of cytoplasm. It is quite a thick layer in some cell types, such as amoeboid cells, for example, and was referred to by early microscopists as

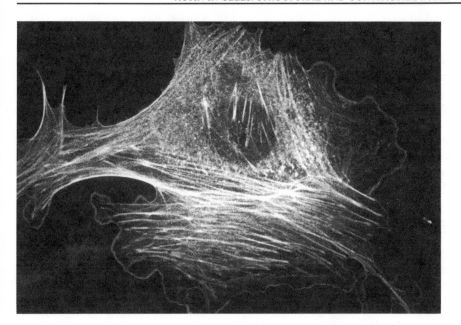

Figure 5.1

Fluorescence image obtained, by confocal light microscopy, from a 3T3 cell fixed and incubated with fluorescent phalloidin, which binds specifically to F-actin. The fluorescent lines crisscrossing the cell are bundles known as stress fibres. × 600.

ectoplasm (as opposed to the granular, organelle-containing, endoplasm). It is thought to be involved in various motile processes (see Sections 6.4 and 11.9). Another probable role is in controlling interactions between the cell membrane and membranes of internal organelles, since bare membranes fuse readily with one another. During secretion by **exocytosis** the layer is probably signalled to move aside, by contracting or disassembling, so that secretory vesicles can fuse with the cell membrane and extrude their contents on the outside of the cell.[23] The details of this process are not yet known.

Actin is also found in a variety of bundles (visible in Figure 5.1), such as stress fibres (Section 5.5) deep within the cytoplasm, or in protrusions (microvilli, microfilopodia and lamellae) from the cell. A significant fraction is soluble, kept in that state by accessory proteins such as profilin (see Section 4.2). The roles of actin in these various forms will be discussed in this chapter and the next.

5.2 Filament polarity in cells[16]

Actin filaments can be oriented in any direction within a cell. In meshworks, the orientation seems to be random. Bundles running through cells often have a bipolar arrangement of filaments. The main exceptions are filaments attached to membranes; experiments to determine their polarities usually indicate that the barbed end (see Section 3.1) is the one attached to the membrane (as in the case of actin filaments attached to Z-lines in striated muscle). This is somewhat surprising, since *in vitro* results (see Section 3.2) suggest that the barbed end is the preferred end for growth of F-actin.

Do these filaments grow by insertion of new monomers between the end of the filament and the membrane? Morphological work on the acrosome of horseshoe crab (*Limulus*) sperm suggests that this is indeed the case. Here stages in the growth of a highly ordered actin-filament bundle on a membrane have been studied. The pointed ends are formed first and are immediately crosslinked to each other to form the fatter near end of the tapered bundle. The thinner end of the bundle is formed last, and contains the barbed ends of the filaments.

The complex of proteins at the membrane end clearly has interesting properties. It may be analogous to a **kinetochore**, a structure that can hold on to the ends of microtubules (see Sections 11.6 & 11.7) but still allow them to grow or shrink. Such a structure presumably binds to sites on the *side* of the polymer and there must be some mechanism for shifting the points of attachment towards the barbed end, possibly involving myosin-like proteins.

5.3 Structural actin

In many situations where actin filaments seem to have a mainly skeletal role, a secondary role may involve interaction with myosin or some other dynamic protein. The following are examples.

5.3.1 Erythrocyte membrane skeleton[12,13,24,25]

Mammalian red blood cells (**erythrocytes**) have no fibrous cytoskeletal structures; shape-determination depends entirely on the membrane and a thin layer of cytoskeletal proteins next to it. The latter retains its shape even after the removal of the lipid membrane with non-ionic detergents. When it is solubilised and run on SDS gels, 75 per cent of it is found to consist of spectrin while about ten other polypeptides, including actin, tropomyosin and ankyrin (see Section 4.4) make up the remainder. Some polypeptides have not yet acquired 'proper' names but are referred to by their positions as **bands on SDS gels**. Proteins closely related to most of these components have been discovered in other cell types, so their properties and interactions are likely to be of general importance.

The structure of the erythrocyte 'membrane skeleton' has been studied in detail by electron microscopy (see Figure 5.2, for example) and found to consist of a chicken-wire-like arrangement of spectrin filaments, with short lengths of F-actin at the 'knots'. Figure 5.3 shows a schematic model of the proposed molecular arrangement. The chain of interaction starts at the membrane with the integral membrane components **glycophorin A** and **band 3 protein** (responsible for transporting ions through the membrane). The peripheral membrane protein, **ankyrin**, links the cytoplasmic domain of band 3 protein to β-spectrin. Actin and **band 4.1 protein** bind to the tails of the spectrin tetramers and both are required to keep the structure intact. Band 4.1 protein apparently strengthens the actin–spectrin attachment in a calcium-dependent manner which involves calmodulin bound to α-spectrin. Structural instability in the presence of Ca^{2+} is probably more important in other cell types, since non-erythroid α-spectrin has a much higher affinity for calmodulin.

Small quantities of myosin are present in red blood cells. It is not known how the molecules interact with the actin, but they may be responsible for a flickering motion of the membrane that appears to be dependent on a supply of energy and probably assists exchange of oxygen and carbon dioxide.

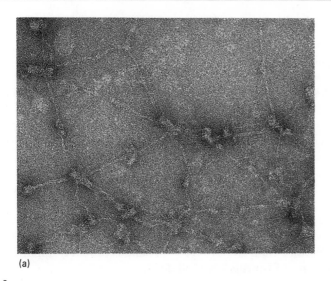

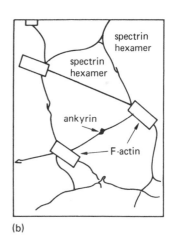

(a) (b)

Figure 5.2

(a) The cytoskeletal framework of a red blood cell, seen after the membrane of a lysed cell has been spread over an electron microscope grid and negatively stained. Many of the protein components can be identified, as indicated in (b). [Micrograph reproduced with permission from Shen, B. W., Josephs, R. & Steck, T. L. (1986). *J. Cell Biol.* **102**, 997–1006.]

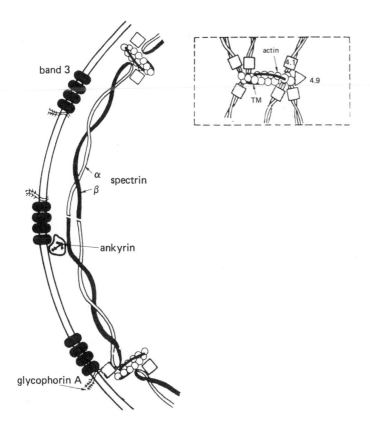

Figure 5.3

Schematic model of the erythrocyte cytoskeletal framework, showing how the spectrin–actin net may be linked to integral membrane protein molecules (the anion transporter protein and glycophorin A) via ankyrin and protein 4.1, respectively. Tropomyosin (TM) and protein 4.9 probably help to stabilise the short actin filaments. [Diagram redrawn from Bennett, V. (1985). *Annu. Rev. Biochem.* **54**, 273–304.]

5.3.2 Intestinal epithelial cells[14]

Vertebrate intestines are lined with multicellular finger-like outpushings, known as **villi**. The surface area is further increased up to twenty-fold by a close-packed array of **microvilli** covering each epithelial cell (Figure 5.4). This array, often called the **intestinal brush border**, is thought to facilitate the uptake of nutrients into the cells.

Each microvillus is 100 nm in diameter and $1-2\ \mu$m long; it is supported by a core of crosslinked actin filaments. Since the microvilli can be isolated in large quantities, the proteins in the core have been studied in detail. The actin bundle is extensively crosslinked by fimbrin and villin (see Section 4.4), and connected to the surrounding membrane by a 110 kD protein related to myosin I (Section 4.3.5). Fimbrin is probably the primary crosslinking agent. Villin, which is also present in the **terminal web** that lies under the microvilli (see Figure 5.4), becomes an actin-filament-severing agent (see Section 4.2) in the presence of raised calcium levels ($>10^{-6}$ M) and is probably important for the rapid destruction of filaments after cell death.

Myosin II and spectrin (sometimes called TW260/240 here) appear to be concentrated in the terminal web. A stress-fibre-like circumferential band of

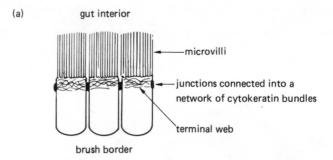

(a)

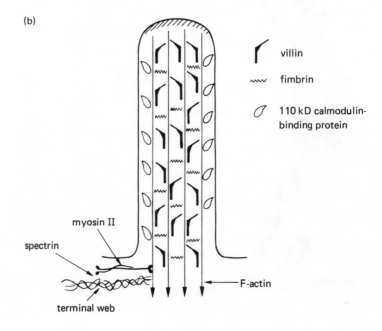

(b)

Figure 5.4

(a) Cells lining a vertebrate intestine form a layer, connected by intercellular junctions (see also Figure 1.1). Their apical surfaces, covered with microvilli, form the so-called brush border. (b) Arrangement of F-actin and associated proteins in a microvillus: each microvillus is supported by a bundle of actin filaments (not to scale). The filaments are crosslinked by fimbrin and villin. A 110 kD myosin-I-like protein links the membrane to the sides of the bundle. Other proteins apparently link the ends of the filaments to the membrane at the microvillar tip. Myosin II and spectrin are associated with other actin filaments in the 'terminal web', at the bases of the microvilli.

filaments, containing actin, tropomyosin, α-actinin and myosin, runs alongside the junctions between neighbouring cells in the epithelium. Also connecting the junctions are strengthening bundles of intermediate filaments (see Section 2.5). The circumferential band is capable of contracting; both it and the terminal web probably have a tension-producing role similar to that of stress fibres.

The brush-border cytoskeleton is not a static structure; in particular, the lengths of the microvilli vary with time. They elongate as the cells migrate towards the tip of the villus and shorten during periods of fasting. Thus, actin filaments in the microvillus cores are apparently able to grow or shrink. Their barbed ends lie at the tip of a microvillus, where they are embedded in a dense 'cap' of protein. This material does not necessarily block the filament ends, as discussed in Section 5.2. It is possible that the filaments are continually growing at the microvillar tip and losing subunits in the terminal web (i.e. treadmilling).

Gut epithelial cells have a relatively short lifetime because of the harsh conditions they experience. They gradually move up from the bases of the villi and are finally shed at the tips. The calcium-activated severing action of villin is thought to be involved in dissolving the cytoskeleton at this time.

5.3.3 Actin bundles in stereocilia and microvilli[16,17,33]

Hair cells in the cochlea of the inner ear contain processes called **stereocilia**. These act as levers, transmitting force to the sensory cell membrane: to function effectively, they have to be stiff. They contain internal bundles of highly crosslinked actin filaments arranged with the same polarity and apparently in perfect axial register. The latter feature seems particularly surprising in view of the fact that the lateral packing arrangement of filaments, as observed in cross-section, appears to be quite random. The molecule responsible for crosslinking actin bundles in stereocilia has not yet been identified, but is probably similar to **fascin**, a 55 kD protein from sea-urchin oocytes (see Section 4.4.6), which is sufficient *in vitro* to crosslink actin filaments in register. The separation between filaments is similar in both cases; 10 nm in fascin bundles and 9 nm in stereocilia.

Sea-urchin eggs produce large numbers of fine microvilli (Figure 5.5) on their surfaces after fertilisation. Each microvillus contains a small polar bundle of actin filaments; the stiff bundle appears to prop up an outpushing of membrane, probably providing a means of packing a large reserve of membrane into a small space in preparation for cell division. Fascin is thought to be responsible for crosslinking these actin filament bundles.

DeRosier and Tilney have analysed the structures of both kinds of bundle by thin sectioning, negative staining and optical diffraction of electron micrographs. Filaments in actin bundles isolated from sea-urchin eggs are packed in approximately hexagonal arrays and the arrangement of crosslinks can be determined directly. Fascin molecules bind to diametrically opposite sides of a filament (Figure 4.13). The other end of each fascin molecule binds to an adjacent filament, automatically putting them into axial register. Further up the actin helix there are monomers that face other neighbouring filaments and new crosslinks can be made. The crosslinking is maximised in a hexagonal arrangement of filaments, with three levels of crosslinks per actin repeat (see Figure 5.6). Electron micrographs often show stripes across the bundles, with an average spacing of one-third of the actin filament repeat (Figure 5.5).

Figure 5.5

Electron micrograph of an actin bundle in one of the numerous microvilli that sprout from the surface of a sea-urchin egg after fertilisation. The egg membrane has been removed and the filaments negatively stained, as for Figure 1.4. The lateral striations due to the presence of fascin have a longitudinal spacing of 12 nm. [Adapted from Spudich, J. A. and Amos, L. A. (1976). *J. Mol. Biol.* **129**, 319–331.]

(a)

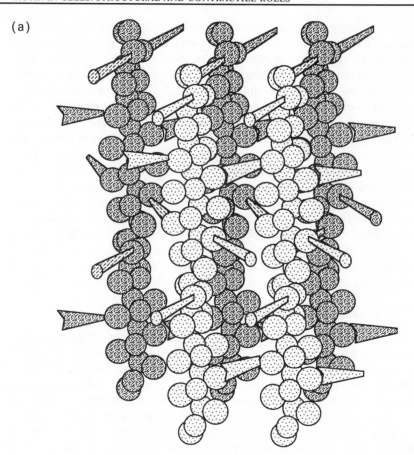

(b)

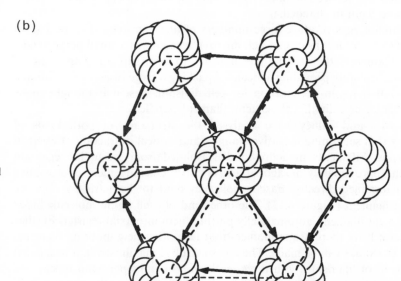

Figure 5.6

Bundling of actin filaments by
fascin into approximate hexagonal
arrays. As the filaments twist
around, fascin molecules can
attach to actin monomers facing
in approximately the right
direction to make crosslinks with
six different neighbours. (a) The
three different levels per
actin-filament crossover repeat.
(b) An end-on view of the
hexagonal arrangement.

In the stereociliary bundles, bundling seems to be random. Although crosslinks must still link diametrically related sites, they do not force the filaments to pack hexagonally; perhaps assembly is too rapid for maximisation of the number of crosslinking molecules that are bound. The result is that crosslinks may lie at any actin monomer level, rather than at only three levels per actin filament repeat, and there are no obvious stripes to the bundles. However, the fact that all crossbridges at a given axial level are oriented at a particular angle to the filaments gives rise to some striking patterns in thin sections through the bundles.

The microvilli of motile cells such as **fibroblasts** (Section 6.2) are supported by randomly packed bundles of actin filaments. It has not been demonstrated directly that all the filaments have the same polarity but this is probably so: it is likely that the bundling is due to proteins similar to those described above.

5.3.4 *Nitella* and *Chara*

Section 10.6 describes the transport of organelles on microtubules, which has been studied particularly in animal cells. However, movement of membranous organelles against actin filaments is clearly also possible (Section 4.3.5). In plant cells Golgi and ER membrane components may perhaps travel along microtubules, but many small vesicles and mobile chloroplasts appear to move against actin bundles.

Cytoplasmic streaming is often studied in the giant algal cells, *Nitella* and *Chara*. Their cell membranes are lined with rows of chloroplasts and these in turn support substantial bundles of actin filaments (Figure 5.7). The cells are sufficiently large to break open and extrude the contents, so the motility has often been studied *in vitro*. More recently, Sheetz and Spudich used opened-out *Nitella* cells to watch the movement of fluorescent beads, coated in myosin fragments, along the actin bundles (see also Section 4.3).[15]

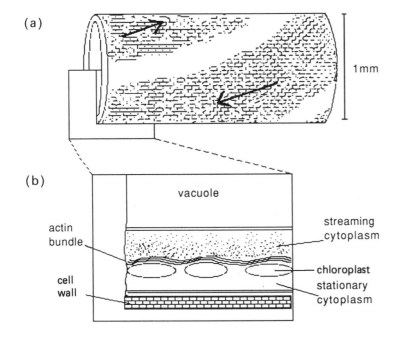

Figure 5.7

(a) A segment of one of the giant cells of the alga *Nitella*. Bands of cytoplasmic streaming are seen close to the cell cortex.
(b) The arrangement of cytoplasmic components in a cross-section. Actin bundles are supported on lines of chloroplasts and neighbouring cytoplasm streams past.

Labelling with myosin fragments has shown that the arrangement of filaments in a bundle is polar and that all bundles in the same zone have the same polarity. The barbed ends of the filaments point towards the direction of streaming. Since the bundles circuit the whole cell, cytoplasmic streaming along them brings material back to the same point. Streaming probably involves membranous organelles and free actin filaments being pulled along the cables by myosin, and fluid being carried along with them.

5.4   Striated muscle as a model for contractile systems

Striated muscle is the best-understood actomyosin system and the concepts developed for it help to provide a basis for understanding motility in non-muscle cells. Because it has a paracrystalline structure, with many parallel filaments moving in unison (see Section 1.6), there is the possibility of following the structural changes by various biophysical methods and of correlating them with force production and biochemical steps in the enzyme mechanism. However, the high degree of structural order is a specialisation for producing force on a macroscopic scale and many features are not needed for ordinary cellular motility. There are many excellent accounts of muscle motility and the variations that occur in different kinds of muscle. Here, we attempt to identify the essential features of the mechanism, which may be conserved in non-muscle systems.

5.4.1   Crossbridge cycles and tension production[1-3]

The myosins of muscle and non-muscle cells are all essentially similar, so the mechanism of producing force and movement is probably common to the actomyosin systems of all eukaryotic organisms. The original model of the crossbridge cycle (based partly on interpretation of EM images of sections through muscle) involved a rod-like crossbridge that bound initially at an angle of 90° to an actin filament and remained attached while rotating to a 45° angle. It was proposed that the rotation would produce a relative sliding of actin and myosin filaments if movement were possible. Alternatively, the rotation could be compensated for by stretching an elastic element (see Figure 5.8), thereby producing tension in the muscle.

The fact that the S-1 subfragment alone is sufficient to produce motility *in vitro* (Section 4.3.3) rules out the alternative model proposed by Harrington in which S-1 remains at a fixed angle whilst the α-helical rod S-2 contracts under the influence of calcium. Yet the evidence for an obvious change in crossbridge angle is very weak. Beads observed travelling along *Nitella* actin cables (Section 5.3.4) show no sign of the rolling that might be expected from a rowing boat with oars only on one side. Thus, instead of regarding the crossbridge as a simple hinged rigid rod, current models of the mechanism focus on the fact that the myosin head consists of a number of structural domains.

Figure 5.8

The concept, introduced by A. F. Huxley, of representing the myosin crossbridge as a rigid rod combined with an elastic element. (a) The head binds to actin. (b) The crossbridge movement during the power stroke is initially taken up by stretching the elastic element. This explains, simply, how muscles can still exert tension even when prevented from shortening.

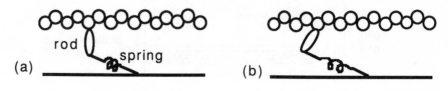

(a)   rod   spring   (b)

The domain that binds directly to actin may indeed remain at a fixed angle, while changes in its interactions with other domains produce the conformational changes that produce sliding. Such changes might be similar to those that appear to occur in allosteric enzymes whose structures are known to atomic level. The elasticity required for tension production can be spread throughout the head.[4]

Muscles are able to maintain tension for long periods even when there is a resistance of actual shortening (isometric contraction). Crossbridges may be arrested at some stage of the conformational cycle; if they are prevented from reaching the endpoint of the stroke they cannot be released by ATP binding. Isometric-tension production also occurs in non-muscle cells, as will be discussed later. In fact, the maintenance of tension under static conditions may be an important function of myosin in non-muscle cells.

5.4.2 Enzymatic activity of myosin[1–3,32]

The essential properties of the myosin enzyme reaction, according to the **Lymn-Taylor model**, are as follows:

(a) The binding of nucleotide (usually ATP) to the myosin head in its final attached state causes it to detach *very rapidly* from actin.

(b) The detached myosin head hydrolyses the nucleotide *fairly rapidly*, but, without help, it releases the products only *slowly* from the hydrolysis site.

(c) Actin catalyses the reaction by increasing the rate of dissociation of the products. This is known as **actin activation** of myosin ATPase. Myosin with bound products readily binds to actin if a site is available; the release of the products is then coupled to the change in conformation described above. The release of the phosphate (P_i) cleaved from the nucleotide is thought to provide the energy for the shearing force.

It is helpful to draw a simple diagram of the various possible kinetic states, as in Figure 5.9. An independent crossbridge with only one actin-binding site must either be unattached (top row of states) or bound to the polymer (bottom row). If the crossbridge is to do work in each cycle, it must go from a state on

Figure 5.9

Kinetic diagram for the interaction between actin and myosin (also for microtubules and kinesin or dynein—see Chapter 8). Any crossbridge cycle must involve crossing over from a (free) state on the top row to a (bound) state on the bottom row at some point, and back again at some other point. There are a number of possible routes. The relevant set of rate constants will favour a particular route and some steps may never take place in practice.

The following conditions seem to be true for all three enzymes:

(a) dissociation of P.E by ATP (steps 1 and 2) is fast;

(b) hydrolysis (step 3) is fast;

(c) unassisted loss of products (step 9) is slow;

(d) binding to polymer (step 4') followed by polymer-activated loss of products (step 5) is faster;

(e) dissociation in the absence of ATP (step 10) is very slow (this produces the rigor state).

Peculiar to *kinesin*:

(f) If AMPPNP is bound instead of ADP + P_i, step 5 is probably blocked, and the head remains attached to the microtubule.

(g) the two heads are more cooperative (*in vitro*) than those of myosin or dynein (see Figure 8.17).

(h) If Mg^{2+} is removed, hydrolysis is inhibited and E remains attached. Perhaps the first head cannot carry out step 5 if the second head is either unable to bind ATP or cannot hydrolyse it.

Peculiar to *dynein*:

(i) If Mg^{2+} is lacking, hydrolysis is inhibited; E cannot carry out step 3 and remains detached. (Myosin can hydrolyse ATP in the absence of Mg^{2+} ['EDTA ATPase activity'] but does not interact with actin under these conditions.)

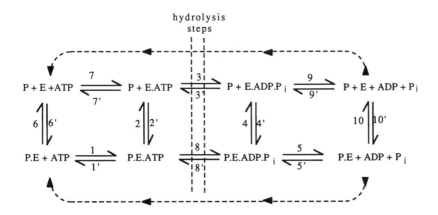

Enzyme E = myosin, kinesin or dynein
Polymer P = F-actin or microtubule
P.E = complex between enzyme and polymer

ATP = adenosine triphosphate
ADP = adenosine diphosphate
P_i = released phosphate

the top line to a state on the bottom line and back again by another route. The normal cycle for myosin (also that of kinesin or dynein interacting with microtubules, discussed in Section 8.8) is thought to follow steps 1, 2, 3, 4′ and 5 (see Figure 5.10). Once you can find your way round this simple diagram, more complex schemes that have been published, with various extra steps, should not be too daunting.

Even the scheme shown here includes steps which are not part of the simple cycle outlined at the beginning of this section. Eisenberg and colleagues have found that every step is actually a reversible equilibrium reaction whose

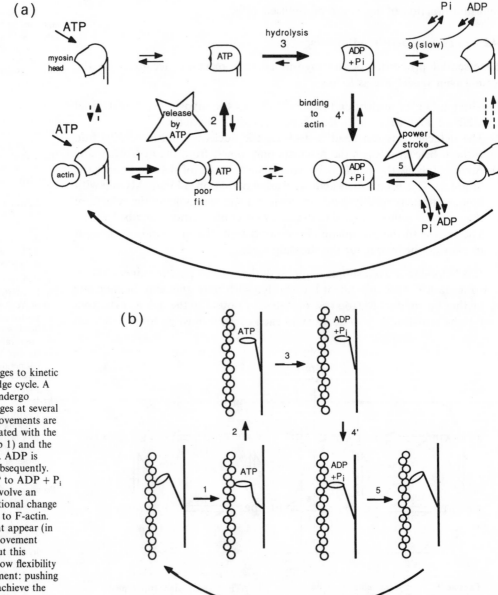

Figure 5.10

(a) Scheme relating conformational changes to kinetic steps of the crossbridge cycle. A myosin head must undergo conformational changes at several stages: the largest movements are thought to be associated with the binding of ATP (step 1) and the release of P_i (step 5). ADP is probably released subsequently. The splitting of ATP to ADP + P_i is also thought to involve an important conformational change that favours binding to F-actin. (b) Steps 1 + 2 might appear (in (a)) to reverse the movement achieved in 4′ + 5 but this diagram illustrates how flexibility biases the net movement: pushing on a rope does not achieve the reverse of pulling on it.

Table 5.1 The range of sliding speeds for myosins	Motor molecule	Speed ($\mu m/s$)
	Rabbit skeletal muscle myosin	2–5
	Nitella myosin	80
	Acanthamoeba myosin I	0.05

net effect depends on the concentrations of the reagents: myosin–ATP rapidly rebinds to and re-dissociates from F-actin, for example, and even the splitting of ATP appears to be reversible. Each step is not the automatic consequence of the previous step taken, as in a mechanical engine: instead, the crossbridge cycle is the most probable outcome of a series of chemical equilibria. ATP hydrolysis produces a cycle in the appropriate direction because the equilibrium between myosin and ATP favours binding whereas that between myosin and $ADP + P_i$ favours dissociation.

5.4.3 Myosin speed and structural organisation

Different myosins do vary considerably with regard to their speed. Highly organised systems like skeletal muscle can achieve great speed by putting many molecules to work in unison, so that large viscous forces can be overcome. The longitudinal organisation of actin and myosin filaments into sarcomeres (Section 1.6) means that the serial sum of many submicroscopic movements produces a macroscopic movement. Non-muscle stress fibres, which are described below, seem to be organised along similar lines to muscle sarcomeres.

More slowly working species of molecule may be better where they must work singly or in small groups, to reduce viscous forces. Single myosin heads may, for example, need to be capable of transporting vesicles along actin cables.

5.4.4 Ca^{2+}-control of striated muscles[31,10]

The interaction between myosin heads and actin filaments is prevented in striated muscle unless specifically activated by calcium ions, which are released from stores in a vesicular system known as the **sarcoplasmic reticulum**. In vertebrate muscle, tropomyosin filaments running alongside both long-pitch helices of an actin filament (see Figure 4.12) inhibit the interaction, either by getting in the way of the heads or affecting the conformation of actin in some way. The position of tropomyosin is controlled by a protein complex called **troponin**, one of whose components, **troponin C**, resembles calmodulin (Section 4.4). When troponin C binds calcium, conformational changes occur in the whole complex (Figure 5.11). The effect is known as **thin-filament regulation**.

Except in vertebrate muscle, myosin activity can be directly controlled by calcium, providing a mechanism known as **thick-filament regulation**. The effect of binding calcium to the myosin head depends on the presence of one of the light chains, the so-called **regulatory light chain** (**RLC**—see Section 4.3.1). Interaction with actin is controlled in this way in the muscles of invertebrates such as scallops. The muscles of some organisms have both thin-filament and thick-filament regulation. Since the interaction with actin is controlled, even in solution, it is thought that conformational changes in the light chain are transmitted to the enzymatic regions of the heavy chain.

Figure 5.11

Scheme for thin-filament control of the interaction between myosin heads (S-1) and F-actin (A), viewed down a filament from the minus end. In the blocked state (a), tropomyosin (TM) lies in the grooves running along F-actin, each molecule interfering with the binding of myosin to seven actin monomers (see Figure 4.8). It is held in place by the troponin complex; troponin-T (T) attaches the complex to TM, troponin-I (I) binds to actin. (b) Ca^{2+} binding to troponin-C (C) alters the conformation of the complex, probably preventing the interaction between TN-I and actin. TM is free to roll over the surface of actin; according to the 'steric blocking hypothesis' such movements may expose sites on actin to which myosin S-1 can bind. [The subunit arrangements shown are modified from Zot, A. S. and Potter, J. D. (1987). *Annu. Rev. Biophys. Biophys. Chem.* **16**, 535–559 to be compatible with Figures 3.1, 3.2 & 4.8.] There may be other equally good schemes but the essential features are likely to be similar.

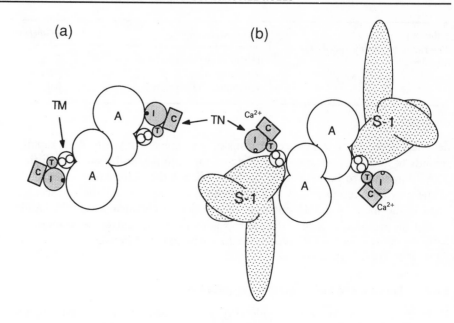

5.5   Smooth-muscle and non-muscle contraction[5,7–10]

Although actin and myosin are the major constituents of smooth-muscle cells, they are not highly organised as in striated muscle. The cytoskeleton can be rearranged to produce contraction in different directions, as required. These processes are thought to be similar to some that occur in non-muscle cells.

5.5.1   Myosin light-chain phosphorylation[7–10]

Even individual thick filaments are labile and it is thought that the assembly and disassembly of myosin into and from filaments is controlled by the same mechanism that regulates actin–myosin interactions (Figure 5.12). The regulatory light chains of smooth muscle and similar non-muscle myosins have a site that can be phosphorylated. If the RLC is *not* phosphorylated, the ATPase activity and any interaction with actin are inhibited in the absence of Ca^{2+}. Furthermore, a site on the tail is able to interact with the junction between the two heads, producing a folded-up molecule (Figures 4.5 & 4.7) that is unable to assemble into filaments.

It is *possible* for the molecule to unfold again without any modification but the tail is much more likely to be released if the RLC is phosphorylated. This modification is carried out, in the presence of calcium ions, by a specific calmodulin-dependent kinase. Unfolded molecules are then able to assemble into filaments as well as being able to interact with actin. Once the RLC is phosphorylated, it is not even necessary for Ca^{2+} to switch on the activity of the head, so it will not matter if the activating pulse of calcium ions is lost whilst the thick filaments are assembling. The equilibrium between myosin monomers and filaments is further controlled by the effects of ATP hydrolysis (see the legend to Figure 5.12). These features make smooth muscle more versatile and responsive to local stimuli than striated muscle.

The filaments formed by smooth-muscle myosin are small ribbons, quite different in structure from the long, helically arranged, thick filaments of striated muscle (Figure 2.6). Each face of a ribbon consists of molecules arranged with the same polarity, with heads projecting from the end. Opposite faces have opposite polarities. In such an arrangement the crossbridges on each side can travel in a particular direction, along any actin filament it meets which has a suitable polarity, and effect the very extensive contractions observed in smooth muscle.

Another important characteristic of smooth muscle, namely the long-term maintenance of tension, seems to be facilitated by reversible crosslinking of actin filaments by proteins such as filamin and caldesmon (see Section 4.4). When activated by calcium, calmodulin binds to caldesmon and inhibits its binding to F-actin. Filamin-binding is not affected by calcium and actin–filamin aggregates are thought to perform long-term structural roles.

5.5.2   Myosin heavy-chain phosphorylation

In invertebrate non-muscle myosins, filament assembly may be controlled by phosphorylation of the myosin *heavy* chain, as described in Section 4.3.2. Whereas light-chain phosphorylation inhibits molecular folding and promotes activity, heavy-chain phosphorylation inhibits the formation of filaments and also lowers their actin-activated ATPase in some unknown manner. It is not clear whether the two opposing activities ever coexist in the same cell.

5.6   Stress fibres in non-muscle cells[6,18,19]

If fibroblasts in tissue culture are confluent (i.e. in contact with each other, so that no cell migration is occurring), staining with an anti-actin antibody produces a pattern of fibres (Figure 5.1). It is different from intermediate filament and microtubule patterns in that the lines are straight and do not emerge from a specific centre. These lines, which are called **stress fibres**, can also be seen under phase contrast. They are connected to the cell membrane at the adhesion plaques described in Section 5.6.2. Such striking bundles of actin are not seen in moving cells, nor do they seem to be evident in cells in normal tissue. But they are worth studying since the intermolecular interactions are probably representative of those found elsewhere, in less well-ordered arrangements in non-muscle cells.

Whereas anti-actin binds all along a stress fibre, anti-myosin binds at regular intervals of about 0.4 μm. Different staining patterns by antibodies to the heads or tails of myosin suggest that myosin in stress fibres is organised into bipolar filaments, as in muscle (Section 4.3.2). Anti-α-actinin and anti-filamin both bind in between the myosin regions. Patterns obtained with anti-tropomyosin are complementary to those obtained with anti-α-actinin. All this suggests a sarcomere-like organisation, as diagrammed in Figure 5.13.

Stress fibres also contain proteins which may regulate their activity, such as caldesmon and myosin light-chain kinase, as in smooth muscle (Section 5.5). If cut free from their attachment points with a laser microbeam in a permeabilised cell, stress fibres can be made to contract when supplied with Mg^{2+} and ATP;

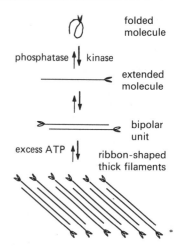

folded molecule

phosphatase | kinase

extended molecule

bipolar unit

excess ATP

ribbon-shaped thick filaments

Figure 5.12

Scheme for the control of smooth-muscle myosin assembly into filaments and of its interaction with actin filaments.[7–9] Two-headed myosin molecules can be in an extended or folded configuration. The extended state is greatly favoured if the regulatory light chains in the heads are phosphorylated by a Ca^{2+}-activated kinase. Folded molecules are inhibited from assembling: they can bind ATP but ATPase activity is inhibited. Extended molecules, bipolar dimers and filaments are in equilibrium with one another. As for F-actin (Section 3.2), each assembly/disassembly step has a critical concentration, which is higher for ATP-myosin than for ADP-myosin. Lack of actin to activate the myosin ATPase may thus make the filaments more likely to disassemble.

Figure 5.13

Probable sarcomere-like organisation of proteins in stress fibres.[18] These bundles occur predominantly in stationary cells, where the components have time to organise into arrangements optimising the interactions between them and balancing the tensile forces. Similar interactions between individual components almost certainly take place in a less organised manner in motile cells.

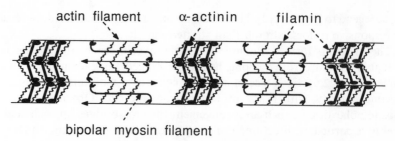

the inter-myosin gaps close up. Such shortening is not observed in living cells with large numbers of obvious stress fibres, as such cells are usually stationary. However, cells do contract whilst moving (see Section 6.4) and microfilament bundles appear to be involved, especially towards the rear.

5.6.1 Tension versus motility[21]

Careful study of cells in culture shows that there are far fewer stress fibres in moving than in stationary cells. There is a correlation between the presence of stress fibres and tension; only flattened stationary cells develop sufficient tension to deform a latex surface. In contrast, protrusive motility seems to require a lack of tension. If small motile clusters of epidermal cells are artificially stretched using a micromanipulated needle, protrusive activity is dramatically reduced and microfilaments line up parallel to the tension. It seems likely that tension is important during the development of an animal, influencing the arrangement of cytoskeletal elements in its cells.

Another interesting result which supports the idea that tension inhibits motility was obtained by the injection of anti-myosin antibodies into stationary chicken fibroblasts.[20] The antibodies were carefully characterised monoclonals, known to inhibit myosin filament formation *in vitro*. After several hours, the stress fibres gradually broke down and finally ruffling motile activity appeared. This experiment also supports evidence that filamentous myosin is not required for motility (see Section 6.4.3).

5.6.2 Adhesion plaques[11,13,35]

Cells with the ability to form tissue make close contacts not only with neighbouring cells (**tight junctions**, included in Figure 1.14) but also with the substrate on which they are growing. In *in vitro* culture, a glass or plastic surface forms the substrate but there are equivalent extracellular substrates *in vivo*, such as **basement membranes**, ends of tendons, and so on. Where the cell membrane makes such close contacts, a dense plaque of material is observed on its cytoplasmic surface (Figure 5.14a). The plaques contain α-actinin and vinculin, which together bind to F-actin (Section 4.4); and talin, which interacts strongly with vinculin.

It has been shown *in vitro* that talin also binds to **integrin**, a 140 kD integral glycoprotein in the membrane, and integrin itself has been localised to focal contacts in fixed cells. The extracellular domain of integrin is a receptor for **fibronectin**, one of the numerous secreted proteins found on the extracellular surface of the membrane next to plaques, with the apparent role of gluing the membrane to the substratum. Fibronectin seems to be a fairly universal adhesion protein; **laminin** is another such molecule. There are also tissue-specific

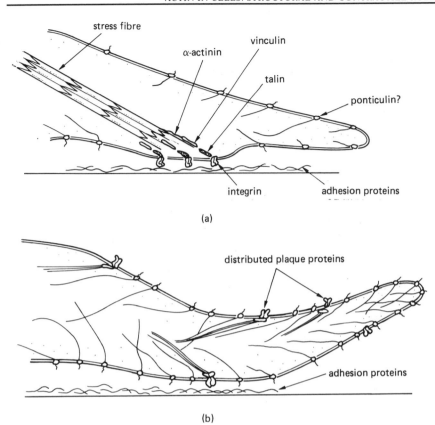

(a)

(b)

Figure 5.14

(a) In a stationary cell, actin filaments associate to form stress fibres; the transmembrane proteins to which they are attached, directly or indirectly, and which also make contact with adhesion proteins stuck to the substrate, will be clustered to form substantial adhesion points at the ends of stress fibres.[35] (b) The cytoskeletal components in a moving cell undergo constant rearrangement and the transmembrane proteins apparently distribute themselves evenly. Stress fibres are uncommon in motile cells (Section 6.2).

cell-adhesion molecules (CAM), such as the **NCAM** found on neurons. The membrane includes integral glycoproteins that are receptors for adhesion molecules, such as a 140 kD fibronectin receptor that has been localised to focal contacts. Binding between talin and this fibronectin receptor has been demonstrated *in vitro*.

5.7   Summary

Actin is a major protein in cytoplasm, often the most abundant protein in the cell. It is an extremely well-conserved protein with little variation in sequence. Yet, thanks to a great variety of associated proteins, it plays a wide range of roles in shape maintenance and in cellular motility. Combined with myosin, it can form powerful contractile systems. Elsewhere, its polymerisation and disassembly are subject to exact control by various accessory proteins. These endow eukaryotic cells with a remarkable plasticity (to be discussed further in Chapter 6).

5.8   Questions

5.1 To what extent do stress fibres and adhesion plaques resemble striated muscle fibrils?

5.2 Outline three mechanisms for controlling the interaction of myosin with actin filaments.

5.3 How is the characteristic shape of the red blood cell produced? What properties of spectrin are particularly relevant?

5.4 Compare the structural requirements for a protein that crosslinks actin filaments into a gel and for one that forms oriented bundles of actin. [A question for biophysicists.]

5.9   Further reading

Reviews

1 Taylor, E. D. (1986). Cell motility. In *J. Cell Sci. Suppl.* **4** (*Prospects in Cell Biology*), 89–102.
   (*A short readable review, centred on the mechanism of myosin.*)

2 Johnson, K. A. (1985). Pathway of the microtubule-dynein ATPase and the structure of dynein: a comparison with actomyosin. *Annu. Rev. Biophys. Biophys. Chem.* **14**, 161–188.
   (*This includes the Lymn-Taylor-Eisenberg kinetic model in a readable form!*)

3 Eisenberg, E. & Hill, T. L. (1985). Muscle contraction and free energy transduction in biological systems. *Science* **227**, 999–1006.
   (*A comparison of myosin–actin interaction with the ATPase that transports Ca^{2+}: worth some effort if you want to understand kinetics.*)

4 Vibert, P. & Cohen, C. (1988). Domains, motions and regulation in the myosin head. *J. Muscle Res. & Cell Motil.* **9**, 296–305.

5 Small, J. V. (1987). The cytoskeleton: selected topics. In *Fibrous Protein Structure*, ed. Squire, J. M. & Vibert, P. J., pp. 357–388. London & San Diego: Academic Press.

6 Burridge, K. (1981). Are stress fibres contractile? *Nature* **294**, 691.

7 Small, J. V. (1988). Myosin filaments on the move. *Nature* **331**, 568–569.

8 Citi, S. & Kendrick-Jones, J. (1987). Regulation of non-muscle myosin structure and function. *Bioessays* **7**, 155–159.

9 Cross, R. A. (1988). Smooth muscle contraction: what is 10 S myosin for? *J. Muscle Res. & Cell Motil.* **9**, 108–110.

10 Kendrick-Jones, J. & Scholey, J. M. (1981). Myosin-linked regulatory systems. *J. Muscle Res. & Cell Motil.* **2**, 347–372.

11 Burridge, K., Fath, K., Kelly, T., Nuckolls, G. & Turner, C. (1988). Focal adhesions: transmembrane junctions between the extracellular matrix and the cytoskeleton. *Annu. Rev. Cell Biol.* **4**, 487–425.

12 Bennett, V. (1985). The membrane skeleton of human erythrocytes and its implications for more complex cells. *Annu. Rev. Biochem.* **54**, 273–304.

13 Brown, S. S. (1985). The linkage of actin to non-erythroid membranes. *Bioessays* **3**, 65–68.

14 Mooseker, M. S. (1985). Organization, chemistry and assembly of the cytoskeletal apparatus of the intestinal brush border. *Annu. Rev. Cell Biol.* **1**, 209–241.

Some original papers

15 Sheetz, M. P. & Spudich, J. A. (1983). Movement of myosin-coated fluorescent beads on actin cables *in vitro*. *Nature* **303**, 31–35.

16 Tilney, L. G., Bonder, E. M. & DeRosier, D. J. (1981). Actin filaments elongate from their membrane-associated ends. *J. Cell Biol.* **90**, 485–494.

17 Tilney, L. G., DeRosier, D. J. & Mulroy, M. J. (1980). The organization of actin filaments in the stereocilia of cochlear hair cells. *J. Cell Biol.* **86**, 244–259.

18 Langanger, G., Moermans, M., Daneels, G., Sobiesszek, A., De Brabander, M. & De Mey, J. (1986). The molecular organization of myosin in stress fibers of cultured cells. *J. Cell Biol.* **102**, 200–209.

19 Byers, H. R. & Fujiwara, K. (1982). Stress fibers in cells *in situ*: immuno-fluorescence visualization with anti-actin, anti-myosin and anti-alpha-actinin. *J. Cell Biol.* **93**, 804–811.

20 Honer, B., Citi, S., Kendrick-Jones, J. & Jockusch, B. M. (1989). Modulation of cellular morphology and locomotory activity by antibodies against myosin. *J. Cell Biol.* **107**, 2181–2190.

21 Kolega, J. (1986). Effects of mechanical tension on protrusive activity and microfilament and intermediate filament organization in an epidermal epithelium moving in culture. *J. Cell Biol.* **102**, 1400–1411.

22 Bray, D., Heath, J. & Moss, D. (1986). The membrane-associated 'cortex' of animal cells: its structure and mechanical properties. *J. Cell Sci. Suppl.* **4**, 71–88.

23 Spudich, A., Wrenn, J. T. & Wessels, N. K. (1988). Unfertilized sea urchin eggs contain a discrete cortical shell of actin that is subdivided into two organizational states. *Cell Motil. & Cytoskel.* **9**, 85–96.

24 Shen, B. W., Josephs, R. & Steck, T. L. (1986). Ultrastructure of the intact skeleton of the human erythrocyte membrane. *J. Cell Biol.* **102**, 997–1006.

25 Byers, T. J. & Branton, D. (1985). Visualization of the protein associations in the erythrocyte membrane skeleton. *Proc. Natl. Acad. Sci., USA* **82**, 6153–6157.

Additional references

26 Irving, M. (1987). Muscle mechanics and probes of the crossbridge cycle. In *Fibrous Protein Structure*, ed. Squire, J. M. & Vibert, P. J. London & San Diego: Academic Press.

27 Cooke, R. (1986). The mechanism of muscle contraction. *CRC Crit. Rev. Biochem.* **21**, 53–118.

28 Bagshaw, C. R. (1982). *Muscle Contraction.* London & New York: Chapman & Hall.

29 Huxley, A. F. (1980). *Reflections on Muscle.* Princeton, N.J.: Princeton University Press.

30 Marston, S. B. & Smith, C. W. J. (1985). The thin filaments of smooth muscles. *J. Muscle Res. & Cell Motil.* **6**, 669–708.

31 Sellers, J. R. & Adelstein, R. S. (1987). Regulation of contractile activity. In *The Enzymes*, ed. Boyer, P. & Krebs, E. G., vol. 18, pp. 381–418. San Diego, CA: Academic Press.

32 Adelstein, R. S. & Eisenberg, E. (1980). Regulation and kinetics of the actin–myosin–ATP interaction. *Annu. Rev. Biochem.* **49**, 921–956.

33 Spudich, J. A. & Amos, L. A. (1976). Structure of actin filament bundles from microvilli of sea urchin eggs. *J. Mol. Biol.* **129**, 319–331.

34 Stossel, T. P. (1984). Contribution of actin to the structure of the cytoplasmic matrix. *J. Cell Biol.* **99**, 15s–21s.

35 Geiger, B. (1989). Cytoskeleton-associated cell contacts. *Curr. Opinion Cell Biol.* **1**, 103–109.

6 ACTIN IN CELLS: ROLES INVOLVING ASSEMBLY

6.1 Polymerising activity[4,6,20]

Actin assembly can be quite dramatic in some cells, especially the specialised 'disposable' kinds such as platelets and sperm. In other, more long-lived, cells the control of polymerisation seems to be more complex and subtle. Studies involving the injection of fluorescently labelled actin into fibroblasts show rapid exchange of labelled actin *monomers* in the motile regions of a cell (the lamellipodia—see Section 6.4), suggesting continuous cycles of assembly and disassembly. Yet in other regions of the cell, where actin bundles (stress fibres—Section 5.6) occur, exchange is very slow.

Injection of fluorescent **phalloidin**, a drug that binds specifically to actin filaments and shows very little affinity for the monomers, gives results indicating that filaments *can* be transferred to different structures within the main body of the cell without having to go through disassembly and reassembly. Thus, filaments labelled in stress fibres in an interphase cell may appear in the contractile ring formed during cell division (Section 11.9). This is surprising, considering the known viscosity of cytoplasm and the pore-size measurements given in Section 1.2. It has been proposed that filaments may 'reptate' (move in a snake-like manner) through the mesh, but this is difficult to reconcile with the apparent stiffness of F-actin (Section 3.1). The mobility of the filaments may be aided by being severed into short lengths and the pieces may be carried along by fluid flow.

6.1.1 *Thyone* acrosomal filament[21]

During fertilisation of an echinoderm egg, the sperm head produces a long thread, known as an **acrosomal process**, which consists of a bundle of actin fibres enclosed by a membrane. The bundle is assembled from a store of actin monomers which are complexed with profilin (**profilactin**—see Section 4.2) until the sperm is activated. It is usually a very rapid process, designed to drive the thread into the egg. Labelling with myosin subfragments has shown that the growing (barbed) ends of the filaments are always at the tip of the process, as in more slowly-growing protrusions (Section 5.3).

The acrosome of the sea cucumber, *Thyone*, is particularly dramatic, extending up to 90 μm in less than 10 s (Figure 6.1a & b). Actin polymerisation was once assumed to drive the elongation process but more recent calculations by Oster have shown that diffusion-limited F-actin assembly is much too slow to account for the elongation.[20] His alternative explanation is that *osmotically-generated hydrostatic forces* drive the extension, and an associated flow of fluid brings actin monomers to the tip. The polymerising bundle of actin shapes and stabilises the surrounding tube of membrane.

How is the hydrostatic pressure produced? There appears to be a correlation between actin assembly and protrusive activity; activated platelets polymerise much of their soluble actin whilst spreading outwards, and the ruffling activity of other motile cells clearly involves actin polymerisation at the edge of the ruffle. Oster has suggested that processes involving actin polymerisation upset the local osmotic pressure equilibrium. When a sperm head is activated, channels in the membrane may open to allow the influx of water to dilute the high concentration of profilactin and begin the protrusion process. Inoué and Tilney

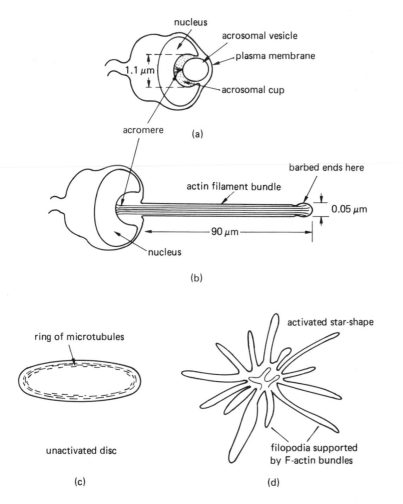

Figure 6.1

(a) & (b) Schematic drawings of the sperm head from *Thyone*. (a) Before the explosion, granular material including the actin–profilin complex is concentrated into a hollow in the side of the nucleus, known as the acrosomal cup (volume $\sim 10^{-13}$ ml). After activation (b), actin assembles into filaments from a filament-organising centre (the acromere) in the middle of the cup. Membrane, originally stored in the form of a large vesicle that fuses with the plasma membrane, is pushed out ahead of the rapidly-assembling filament bundle. (c) & (d) Platelet activation. Initially (c) the cell shape is determined by very long microtubules, coiled into a round bundle. The microtubules break down when the cell is activated. (d) The cell becomes star-shaped, with long membranous protrusions, known as filopodia, supported by bundles of actin filaments that assemble on activation.

found that the volume of the acrosomal region nearly doubles within 70 ms of the induction of protrusion and there is a precipitous drop in the refractive index. This is direct evidence of a decrease in concentration of solids in the cytoplasm.

Conditions at the growing tip cause actin and profilin to dissociate (see Section 4.2.9), increasing the number of molecules at the tip. This may cause a further influx of water to be driven osmotically into the acrosomal tip. It is likely that fresh profilactin subunits are continuously brought to the tip by fluid flow brought about by the large influx of water. As the membrane is driven out and actin monomers are released, actin filaments are able to elongate without hindrance. Actin polymerisation displaces water molecules bound to the monomers (see Section 7.3.1 on microtubule assembly, where the effect is even greater), which may also help to generate space around the filament end.

6.1.2   Platelet activation[24]

A blood platelet is a small white blood cell which is produced by the subdivision of a larger parent cell: it lacks a nucleus. Before activation the platelet has a discoidal shape maintained by a ring-shaped marginal bundle of microtubules. Various external stimuli, including the presence of thrombin, cause a dramatic change (Figure 6.1c & d); the cell assumes a stellate shape and becomes motile. Profilin seems to be a major participant in this transformation.

Approximately half the total actin in inactive platelets is thought to be bound to profilin as a soluble 1:1 profilactin complex; after activation, the complex apparently dissociates, leaving the actin free to polymerise. Other actin-binding proteins present, such as spectrin, actin-binding protein, tropomyosin and myosin, can then play a role in organising the filaments for motile activities. Like the echinoderm acrosomal body, a blood platelet needs to be activated only once, so the entire change does not need to be reversible. But profilin probably continues to play a role in the motile activity, as elsewhere (Section 6.5).

6.2   Fibroblast motility[1-8]

The motility of cells such as fibroblasts (see Figures 1.7 & 6.2 for their general shape) has been studied in detail in time-lapse films. The process is most obvious when a cell is spreading out on a new surface. A cell growing in a tissue-culture dish will round up and detach from the substrate when treated with trypsin. If it is then allowed to resettle, a thin membrane lamella begins to flow out over the surface in all directions, gradually expanding beyond the still rounded cell body. **Microspikes** (or **microfilopodia**) develop around the cell periphery (Figure 6.3), each starting suddenly as a small protrusion and slowly elongating into a spike. Stress fibres (Section 5.6) are not seen anywhere at this stage. After a time, material gradually moves out of the centre until the cell eventually assumes the typical fibroblast shape. Motile activity may continue but becomes polarised so that the cell moves across the surface which is supporting it.

6.2.1   Translocation[2,4,6,11,27]

In translocating cells, the cell periphery is differentiated, with only one or a few major sections engaged in obvious motility. The **leading edge**, or **lamella**, does

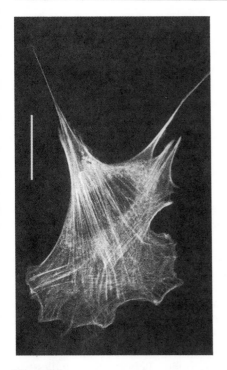

Figure 6.2

Tissue-culture cell with the morphology typical of a moving cell; F-actin has been stained with fluorescent phalloidin. The leading lamella is at the bottom, rearwards retraction fibres are at the top of the picture. The bar represents 50 μm.

Figure 6.3

Actin filaments in the leading lamella of a fibroblast, shown negatively stained (as in Figures 1.4 & 5.5). The density of actin filaments is so great, compared with the erythrocyte cytoskeleton (Figure 5.2), that it is difficult to identify any associated proteins without antibody labelling. The microspike bundles are embedded in the more randomly oriented peripheral weave of filaments. × 100 000. [Micrograph provided by Dr Victor Small.]

not move forwards steadily but shows periods of protrusion followed by periods of retraction. During protrusion the membrane may form folds on the upper surface and both folds and spikes may extend upwards away from the substrate. This behaviour is known as **ruffling**.

Ruffling can be an extremely complex process. Spikes quite often raise into the medium, swing around for a while and attach either to the substrate or to the surface of neighbouring cells, or they may fold back over the advancing lamella and disappear. The cell edge may flow out between adjacent spikes. Parts of the lamella itself may detach from the surface and wave about in the medium. Loose spikes and membrane ruffles often move towards the cell centre, gradually decreasing in size.

If particles in the medium happen to attach to the exposed cell surface, the cell moves them in a straight line *backwards* relative to the direction in which it is itself advancing. As detailed in Section 6.4, video-enhanced images even suggest that waves of material just inside the membrane of the lamella are moving continually backwards and that particles on the external surface are carried along with them. No regions of surface near the front of the cell appear to move forwards. Nor are there signs of forward movement of anything within the cytoplasm of the lamella. Nonetheless, the leading edge of the membrane moves forwards, whilst the rear of the cell is pulled out into a long tail until the tension produced suddenly appears to reach a limit and the tail is retracted towards the cell centre.

Studies of cells microinjected with fluorescent actin show evidence of actin polymerisation close to the leading edge. Treatment with the drug cytochalasin, which blocks the elongation of actin filaments and severs them, inhibits the motility of fibroblasts and similar cells.

6.2.2   Cytoskeleton arrangements[4,22]

Not surprisingly, actin is the most obvious cytoskeletal component of the thin layer of cytoplasm within the leading lamella (Figure 6.3). Within the body of the cell, the distribution of actin is different and apparently less organised than in stationary cells. In particular, there are considerably fewer stress fibres.

Electron microscopy reveals that a microspike consists of a bundle of microfilaments, surrounded by membrane. The bundles are superficially similar to stress fibres (see Figure 6.3) but there is no evidence for periodic arrangements of myosin in them. The actin filaments probably have a polar arrangement, with the plus end at the tip, unlike the bipolar arrangement of stress fibres. Where microspikes protrude from the edges of the ruffles, the microfilament bundles are found to extend deep inside the ruffles. Between the bundles is a more random network of individual filaments, of a more labile nature (much less easy to fix). This network is sometimes known as the **peripheral weave** of actin filaments. Immunofluorescent labelling indicates the presence of the filament crosslinking protein filamin (Section 4.4.4) in these regions.

In the rest of the cell there may be bundles (potential stress fibres) and sheets of microfilaments close to the upper and lower layers of cell membrane. Spectrin is one of the actin-associated proteins to be localised in these parts of the cell. The bundles often converge to focal points (**adhesion plaques**—see Figure 5.14a) at the base of the leading lamella. Near the rear end of the cell, fibres and sheaths of bundled microfilaments taper off into one or several retraction fibres.

Forward progress depends upon the protrusions making new points of attachment to the substrate. Labelling with specific antibodies has shown that potential adhesion plaques, containing vinculin and α-actinin (Section 5.6), exist before contact is made. Actin also appears to become concentrated at such points before they attach. After contact is made, vinculin accumulates at the new plaque. There is a correlation between contact sites and free ends of microtubules. Microtubules seem to be involved in stabilising the plaque before new stress fibres assemble on to it. As the cell moves forwards, it moves over these stationary adhesion plaques, eventually losing contact by retraction at the rear, as described above, often leaving behind a **footprint** containing traces of all the adhesion-plaque proteins.

6.2.3   Motile behaviour in other crawling cells[2,6]

The details of motility vary in different types of crawling cells but there are sufficient similarities to suggest that the basic mechanisms may be much the same in most cases. Macrophages and other white blood cells tend to have a large proportion of their edges undergoing ruffling, giving them a more symmetrical shape than fibroblasts and often resulting in no net translocation of the cell. Ruffling may be more vigorous than in fibroblasts and waves of density moving through the leading lamella often appear to accumulate as fibrous bundles or 'arcs' at the base of the lamella. The arcs may move back over the body of the cell, sometimes producing a construction not unlike a transient cleavage furrow (Section 11.9).

Some amoebae, such as *Dictyostelium* and *Mayorella*, behave very like fibroblasts. Others show interesting variations. *Acanthamoeba* produces very long thin filopods instead of ruffles and short protrusive spikes. *Amoeba proteus* and *A. chaos* produce very large extensions, called **lobopodia**. Often only the tips of the extensions contact the substrate, so the cells appear to 'walk' over it. Shelled amoebae, such as *Difflugia*, send out forward protrusions and pull their shells after them. **Syneresis** occurs to one side of the protrusion whilst the pseudopod undergoes contraction. Strangest of all is the *Limax* type of amoeba, which has the appearance of a sausage, elongated in the direction of motion and charging along at $100 \, \mu m/min$—no need for time-lapse filming in this case!

Growing neurons put out very long extensions known as **neurites** (potential axons or dendrites). Along most of the extension there is no obvious motile activity, but the growing end spreads out into a large cone-shaped appendage known as the **growth cone**, remarkably like a fibroblast in its movement. Some interesting studies on growth cones are described in Section 6.4.

Lymphocytes tend to maintain a rounded shape and do not adhere well to a substrate. Their surfaces do, however, show a type of motility, as demonstrated by the phenomenon known as **capping**: this also is discussed in Section 6.4.

6.3 The cell membrane and the locomotory mechanism

Various theories have been proposed to account for the observed movements. We shall discuss only the current more likely mechanisms. Fibroblast movement, as described above, shows features which cannot be explained by simple contraction or sliding-filament models. All current models require the membrane to play a key role in motility. It is an area of cell biology where a possible consensus of opinion is only just beginning to emerge.

6.3.1 Membrane recycling and lipid-flow theory[7–9,27]

To account for the observed behaviour, Abercrombie and his collaborators first suggested in 1970 that the cell membrane of a moving cell moves backwards in a continuous flow from the leading edge to the rear and that external particles are carried along by this flow. They proposed that membrane material (lipids plus integral membrane proteins) is internalised at the back of the cell, transported through the interior of the cell to the front and reinserted there into the cell membrane. Experiments carried out with ligands such as **concanavalin A** and **ferritin**, which bind to specific membrane receptors (domains of membrane glycoproteins that are exposed on the exterior surface), have indeed shown that recycled membrane proteins are inserted close to—though not necessarily right at—the leading edge. Membrane proteins involved in making contact with the substrate are included in this group.

The **lipid-flow theory** proposed by Bretscher is a modification of Abercrombie's theory, in which it is suggested that lipids, together with selected membrane proteins, are taken into the cell's interior at points *over the entire surface* rather than over a limited region of the rear surface. Electron-microscopic evidence certainly suggests that receptor proteins in the membrane are internalised over a wide area.

Membrane is thought to be drawn into the cell with the help of the 180 kD cytoplasmic protein **clathrin**, which assembles into a network on the inner surface of the cell membrane and deforms it so as to form small pits in the outer surface. These 'coated pits' are believed to be able to filter integral membrane proteins in some way, so that some are trapped in the pits and others are excluded. The pits deepen and eventually pinch off as small 'coated vesicles', in a process known as **endocytosis** or **pinocytosis** (see Figure 6.4). Figure 6.5 shows small vesicles with and without clathrin coats. The way in which clathrin trimers are thought to interact to assemble the coats is summarised in Figure 6.6.

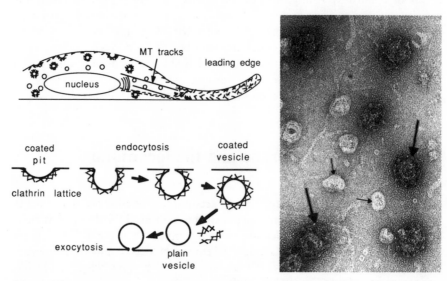

Figure 6.4

Scheme for membrane recycling in animal cells: the protein clathrin assembles on the cytoplasmic side of the plasma membrane,[9] producing coated pits which pinch off to form coated vesicles. The vesicles soon lose their clathrin coats and fuse with internal membrane systems. Eventually, vesicles or tubules of membrane are transported along microtubules to the cell periphery. In a moving cell, the microtubules are directed predominantly towards the base of the leading lamella. Fusion of vesicles with the plasma membrane here (exocytosis) will provide material for further protrusion of the lamella.

Figure 6.5

Coated (large arrows) and uncoated vesicles (small arrows) as seen by electron microscopy in negative strain. The densely-staining coats consist of a polyhedral cage of clathrin and various associated proteins. The small vesicles are pinched off by the coat material either from the plasma membrane or from internal membrane systems such as the Golgi.

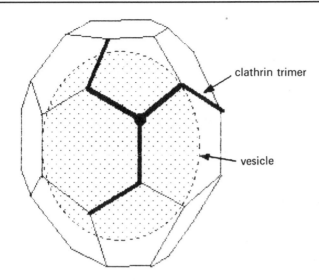

clathrin trimer

vesicle

Figure 6.6
Clathrin trimers form polyhedral cages in a variety of sizes.[9] The arrangement deduced for one of the smallest and most regular cages is shown here. The cage size can be enlarged indefinitely by insertion of additional six-sided faces, keeping essentially the same intermolecular interactions.

6.3.2   Intracellular route for membrane

The vesicles, having rapidly lost their clathrin coats once they are in the cytoplasm, are thought to coalesce with the cell's lysosomal membrane system, whilst any substances trapped inside are processed. Finally the membrane components themselves (lipids and any recyclable membrane proteins) are returned to the cell membrane. They may travel to the surface as individual vesicles which then fuse with the cell membrane by the process known as **exocytosis**. Alternatively, continuous membrane systems within the cell may periodically fuse with the cell membrane and transfer membrane components.

In either case, recycled and new material to be inserted in the cell membrane is probably directed along microtubules, many of which end at the base of the leading lamella (see evidence in Section 10.6 for vesicle movement along microtubules in animal cells). This might explain why the arrangement of microtubules is often correlated with the position of the leading edge in a motile cell (see Sections 6.2.2 & 10.1.1). Some cells treated with anti-microtubule drugs, such as **colchicine** (Section 7.4.6), lose their leading edge and ruffling activity is observed all round the periphery. However, it is evident that microtubules do not normally extend *into* the lamella, only to its base (see Section 6.4.1). If vesicles did move up to the leading edge, they would need an alternative transport mechanism.

6.3.3   Could membrane flow drive motility?[2,3,7,12]

The theory that motility is *driven* by membrane recycling[7] requires that (a) forward progress of the leading edge depend upon the rate at which membrane can be recycled; (b) vesicles be reinserted into the cell membrane right at the leading edge, in order to produce the observed extension; (c) backward transport be entirely accounted for by a flow of molecules diffusing over the cell surface to replace material taken inside by endocytosis: large particles observed moving on the membrane surface must be carried along in the lipid current. The extent to which these requirements are met is as follows.

(a) Calculations show that the rate at which fibroblasts move (around 1 mm per day) is consistent with the average time that cell surface receptors spend in a cycle of internalisation and reappearance on the surface. However, there are types of cell that move much faster than fibroblasts (Section 6.2.3) and others that move despite carrying out very little endocytosis. Polymorphonuclear leucocytes (a kind of white blood cell) can move twice their diameter in a minute without taking up significant amounts of external fluid. On the other hand, membrane recycling is known to occur actively in most non-motile cells, as they take in nutrients by pinocytosis from the external medium. No ruffling activity is produced by reinsertion of lipid and proteins into the cell membrane.

(b) Vesicles have never been observed within the leading lamella, either by electron microscopy or by the latest video-enhanced light-microscopy techniques that have been proved capable of detecting moving objects of this size. It could be argued that vesicles were present but fused with the membrane during fixation for electron microscopy. However, with light microscopy living cells can be observed whilst still moving.

(c) The membrane-flow theory seems inadequate to explain, simply in terms of diffusion, the backwards movement over the cell body of spikes, ruffles and large exogenous particles protruding into the surrounding medium, where viscous forces are always large on a microscopic scale. Other observations described below also indicate that it is unlikely that membrane flow is the driving mechanism involved in fibroblast-like motility.

6.4　Role for the actin cytoskeleton in locomotion[2-4]

Other theories feature a direct involvement of the cytoskeletal scaffold underlying the membrane. There is good evidence for structural connections between molecules situated on the external surfaces of cell membranes and the cytoskeleton. In the case of *Dictyostelium*, transmembrane proteins that bind directly to actin have been identified. In other cells, indirect associations between actin and various membrane proteins are well documented (Section 4.5). Details of how such processes might operate are gradually emerging from a series of interesting new observations.

6.4.1　Two-headed myosin and ruffling in amoebae[28,29]

Experiments to manipulate the expression of myosin II in *Dictyostelium* have suggested that thick-filament-forming myosin is not essential for amoeboid motility. Cells were transformed either by introducing a gene that produced **antisense RNA** (Appendix A.6), complementary to the mRNA encoding the myosin II heavy chain (with which it would hybridise and would thereby inactivate), or by replacing the normal gene with a plasmid encoding a defective product. Even though the cells produced little or no myosin II heavy chain, they were able to adhere to surfaces and exhibit amoeboid locomotion, including extension and retraction of pseudopods and filopodia, ruffling and some phagocytosis, though the rate at which these cells advanced was considerably slower than normal (one-half to one-third).

Thus, either contractile activity is not essential for movement, or myosin I is sufficient for the purpose. The behaviour of cells totally lacking myosin I may be known soon. Immunofluorescent labelling of normal cells has indicated that myosin I is concentrated in leading lamellae, whilst myosin II is dominant in the rest of the cell cortex.

Cytokinesis was prevented in the modified amoebae, so the cells grew very large and multinucleated. The latter result supports findings that cell division is inhibited by injection of anti-myosin antibodies into sea-urchin eggs. Myosin II is highly concentrated in the new contractile ring that pinches off two daughter cells during cell division (Section 11.9) and the ability to form filaments may be essential for its function here.

4.6.2   Patching and capping[18]

The phenomenon known as **capping** is an interesting expression of cell motility. When a motile cell such as a lymphocyte is exposed to antibodies or lectins that bind to proteins on its external surface, these become crosslinked to form dense **patches**. The patches do not move if the cells are kept cold and hence incapable of motility. But, after warming, they slowly migrate to the rear of the cell (or to a position over one pole if the cell is detached and thus not actively moving) to form a **cap**. Electron microscopy has shown that microtubules and microfilaments become concentrated beneath the cap. Eventually, the cell may engulf the crosslinked material in a large-scale process known as **phagocytosis**.

Capping, which can be detected on the surfaces of normal *Dictyostelium* cells, did not take place on the modified cells described in Section 6.4.1. This helps to explain the reduction in the locomotion rate of the modified cells; myosin II may be responsible for the continuation of waves of contraction behind the leading lamella. Myosin II filaments probably also make an important contribution to forward movement within the retraction fibres at the rear of the cell.

6.4.3   Observations on nerve growth cones

Recent observations of nerve growth cones by Forscher and Smith[10] provide direct visual support for the role of the cytoskeleton in ruffling. A growing neuron is simpler to study than a whole fibroblast, since the cell body remains behind in one place, while the axon extends, and obvious motility is restricted to the growth cone. The behaviour is very like the front part of a fibroblast, with the same kinds of protrusion and ruffling, and backwardly-moving waves of density within the cell cortex (see Section 6.2.1). A growth cone provides the opportunity of studying leading lamellar behaviour without complications introduced by changes in the rest of the cell. (Transport of material along the axon to the growth cone from the cell body, which provides any new components needed, will be considered in Sections 10.4 & 10.5.) It should be mentioned that the results obtained have been confirmed by observations made by Taylor's group on fibroblasts.[11]

Video-enhanced microscopy shows waves of density moving back over the lamella, beginning at the leading edge and sweeping back to the base of the cone. If particles such as polystyrene beads in the external medium stick to the outer surface, they also move back in straight lines at the same rate as the cytoplasmic waves. When cells were treated with cytochalasin (Section 3.3), material already in the lamella continued to move back but left behind at the leading edge a clear zone that gradually widened until the entire lamella appeared clear (Figure 6.7). Some particles attached to the outer surface became

Figure 6.7

Video-enhanced light micrographs of growth cones produced by cultured *Aplysia* neurons, using DIC optics (see Appendix A.1.4). (a) Under normal conditions: the clear leading lamella contains actin bundles that run into the microvilli protruding from the leading edge. Above is the granular endoplasm, obscuring the ends of the axonal microtubules. (b) After 1.5 min in 5 μM cytochalasin B (CB); the arrow indicates the distal edge of the receding lamellar actin network. (c) The edge has receded completely after 2.5 min. (d) 1 min after removal of CB, the first signs of actin network assembly appear as a dense swelling along the growth cone margin (arrows). (e) The DIC image after 1.5 min recovery; compared with (f), the distribution of actin as revealed by indirect immunofluorescence. The distribution would have returned completely to normal after a few more minutes. [Reproduced with permission from Forscher, P. and Smith, S. J. (1988). *J. Cell Biol.* **107**, 1505–1516.]

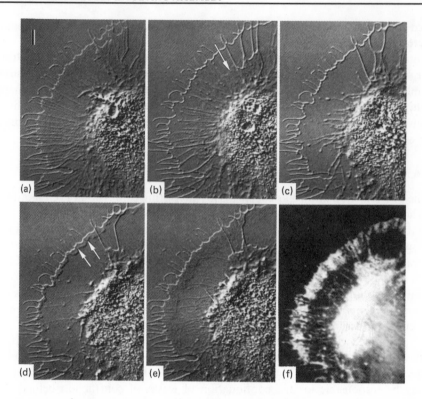

immobilised and appeared unable to diffuse. When cytochalasin was washed out of the medium, phase-dense material appeared again in the lamellar cytoplasm, originating as a layer at the leading edge and gradually expanding back into the cone until normal movement was achieved.

The normal behaviour is interpreted as resulting from polymerisation of actin at the leading edge and drawing of the resulting net into the base of the cone. Cytochalasin is thought to stop assembly, although contraction at the base continues. Particles on the external surface appeared to move only via connections through the membrane to the moving net. When the lamella was clear after cytochalasin treatment, microtubules were seen to move into the space, suggesting that normally they were excluded by the dense meshwork of filaments.

6.4.4 Analysis of glycoprotein movements[12,13]

Sheetz and colleagues have used video sequences for detailed analysis of the movements of particles attached to the outer surfaces of cells. Gold particles coated with the lectin concanavalin A were allowed to bind to the extracellular domains of glycoproteins in the membranes of macrophages. They were found to move in two distinct ways, either moving unidirectionally towards the nucleus, at speeds of 0.1–0.3 μm/s, or else diffusing randomly. There was no net drift in the random movements that could be ascribed to bulk flow of membrane.

Nor were there any random *sideways* excursions by the particles undergoing directed movement; such movement is evidently due to direct connections between the cytoplasmic domains of the glycoproteins and components of the moving cytoskeleton on the other side of the membrane. Individual particles were seen to change from one type of movement to another, showing that the connections are not permanent. When disconnected from the moving cytoskeleton, the particles moved freely, but only over short distances. The behaviour suggests the glycoproteins are constrained within 50–100 nm wide 'corrals' formed by some component of the submembrane cytoskeleton.

Additional observations on **keratocytes**, small motile cells from fish epidermis, showed particles sometimes moving at greater speeds (up to 20 μm/s) *towards* the leading edge, in addition to the more commonly observed rearwards movement. Sheetz and colleagues suggest that, whilst the rearwards movements are probably coupled to those of membrane-associated F-actin, the forwards movements might be due to linkage between glycoproteins and membrane-associated myosin I (see Figure 6.8). Such a process would provide an alternative explanation for the aggregation of newly-recycled proteins at the leading edges of cells, which has previously been cited as evidence for the direct reinsertion of membrane here.

6.4.5 Membrane reinsertion

Exactly where vesicles re-fuse with the cell membrane is unknown. In the case of motile cells, the process may occur preferentially towards the front of the

Figure 6.8

(a) Diagram of a leading lamella, as in a section cut normal to the substrate; the scale is greatly expanded in the vertical direction. Within the protrusion is a network of short actin filaments crosslinked together by passive crosslinkers, such as filamin. Myosin I acts between the network and the membrane and both myosins I & II may act within the network. The network is continually drawn towards the cell centre,[10,11] while actin monomers add to the plus ends of the filaments, many of which are associated with the plasma membrane. Hydrostatic forces probably cause the membrane to protrude slightly beyond the filament network, allowing space for monomer addition. There may be specialised capping proteins to facilitate assembly close to the membrane (see Figure 6.9). Lipids are probably recruited by diffusion from membrane adjoining the lamellar base. Membrane proteins are also free to diffuse from the rear to the leading edge, where they can initiate the assembly of new filaments. Some are small complexes of potential adhesion plaque proteins, which occasionally form new focal contacts.[14] (b) Particles on the outer surface stick to the extracellular domains of membrane proteins. Many of the latter are connected through their cytoplasmic domains to the backwardly-moving cytoskeletal network.[10–12] Others may associate with myosin I molecules bound to the cytoplasmic surface of the membrane and travel rapidly forwards when the molecules are pushing against a filament.[13]

(a)

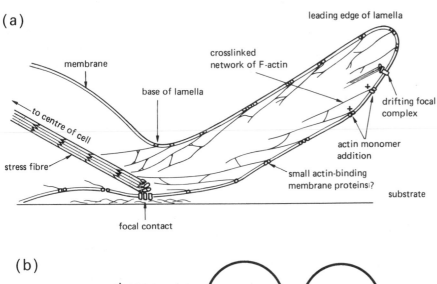

leading edge of lamella

crosslinked network of F-actin

membrane

base of lamella

to centre of cell

drifting focal complex

actin monomer addition

stress fibre

small actin-binding membrane proteins?

substrate

focal contact

(b)

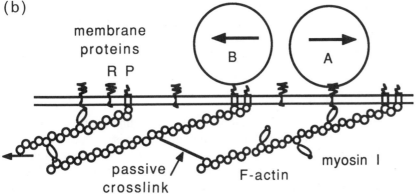

membrane proteins
R P

B

A

passive crosslink

F-actin

myosin I

cell, perhaps at the base of the leading lamella, where most microtubules are directed. Sites favourable for membrane fusion might also initiate the formation of microspikes. Experimentally, substances such as **glycerol** or **polyethylene glycol**, whose molecules are able to insert into membranes, are known to encourage fusion of separate membranes. It is thought that such molecules introduce point defects into the lipid bilayers. In a cell, a similar point defect may be introduced by cleavage of a lipid molecule in the membrane or by insertion of an extra cone-shaped lipid.[1] If such a site existed at the base of a microspike, a whole series of vesicles might fuse with it, thus assisting its growth. Lipids inserted near the base of a lamella would readily diffuse towards the leading edge during protrusive activity.

6.4.6 Contractile and protrusive forces[1-4]

The force for forward movement over a substrate must be applied to the points of attachment. There are two obvious kinds of force, *pulling* and *pushing*. Both may play a part.

Actin bundles ending at an attachment plaque behind the leading lamella could contract and exert a *pulling* force on the *rearward* part of the cell. This does seem to occur, though bundle contraction is probably not essential. Its main role must be to help pull the nucleus and the rear of the cell forwards towards the leading lamella, after a new contact has been established by the ruffling membrane.

Ruffling locomotion still occurs in small segments of cell experimentally severed from the main body. Similarly, nerve growth cones appear to move almost autonomously. Thus, the main motile activity must depend on the constituents of the leading lamella and the adjacent cytoplasm. Waves of contraction by the actin network within the leading lamella and in the cortical layer behind it may produce a *pushing* force on the *leading edge*. In addition, osmotic forces associated with actin polymerisation may also operate, as proposed for the protrusive behaviour of *Thyone* sperm (Section 6.1).

6.4.7 Osmotic forces[1,20]

Oster has suggested a mechanism for producing an influx of water into the cytoplasm near the front of the cell, analogous to that proposed for the *Limulus* acrosome (Section 6.1.1), which involves actin-filament polymerisation close to the membrane. A leading lamella is different in that an isotropic network of actin filaments is formed, in addition to numerous anisotropic bundles (Figures 6.6 & 6.9). The precise structures assembled in any cell type will depend on the bundling and other crosslinking proteins present. The actin monomers used to assemble these structures may be recruited from actin–profilin, actin–gelsolin, actin–ADF or other such complexes.

6.4.8 Possible role of inositol lipids[1,3]

A set of reactions that may be important in controlling protrusive activity is the **inositol-lipid pathway**, which is currently under intensive study but can only be mentioned briefly in the present context. *In vitro* experiments show that micelles of the lipid **phosphatidylinositol 4,5-bisphosphate (PIP$_2$)** can dissociate otherwise high-affinity complexes of actin with gelsolin or profilin, releasing

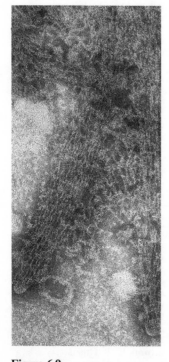

Figure 6.9

Actin in demembranated cell protrusions, from a cell prepared as in Figure 6.3. The composition of the material at the tips of the protrusions is unknown.

actin monomers which can then polymerise on the cytoplasmic surface of the membrane (see Section 4.2). Presumably PIP_2 does not occur as micelles *in vivo*, but there may be sites next to membrane proteins where this lipid congregates to produce a concentrated pool.

The products resulting from cleavage of PIP_2 on the cytoplasmic surface by a phospholipase (activated via a G-protein in the membrane) may be important in subsequent events required for motility (see also Section 1.5). Activation of the G-protein may be triggered by the binding of a ligand to a receptor on the outer surface. For example, chemoattractants that stimulate movement in amoebae and stimulants that activate platelets are known to have an effect on PIP_2 metabolism. In other motile cells, attachment of an adhesion-plaque precursor to the substrate may provide the signal for a cascade of reactions.

The cleavage products of PIP_2 are **diacylglycerol (DG)**, which stays in the membrane (making a useful point defect), and **inositol trisphosphate (IP_3)**, which is released into the cytoplasm. IP_3 can trigger calcium release from internal stores. Calcium can activate gelsolin and promote solation, as described above. Intact PIP_2 inhibits gelsolin-severing activity, presumably protecting actin filaments close to the membrane from attack, whilst filaments further inside the cell are being dismantled.

The released IP_3 and calcium ions may also contribute a little to the osmotic pressure on the protruding membrane. Finally, calcium is also a promotor of membrane fusion since it binds to and shields negative charges on the lipid head groups. The presence of both calcium ions and point defects in the membrane should encourage fusion of vesicles close to motile regions of the cell.

6.5   Possible locomotion cycles

6.5.1   Lamellar activity[1,3–5,10–16,22]

The evidence summarised above seems to point to actin polymerisation in association with protrusive forces on the membrane as the primary source of crawling activity in leading lamellae (Figure 6.8). As in the *Thyone* acrosome, dissociation of actin–profilin and other actin-monomer-containing complexes at the membrane may contribute to protrusive forces, though other processes may be involved. Continuous assembly of the actin filament network into the advancing protrusion will tend to maintain the new position of the leading edge. Depolymerisation at the rear of the lamella, promoted by gelsolin or similar proteins, may supply the actin monomers required at the front. This scheme could be described as actin treadmilling (Section 3.2.4), amplified by an array of associated proteins.

The observed behaviour, with waves of protrusion followed by retraction, seems to require an additional force, namely contraction. An actin filament network initially assembled at the front of a lamella is pulled towards the base of the lamella, where there appear to be more firmly anchored bundles of filaments. The position of the leading edge may depend on the balance between protrusion coupled with polymerisation and retraction due to contractile forces. For some reason, conditions oscillate in time, perhaps because of cyclic changes in the concentrations of activators such as calcium ions, which are discussed in the next section.

6.5.2   Gelation–solation and contraction of bulk cytoplasm[2,6,18]

Behind the leading lamella of a motile cell such as a fibroblast or an amoeba, localised domains of cytoplasm appear to contract in much the same way as a smooth-muscle cell. But the concentration of myosin in non-muscle cells is much lower. Other crosslinking proteins (gelation factors and bundling proteins) probably play important roles in promoting or inhibiting contraction.

A rigid gel crosslinked by calcium-insensitive proteins may inhibit contraction; the actin filament meshwork in the leading lamella, for example, may be maintained in its extended conformation by proteins such as filamin. Less extensive crosslinking should assist myosin in pulling actin filaments together. Furthermore, Oster has suggested that a local drop in pH due to the ATPase activity of myosin could promote shrinkage of a crosslinked actin gel. Thus, after the activation of myosin, the synergistic effects of actomyosin contraction and actin gelation in a particular region could expel solvent from it. Expulsion of water (**syneresis**) has in fact been observed by light microscopy towards the front ends of fast-moving amoebae. The flow of fluid might also carry forwards, towards the leading lamella, any actin monomers released at the rear.

Calcium-sensitive crosslinking proteins such as α-actinin may contribute to the reverse effect. A rise in the concentration of free calcium ions which would activate myosin kinase and gelsolin would also lead to dissociation of the α-actinin crosslinks and allow filaments to slide relative to one another. Once the crosslinks are fully detached and the filaments have been severed into short lengths, relatively low concentrations of myosin are probably unable to maintain a contracted state. Then the raised concentrations of free protein aggregates and ions may cause the now solated gel to absorb water and swell. Contraction of cytoplasm towards the rear of this region may ensure that pressure caused by the swelling (referred to as **gel osmotic pressure**) is exerted in the forwards direction.

Thus a wave of calcium-induced contraction, starting from the base of the lamella, could be followed automatically by a wave of solation which would take up solvent expelled further towards the rear of the cell, as the contraction wave continues to move backwards. Contraction at the rear combined with solation near the front would tend to produce a forwards protrusive force on the leading edge. Further protrusion and a new attachment to the substrate could stimulate a new wave of cortical contraction, followed by another solation wave.

Experimental evidence for such a cycle is lacking. Some reports even question any role for calcium as a regulator of non-muscle motility. It does seem unlikely that calcium ions introduced at the leading edge diffuse rearwards as a wave; any rise in Ca^{2+} concentration may be short-lived and highly localised. But the message may be transmitted in other ways, leading to the release of small amounts of calcium from intracellular stores where and when they are required.

6.5.3   Cytoplasmic streaming

Streams of cytoplasm are observed during the movement of the faster types of amoebae, presumably as a result of large-scale solation of previously gelled actin aggregates. The multinucleate slime mould *Physarum* exhibits remarkable cycles of cytoplasmic flow in alternate directions along its branching strands.[25] There are no obvious actin cables such as are found in *Nitella*. Nor can the alternation in direction of flow be easily explained in terms of components merely sliding over fixed filaments. It seems more likely that high concentrations of actin and suitable associated proteins take part in waves of solation and gelation which alternate in direction.

6.6   The amoeboid sperm of nematodes[23]

Nematode sperm are unusual in not moving by means of flagella (Chapter 9) but instead by crawling in much the same fashion as an amoeba. They extend pseudopods with which they pull themselves over substrates at rates exceeding $70\,\mu m/min$. Even more unusual is the fact that the cells contain only traces of actin and tubulin, and no detectable myosin. Instead, the main pseudopod protein belongs to a 14 kD family, known as **major sperm protein** (**MSP**). This protein apparently assembles into straight filaments, only 2 nm in diameter. The filaments form bundles and networks analogous to those made by F-actin in more conventional motile cells.

The fibre bundles project into membrane protrusions called **villipodia**, whose tips can attach to the substrate. During crawling, the entire cytoskeleton and the villipodia move continuously back towards the base of the pseudopod at the same speed as the cell moves forwards. The rearwards progress of the cytoskeleton still continues even if the cell is not attached to the substrate by its pseudopod. Villipodia and associated cytoskeletal complexes disappear at the junction between the pseudopod and the cell body, presumably being taken into the interior by some method of endocytosis. It is very probable that forward extension of the pseudopod is associated with rapid assembly of MSP filaments at the leading edge.

MSP evidently substitutes for actin in nematode sperm. It remains to be seen whether there is also a substitute for myosin. Comparisons between this system and motile cells with actin-based cytoskeletons should provide valuable insight into the fundamental processes required for amoeboid motility.

6.7   Summary

The cell membrane is not just an impermeable bag separating the cytoplasm from the external medium; it plays an active role in many processes, especially those of cell motility. In most cells, actin is most highly concentrated in a layer of cytoplasm next to the membrane; the assembly and disassembly of actin, as well as contraction, are controlled by events at the membrane. It probably provides a catalytic surface where soluble actin-containing complexes are split; actin assembly at the leading edge probably controls the production of osmotic forces that seem to be required for protrusive activity. In addition, signals originating at the membrane probably lead to localised release of calcium ions which directly or indirectly activate both F-actin-severing proteins and myosins.

Almost certainly, a combination of different processes produces cell crawling. They include assembly at the leading edge, severing and disassembly towards the rear, gelation and solation, contraction and relative sliding of filaments, osmotic protrusion, and recycling of glycoproteins. The net effect of the wide range of actin-associated proteins varies at different locations in the cell according to the precise local conditions, which may be governed by cyclic processes.

6.8   Questions

6.1 Why does the sperm acrosome of *Thyone* provide an experimentally useful model for the formation of microvilli, microspikes and other thin protrusions in more complex cells?

6.2 Where in the cell do the following processes tend to operate: contraction of actin bundles; gelation; contraction and solation of actin networks; assembly of actin filaments from monomers; and osmotically-driven protrusion?

6.3 How might particles be carried back over the surface of a ruffling cell and finally phagocytosed?

6.4 Might adhesion-plaque formation be regarded as a form of frustrated phagocytosis?

6.9   Further reading

Reviews

1 Oster, G. (1988). Biophysics of the leading lamella. *Cell Motil. & Cytoskel.* **10**, 164–171.

(*Protrusion theory: physics for biologists, without any equations.*)

2 Bray, D. & White, J. G. (1988). Cortical flow in animal cells. *Science* **239**, 883–888.

(*The ubiquity of actin-associated contractile forces.*)

3 Zigmond, S. H. (1989). Cell locomotion and chemotaxis. *Curr. Opinion Cell Biol.* **1**, 80–86.

(*A succinct summary of recent findings.*)

4 Small, J. V. (1989). Microfilament-based motility in non-muscle cells. *Curr. Opinion Cell Biol.* **1**, 75–79.

(*More emphasis on structure.*)

5 Small, J. V., Rinnerthaler, G. & Hinssen, H. (1982). Organization of actin meshworks in cultured cells: the leading edge. *Cold Spring Harbor Symp. Quant. Biol.* **46**, 599–611.

6 Bershadsky, A. D. & Vasiliev, J. M. (1988). *Cytoskeleton.* New York & London: Plenum Press.

7 Bretscher, M. S. (1987). How animal cells move. *Sci. Am.* **257**(6), 44–50.

(*Motility without the aid of the cytoskeleton?*)

8 Abercrombie, M. (1980). The crawling movement of metazoan cells. *Proc. Roy. Soc. Lond.* (*Biol.*) **207**, 129–147.

(*Review of original observations.*)

9 Pearse, B. M. F. & Crowther, R. A. (1987). Structure and assembly of coated vesicles. *Annu. Rev. Biophys. Biophys. Chem.* **16**, 49–68.

Some original papers

10 Forscher, P. & Smith, S. J. (1988). Actions of cytochalasins on the organization of actin filaments and microtubules in a neuronal growth cone. *J. Cell Biol.* **107**, 1505–1516.

11 Fisher, G. W., Conrad, P. A., DeBiasio, R. L. & Taylor, D. L. (1988). Centripetal transport of cytoplasm, actin and the cell surface in lamellipodia of fibroblasts. *Cell Motil. & Cytoskel.* **11**, 235–247.

12 Sheetz, M. P., Turney, S., Qian, H. & Elson, E. L. (1989). Nanometre-level analysis demonstrates that lipid flow does not drive membrane glycoprotein movements. *Nature* **340**, 284–288.

13 Kucik, D. F., Elson, E. L. & Sheetz, M. P. (1989). Forward transport of glycoproteins on leading lamellipodia in locomoting cells. *Nature* **340**, 315–317.

14 Izzard, C. S. (1988). A precursor of the focal contact in cultured fibroblasts. *Cell Motil. & Cytoskel.* **10**, 137–142.

15 Fukui, Y., Lynch, T. J., Brzeska, H. & Korn, E. D. (1989). Myosin I is located at the leading edges of locomoting *Dictyostelium* amoebae. *Nature* **341**, 328–331.

16 Euteneuer, U. & Schliwa, M. (1984). Persistent, directional motility of cells and cytoplasmic fragments in the absence of microtubules. *Nature* **310**, 58–61.

17 Honer, B., Citi, S., Kendrick-Jones, J. & Jockusch, B. M. (1988). Modulation of cellular morphology by antibodies against myosin. *J. Cell Biol.* **107**, 2181–2189.

18 Pasternak, C., Spudich, J. & Elson, E. (1989). Capping of surface receptors and concomitant cortical tension are generated by conventional myosin. *Nature* **341**, 549–551.

19 Lassing, I. & Lindberg, U. (1985). Specific interaction between phosphatidyl-inositol 4,5-bisphosphate and profilactin. *Nature* **314**, 472–474.

20 Oster, G. & Perelson, A. S. (1987). The physics of cell motility. *J. Cell Sci. Suppl.* **8**, 35–54.
  (*Details of osmotic theory, including the* Thyone *acrosome. Includes some mathematics but it is not essential to follow the equations.*)

21 Tilney, L. G. & Inoué, S. (1982). Acrosomal reaction of *Thyone* sperm. II: The kinetics and possible mechanism of acrosomal process elongation. *J. Cell Biol.* **93**, 820–827.

22 Wang, Y. (1985). Exchange of actin subunits at the leading edge of living fibroblasts: possible role of treadmilling. *J. Cell Biol.* **101**, 597–602.

23 Sepsenwol, S., Ris, H. & Roberts, T. M. (1989). A unique cytoskeleton associated with crawling in the amoeboid sperm of the nematode, *Ascaris suum. J. Cell Biol.* **108**, 55–66.

Additional references

24 Pollard, T. D., Fujiwara, K., Handlin, R. & Weiss, G. (1977). Contractile proteins in platelet activation and contraction. *Ann. N.Y. Acad. Sci., USA* **283**, 218–236.

25 Kamiya, N., Allen, R. D. & Yoshimoto, Y. (1988). Dynamic organization of *Physarum* plasmodium. *Cell Motil. & Cytoskel.* **10**, 107–116.

26 Jacobsen, K., Ishihara, A. & Inman, R. (1987). Lateral diffusion of proteins in membranes. *Annu. Rev. Biophys.* **49**, 1673–1675.

27 Abercrombie, M., Heaysman, J. E. M. & Pegrum, S. M. (1972). The locomotion of fibroblasts in culture. IV: Electron microscopy of the leading lamella. *Exptl. Cell Res.* **67**, 359–367.

28 Knecht, D. A. & Loomis, W. F. (1987). Antisense RNA inactivation of myosin heavy chain gene expression in *Dictyostelium discoideum. Science* **236**, 1081–1086.

29 De Lozanne, A. & Spudich, J. A. (1987). Disruption of the *Dictyostelium* myosin heavy chain gene by homologous recombination. *Science* **236**, 1086–1091.

PROPERTIES OF 7 TUBULIN

As described in Chapter 1, almost all eukaryotic cells have microtubules; most are single cylinders, but doublet tubules occur in flagella and triplets in basal bodies and centrioles. These complex tubules are described in more detail in Chapter 9.

Every kind of microtubule so far investigated has been found to be composed principally (but not entirely) of a highly conserved protein named **tubulin**. On SDS polyacrylamide gels, the protein runs as two closely spaced bands, apparently at about 55 kD. The reason for this is that tubulin occurs as equal amounts of two similar but distinguishable types of monomer. The faster of the two bands is known as **β-tubulin**, the slower as **α-tubulin**. Undenatured protein molecules are found to be in the form of **heterodimers** of α- and β-tubulin, with a sedimentation coefficient of 6S.[14–16]

7.1 Microtubule structure[13,14,17,23]

Analysis of electron-microscope images of negatively-stained flagellar microtubules (see examples in Figure 7.1) by optical diffraction gave the first indication that the tubulin was present as dimers, with an axial spacing of 8 nm, arranged in longitudinal rows called **protofilaments**.[13,17] However, the two subunits comprising each dimer are so similar in structure, at the level shown by the electron microscope, that most micrographs of cytoplasmic microtubules show only the 4 nm axial spacing of tubulin monomers. The monomers are arranged in protofilaments in a polar fashion (see Figures 7.2 & 7.3), giving the whole microtubule a longitudinal polarity, like the actin filament.

7.1.1 The tubulin lattice in a microtubule

Most native microtubules consist of 13 protofilaments, though there are exceptions, with 7, 9, 12 or 15 protofilaments; they can be counted in cross-sectional views present in thin sections, especially if the fixation buffer includes tannic acid. The latter binds to the outer surfaces of tubulin and takes

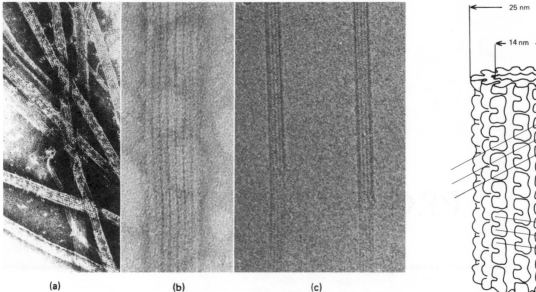

(a)   (b)   (c)

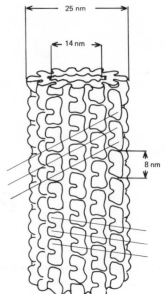

Figure 7.1

EM images of microtubules: (a) disintegrating flagellar microtubules, and (b) a reassembled brain microtubule, both in negative stain;[17] (c) repolymerised pure brain tubulin, frozen in a thin layer of amorphous ice, which appears *less* electron-dense than protein. The longitudinal protofilaments can be seen in all three cases, though they are partially obscured in (a) by associated material. Superposition of the back and front of a tubule is seen most clearly in ice, which produces less distortion than negative stain. A slow twist of the protofilaments around the tubule shows up in the right-hand example in (c) as an axial variation of the superposition pattern. (a) × 120 000, (b) × 500 000, (c) × 200 000. [(c) Photographed by Dr Nigel Unwin.]

Figure 7.2

Diagram showing the main structural features of a microtubule.[14] Dumbbell-shaped units represent 6 S tubulin heterodimers into which microtubules disassemble. They are arranged end to end, in a polar fashion, to form longitudinal protofilaments. Helical lines can be drawn to follow neighbouring subunits in various other directions defined by the helical surface lattice; in this case the details of the lattice are those found for a flagellar A-lattice (shown opened out in Figure 7.4a).

up stain, thus helping to delineate the structure (see Figures 7.8 & 9.1 for examples). Microtubules reassembled *in vitro* from purified protein often include a significant proportion of 14- and 15-protofilament tubes, especially after several cycles of assembly and disassembly. Control of protofilament number may be due to some cellular component that tends to be lost during tubulin purification.

Whether there are 13 or more protofilaments, they associate side by side with the monomer subunits slightly staggered so that any line joining equivalent points round a closed tubule forms a shallow helix, joining every third subunit along a protofilament. Thus, three independent parallel helical lines can be drawn through the monomers at this angle; the group is described as a **3-start helical family**. Other, steeper, families of helical lines can also be drawn through the subunits (see Figure 7.4).

For lateral associations, it does not seem to be critically important whether a given monomer binds to an α-tubulin or a β-tubulin site. In the case of flagellar microtubules, there is evidence for both modes of association.[17] In the complete

Figure 7.3

Model of a microtubule calculated by 3-D image-reconstruction methods from electron micrographs of specimens in negative stain. The protofilament shape shown here was actually derived from sheet specimens such as those shown in Figure 7.5 but the protofilaments are arranged as in a flagellar A-tubule. The most obvious features of the outer surface are the longitudinal grooves between protofilaments; the shallow 3-start helical grooves show up most clearly on the inner surface.

13-protofilament **A-tubule** of an outer doublet microtubule, optical diffraction patterns suggest that all lateral associations are between unlike tubulin monomers. This is due to an approximately half-staggered arrangement of tubulin heterodimers, referred to as the **A-lattice** (Figures 7.4a & 7.2), which has full helical symmetry. In the incomplete, 10-protofilament, **B-tubule** attached to the side of the A-tubule, the heterodimers appear not to be staggered; tubulin monomers associate laterally with identical monomers. This unstaggered arrangement of dimers is referred to as the **B-lattice**. It cannot have full helical symmetry, even in a closed tubule, since one of the three helical lines of monomers remains unpaired into dimers unless there is a glitch, or seam, in the lattice between two of the protofilaments (see Figure 7.4b).

There is almost no information about the dimer arrangement in microtubules other than those from flagella, since the tertiary structures of α- and β-tubulin subunits are so similar. Analysis of images of reassembled microtubules has suggested they may form using a mixture of the two modes of lateral association. On the other hand, associated proteins apparently tend to bind in a manner suggestive of an underlying A-lattice (see Section 8.2.1). Data obtained by X-ray diffraction of native cytoplasmic microtubules is inconclusive.[24]

Reassembled tubules mostly show a slight twist, so that the 'longitudinal' protofilaments actually follow a very long helical path around the microtubule. The twisting has been correlated with the presence of more than 13

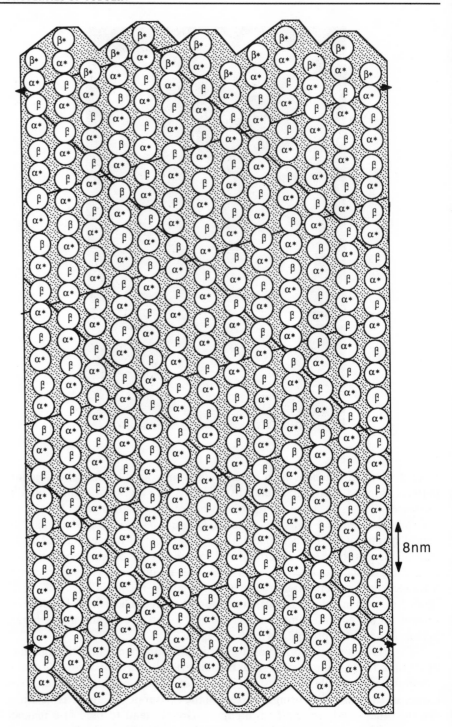

8nm

Model 7.1

A model of a microtubule. Photocopy this page (enlarging it, if possible) and cut around the boundary, leaving a white strip along one side for gluing. When joining the two sides to make a cylinder, the arrows at the sides should line up.

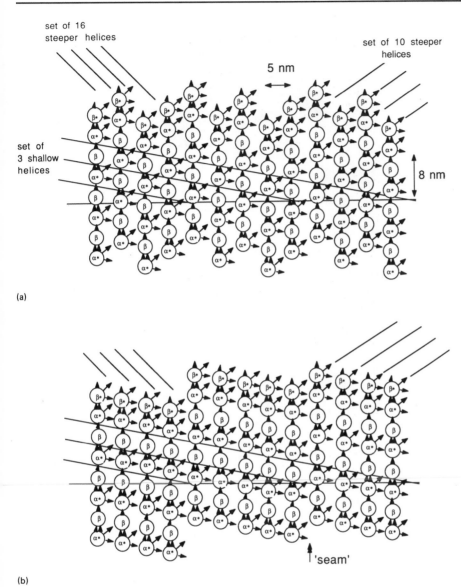

set of 16
steeper helices

set of 10 steeper
helices

5 nm

set of
3 shallow
helices

8 nm

(a)

(b)

'seam'

Figure 7.4

The lattice of subunits in an opened-out microtubule, as seen from the outside surface.[14,17] Each circle represents a 50 kD tubulin monomer. The monomer lattice seems to be the same in all microtubules, producing a 'family' of three shallow helices running in parallel. The way α-β heterodimers line up is variable: in (a) they are in a staggered arrangement, as in the A-tubule of an axonemal doublet microtubule; in (b) they are unstaggered, as in a B-tubule. The arrangement in cytoplasmic singlet microtubules is uncertain (see text). For a 13-protofilament A-lattice, though not for the B-lattice, the 10 right-handed monomer helices pair to give 5 dimer helices. The 3 shallow monomer helices cannot pair consistently in the B-lattice, so a complete tubule with such an arrangement would always contain a dislocation, or seam.

protofilaments. The 13-protofilament A-tubules of flagellar doublet tubules (Section 9.1) do not twist relative to the attached B-tubules.

7.1.2 Zinc-induced tubulin sheets

If microtubule protein is assembled in a buffer containing zinc ions in place of magnesium, the lateral bonding of protofilaments is greatly disturbed. Each protofilament associates in an upside-down orientation relative to its neighbours (see Figure 7.5c). The polymers formed do not close to form tubules, but may keep on growing to large sizes (Figure 7.5a). The assembled sheets are essentially *two-dimensional crystals* and, as such, useful for three-dimensional structure analysis of electron-microscope images by computer methods. These sheets,

Figure 7.5

Extended sheets of tubulin protofilaments induced by replacing magnesium ions in the assembly medium with zinc. (a) An EM image of a rolled-up sheet, contrasted with negative stain (scale bar 0.5 μm). (b) A closer view of protofilaments in a sheet (scale bar 50 nm). (c) A computer-processed image of protofilaments in such sheets,[17] × 4 500 000. Notice that the orientation of neighbouring protofilaments is reversed, whereas all protofilaments have the same polarity in a normal microtubule. The pear-shaped units correspond roughly to individual tubulin monomers; this is a projected view of all levels in the 3-D structure superimposed. Calculation of a 3-D model (Figure 7.3) required data from a large number of tilted views, each showing the density projected in a different direction.

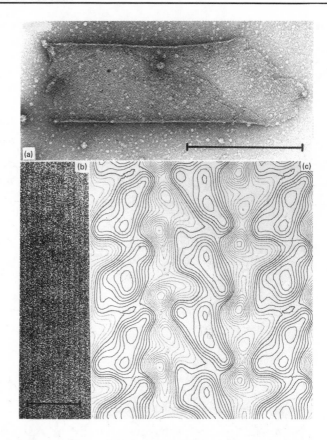

which allow information from a large number of subunits to be accurately combined, have yielded the best structural information on tubulin protofilaments to date. The microtubule model shown in Figure 7.3 was assembled from the resulting three-dimensional protofilament images.[17]

7.2 Molecular structure

Little is known about the internal structure of the tubulin dimer. Spectroscopic data from **circular dichroism (CD)** suggest merely that the percentages of each tubulin polypeptide assuming α-helical or β-sheet conformations are typical of globular proteins.[14] Predictions based on sequence data are in agreement with this. Tubulin crystals suitable for X-ray crystallographic analysis to atomic resolution have not been reported.

7.2.1 Monomer domains

X-ray diffraction data have, however, been obtained from aligned preparations of whole microtubules. The most recent data obtained by Beese and colleagues are sufficiently accurate to refine the model in Figure 7.3 to show 2 nm details fairly reliably.[23] In the refined model (Figure 7.7), each monomer subunit divides into two main globular domains and a smaller rod-like extension. The globular domains of the monomers in a protofilament lie on a continuous helix along the protofilament.

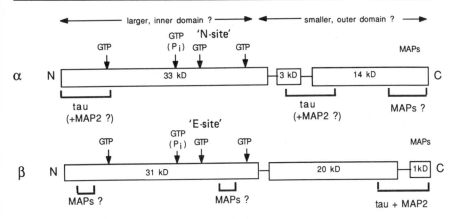

Figure 7.6

One-dimensional maps of the two monomers of tubulin, showing the most proteolytically-sensitive points.[25] Exchangeable ('E-site') GTP, hydrolysed during microtubule assembly, is bound to the N-terminal domain of β-tubulin. Non-exchangeable ('N-site'), non-hydrolysable GTP binds to related sites on α-tubulin.[26] Regions where the microtubule-associated proteins MAP 2 and tau appear to bind are indicated. Chemical crosslinking of *tubulin dimers* suggests the N-terminal domain of α-tubulin binds to the C-terminal domain of β-tubulin, 'burying' the N-site GTP. If *microtubules* are crosslinked, the N-domain of β-tubulin is associated with the C-domain of α-tubulin.

The connections between the domains have the appearance of narrow hinges, suggesting that the monomer may be a flexible molecule, like the actin monomer with its two domains. Eagles and his colleagues have compared X-ray diffraction data from hydrated and dehydrated microtubule specimens and found evidence for reversible conformational changes;[24] it is probable that these changes are due to relative movement of the domains and may be indicative of hydrophobic interactions (see Section 7.3) between molecules or domains. The submolecular domains probably move during assembly, as is thought to occur in the case of actin.

7.2.2 Tubulin sequences and post-translational modifications[10,18,33]

The amino-acid sequences of tubulins have been determined from a range of species. Comparison of the sequence of α-tubulin with that of β-tubulin from any one species shows that they have a 40–55 per cent homology, suggesting that they originally evolved from a single protein, presumably by gene duplication. Each has a true molecular mass of only 50 kD. The C-terminus of each subunit is rich in glutamic acid and is thus very acidic, the β-subunit being slightly more so. This feature may explain their slightly anomalous electrophoretic mobilities on SDS polyacrylamide gels and also the difference in mobility, despite their roughly equivalent size.

If the sequences of either subunit in different species are compared, the homology is considerably higher than between α- and β-tubulin from the *same* species, suggesting that the heterodimer is a very ancient feature. Differences between the two tubulin subunits have been strongly conserved and must be functionally important. The dimers from different species are so similar that co-assembly is possible, even between tubulins isolated from species as different as yeasts and mammals. This conservation of the structural interactions between subunits of the microtubule lattice obviously places strong constraints on the sequences. Probably interactions with other proteins, especially the **microtubule-associated proteins** (MAPs—see Sections 8.1 & 7.6.7), are also conserved.

Most species have more than one gene for each type of tubulin monomer; the expression of these genes depends on the cell type and the stage of development of the organism. Equivalent isotypes from different species are so highly conserved as to suggest that the isotypic variations may be important, either for the binding of specific associated proteins or for subtle changes in tubulin characteristics. The complexity of tubulin is further increased by various post-translational modifications: for example, the β-subunit can be phosphorylated; the α-subunit can be acetylated; and a tyrosine residue at the C-terminus of the α-subunit can be removed and subsequently replaced. None of the biological functions of these modifications is yet quite clear.

7.2.3 Regulation of tubulin synthesis[9,10]

Translation of tubulin mRNA is regulated by a feedback mechanism that depends on the concentration of soluble tubulin. Alteration of the latter by drugs that induce disassembly or by microinjection of additional tubulin causes repression of new tubulin synthesis. Recent research involving the introduction of altered tubulin mRNAs into cells has shown that the message must include coding for the first four amino acids of β-tubulin if its synthesis is to be regulated. The theory is that free tubulin binds to any nascent β-tubulin polypeptides on ribosomes, blocking further elongation and causing degradation of the mRNA. The synthesis of α-tubulin may be controlled in a related manner.

7.2.4 Structural data from proteolysis and antibody labelling[25,30]

Each tubulin monomer can be cleaved proteolytically into two unequally-sized fragments, the larger one containing the N-terminus, the smaller one the C-terminus. A useful difference is that trypsin preferentially cuts α-tubulin at arginine 339, while chymotrypsin attacks β-tubulin at tyrosine 281, producing fragments with distinguishable sizes. Chemical crosslinking studies have shown that the C-terminal domain of β-tubulin binds to the N-terminal domain of α-tubulin to form the heterodimer. Dimers then associate longitudinally by an interaction between the N-terminal domain of β-tubulin and the C-terminal domain of α-tubulin. Polypeptides that have been cleaved in two can still form curved sheets but the structure does not close to form a complete microtubule. Curvature control apparently requires the protein to be intact.

Labelling of microtubules with monoclonal antibodies to specific sections of sequence has shown that the C-terminal domain is readily accessible on the outer surface; the N-terminal domain apparently cannot be labelled unless the microtubules are denatured by fixation (e.g. in formaldehyde, which does not preserve the lattice structure as seen by electron microscopy, although the tubules appear continuous by immunofluorescence). These results, illustrated in Figures 7.6 and 7.7, suggest that the large N-terminal domain may lie more towards the inside of the cylinder. The scheme in Figure 7.7, derived from EM and X-ray data, apparently has domains of a similar size on the inner and outer surfaces. However, the data may not be adequate to represent their relative sizes very accurately. Alternatively, there may not be a simple correspondence between proteolytic fragments and apparent structural domains.

7.2.5 GTP-binding sites[25,26]

Each tubulin monomer has a binding site for a guanine nucleotide in its N-terminal domain (see Figure 7.7). GTP or GDP bound to the site on β-tubulin (the so-called **E-site** of a dimer) can exchange with nucleotide in the medium. GTP is hydrolysed here to GDP during polymerisation. The properties of the E-site are very similar to those of the ATP binding site on the actin monomer

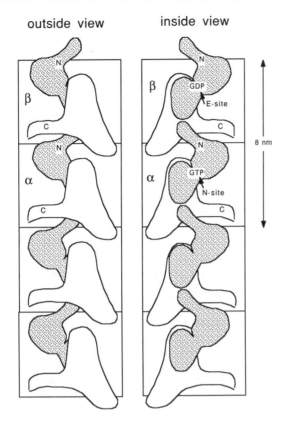

outside view   inside view

Figure 7.7

A speculative scheme for the domain structure of tubulin protofilaments, combining the best available structural data (the model shown in Figure 7.3 after refinement against X-ray diffraction patterns of aligned microtubule samples[23]) with the information summarised in Figure 7.6. Antibodies to the C-terminus label the outer surfaces of microtubules, while the N-terminal domain is less accessible. The suggestion is that each monomer consists of an exposed C-terminal domain (shown in white) and an inner N-terminal domain (shown shaded), with nucleotide bound in the cleft between its two lobes. In the model of Figure 7.3 the inner surfaces of α and β monomers appear slightly different, probably reflecting a structural difference between the E-site and the N-site (see Figure 7.6) where nucleotides bind, but the X-ray data were not adequate to refine the difference.

(Section 3.2). Divalent cations such as Mg^{2+} bind here and are required for assembly and GTP hydrolysis. The α-subunit, on the other hand, contains a non-exchangeable site (the N-site of the dimer) where GTP alone is bound and cannot be removed except by denaturing the protein. It is not hydrolysed during assembly and probably serves an important structural role.

7.3   The assembly of tubulin[13-22]

The most characteristic property of tubulin is, of course, its ability to self-assemble into microtubules. Polymerisation of a solution of purified microtubule protein *in vitro* can be initiated simply by warming it in a buffer of suitable composition.

7.3.1   Tubulin assembly: driven by entropy

Tubulin assembly provides a striking example of a reversible association between protein molecules that is favoured at warmer temperatures. The surface of a protein that polymerises is thought to include non-polar (hydrophobic) regions that tend to be driven together because water molecules interact more readily with other water molecules than with these regions. But water-soluble protein molecules such as tubulin also have charged regions on their surface which tend to *inhibit* the aggregation. The very acidic C-terminal end of tubulin polypeptides is important in this respect; if it is specifically removed by proteolysis, the subunits assemble more readily.[29]

At low temperatures, the repelling force dominates and the protein molecules are dissociated. Water molecules produce a balance of intermolecular forces near a hydrophobic region by ordering themselves into a closed polyhedral cage around it. In the warm, the fine balance between charge repulsion and hydrophobic attraction can be reversed. A higher temperature makes the water molecules more disordered (there is an increase in entropy) and it may no longer be energetically favourable for the water to remain close to the hydrophobic regions of the protein. Instead, the tubulin molecules are driven together into a polymer.

During polymerisation there is an increase in the total volume; water molecules move around forming transient polyhedral arrangements but hydrophobic bits of protein no longer fit neatly into spaces between polyhedra. Changes other than warming that interfere with the properties of water, reducing the freedom of the molecules to arrange themselves favourably around the hydrophobic regions, also alter the balance of forces. Thus, for example, the addition of glycerol or substitution of D_2O for H_2O both hyperstabilise microtubules. Hydrostatic pressure (above 100 bar), on the other hand, tends to destabilise them; the decrease in total volume caused by the pressure makes the ordered arrangement of water molecules around the hydrophobic regions more favourable again, even at higher temperatures.

7.3.2 Purification of microtubules by temperature cycles

Methods of purifying assembly-competent tubulin make use of the fact that cytoplasmic microtubules are cold-labile: if a cell homogenate is chilled on ice, most of the microtubules are depolymerised and the protein is left in the supernatant when the homogenate is centrifuged. The cell membrane, organelles, intermediate filaments and most of the actin are lost in the pellet. The supernatant fraction is then warmed, usually after the addition of GTP, and microtubules self-assemble. They can be isolated by centrifuging again. If the pellet is resuspended in fresh buffer and cooled to 4 °C, the microtubules dissolve and more contaminants can be removed by centrifuging again. This cycle may be repeated several times until the microtubules are relatively pure. A disadvantage of a large number of such cycles is that the final yield may be very poor as some tubulin is lost at each stage; not all of the microtubules depolymerise at each cold step and unpolymerised tubulin (at the critical concentration) is left in each warm supernatant.

Other proteins, called **microtubule-associated proteins** (MAPs), copurify with tubulin in a quantitative way throughout this procedure. They and their properties are discussed in detail in the next chapter. One property worth mentioning at this point is that many of them appear to bind to the C-terminus of tubulin, neutralising the negative charges and enhancing assembly.

7.4 Assembly initiation, elongation and depolymerisation[14–16,19–22]

Polymerisation of microtubules *in vitro* shows broadly similar characteristics to actin-filament assembly (see Section 3.2), with an initial lag phase followed by a first-order exponential elongation (see Figure 7.12a). As for actin, assembly can be accelerated by adding fragments of pre-assembled microtubules (or flagellar axonemes) as 'seeds' and the consequent elongation requires a lower critical concentration than that for initiation of polymers.

7.4.1   Quantitative assay of assembly and disassembly[37]

Measurement of the turbidity (scattering of visible light) is currently the most suitable way of following microtubule assembly in bulk (as in Figure 7.12a). Apparently, the turbidity produced by a solution of rod-like molecules whose length is longer than the wavelength of the light used (330–350 nm) is independent of their length. Instead it is just a function of the mass of assembled material (though, unfortunately, not linearly related). A fluorimetric method has not yet been devised: labelling with any fluorescent compound seems to affect the properties of tubulin. Viscometry is also not very suitable as it introduces shearing forces that readily break microtubules.

Assembly assay by electron microscopy is possible provided the sample is carefully sprayed onto support films as small droplets (so that the amount of polymer on the support does not depend on a representative number of particles in a larger sample happening to stick to it). This method is very time-consuming and tedious (as it involves measuring microtubule lengths).

New methods of watching the assembly of individual microtubules by video-enhanced dark-field or interference light microscopy are also time-consuming if a statistically valid number of microtubules is to be studied. However, the advantages of being able to distinguish the behaviour of different ends and of observing directly the changes with time make this the most valuable of assays (see Section 7.6).

7.4.2   Polymorphic assembly of pure tubulin[6–8,17,36]

Pure tubulin can be obtained from whole 'microtubule protein' by gel filtration (6 S-tubulin) or ion-exchange chromatography (PC-tubulin or DEAE-tubulin) which separate it from MAPs due to differences in size or charge respectively. Once purified, it assembles more slowly than with MAPs and aberrant forms, such as twisted ribbons and curved sheets of protofilaments, are common. Two or more sheets may join with reversed polarity (with bonds between sheets like those found throughout the Zn^{2+}-induced sheets) to give an S-shaped profile in cross-section (see Figure 7.8). Sheets may even attach by their long edges to the *outside surface* of microtubules or other sheets, to form what, in cross-section, looks like a circle with 'hooks' attached to it. A complete tubule with one sheet attached to it (Figure 7.9) resembles part of a flagellar doublet, as described in Chapter 9. It seems that initiation of new microtubules may be so difficult that assembly can often proceed more easily by sideways association with existing tubes.

7.4.3   Properties of lateral bonding[17,6]

All aberrant assembly seems to involve 'mistakes' in the lateral bonds; *longitudinal* protofilament formation is always normal. Each monomer can associate laterally with like or unlike monomers to produce the A- or B-lattices (see Section 7.1.1), both of which consist of protofilaments oriented all one

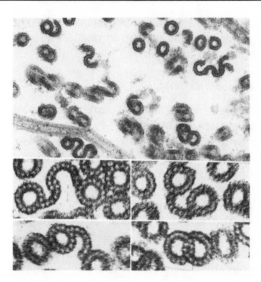 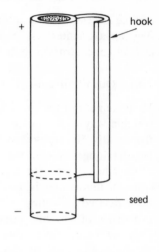

Figure 7.8

Polymorphic forms of pure tubulin seen in cross-section after tannic-acid fixation. A variety of types of lateral association between protofilaments produce S-shapes from partial tubules as well as whole tubules decorated with hooks.[6,14] The polymers were assembled *in vitro* so their polar orientations are random. Top panel × 120 000, others × 300 000. [Reproduced with permission from Burton, P. R. & Himes, R. H. (1978). *J. Cell Biol.* **77**, 120–133.]

Figure 7.9

The assembly of a tubulin 'hook' onto the side of a normal microtubule involves interactions between one side of the first protofilament in the hook and the outer surface of a protofilament in the tubule.[7,8] Individual interactions may not be highly specific but cooperativity resulting from the matching periodicity may promote associations that are weak individually.

way. The association between tubulin protofilaments with zinc ions bound to them into sheets (Section 7.1.2) apparently involves similar regions of the surface, but alternate protofilaments are rotated by 180° about the bond direction. Finally, hooks involve tubulin–tubulin bonding sites on the outer surface of a protofilament. All these associations presumably involve a combination of hydrophobic and ionic forces (see Section 7.3.1). The variations suggest that interprotofilament bonding may be largely the result of relatively unspecific hydrophobic and ionic interactions, which match up because of the periodic spacing of monomers along each protofilament. However, under normal conditions some features of the surface of the tubulin dimer must favour the particular associations that produce a closed microtubule.

7.4.4 Initiation of assembly *in vitro*[20-2]

Formation of a nucleus during the lag phase of microtubule assembly appears to involve assembling a small sheet of protofilaments. (The mechanism appears to be similar whether or not the solution also includes the MAPs that survive assembly–disassembly cycles; only the protein concentration required is different.) It is not clear whether there is a precise point at which nucleation turns into elongation. Presumably the polymer becomes gradually more stable as it grows; it should be maximally stable once the sheet has grown wide enough to close into a cylinder.

There is evidence, both from the 3-D structure and the predominant occurrence of longitudinal protofilaments, that axial bonds between dimers are significantly stronger than lateral bonds. Lateral association only becomes stable when there are a sufficient number of bonds working in concert, holding two protofilaments together. But individual protofilaments do not appear to be capable of assembling to any significant length. Extended ribbons of three protofilaments, in the presence of MAPs, are the narrowest that have been identified under conditions of growth.

7.4.5   Tubulin rings and oligomers[21,22]

Tubulin from microtubules disassembled by chilling or exposure to calcium ions is known to form rings or coils (Figure 7.10). Measurement of the subunit dimensions suggests that the rings are equivalent to rolled-up single protofilaments, or small groups of protofilaments. Electron microscopy of purified preparations sometimes shows microtubules with rings or coils of protofilaments apparently peeling from their ends, indicating that these aggregates may be formed directly from the polymers. But it is known that rings can also be formed by assembly from dimers, under non-polymerising conditions.

Figure 7.10

Tubulin rings formed at low temperature, when assembly into microtubules is inhibited.[20-2] The rings shown here have been shadowed with platinum, which reveals the monomer periodicity along these rolled-up protofilaments. × 160000. [Micrograph by Dr John Heuser.]

The dimers in a coiled protofilament must have a different conformation from those in a straight one. The curly conformation is favoured by the presence of GDP rather than GTP in the E-site. The resulting rings can be quite stable under some conditions, especially if a high level of divalent ions (e.g. Mg^{2+} or Ca^{2+}) is present.

The effect of calcium ions on microtubules is still somewhat mysterious. Micromolar Ca^{2+} levels seem sufficient to cause depolymerisation *in vivo*, but highly purified protein requires millimolar levels for an observable effect *in vitro*. Presumably in these circumstances Ca^{2+} displaces Mg^{2+} and alters the conformational stability of the protein. It is assumed that some additional factor increases the sensitivity *in vivo*.

When a solution of rings is warmed (in the absence of Ca^{2+}), there is evidence from time-resolved light- and X-ray-scattering studies that the coils of protofilaments break down into smaller oligomers, which may then, with fresh supplies of GTP, take part directly in the reassembly of microtubules (especially if assisted by MAPs) or disassemble further into dimers before adding on to microtubule ends.

7.4.6　Drugs affecting tubulin assembly[13–16,32]

Specific inhibitors of polymerisation include **colchicine**, **colcemid**, **nocodazole** and **benomyl** (the last of which is specific to fungal tubulin and therefore useful as an antifungal agent). The binding of labelled colchicine (an alkaloid produced by the autumn crocus, *Colchicum*) was used as an assay for tubulin in early attempts at purifying the protein. Tubulin loses its colchicine-binding activity at the same time as becoming assembly-incompetent.

One drug molecule binds to each dimer, but the drugs block polymerisation totally at a much lower stoichiometry. It is thought that one or a few poisoned dimers are enough to block assembly at a microtubule end.

Vinblastine (from a plant of the genus *Vinca*) disassembles microtubules into helically coiled protofilaments, similar in form to the rings produced by cold disassembly. It is not clear whether it does this by displacing lateral bonding between protofilaments or by directly perturbing the conformations of the subunits. The helices readily form crystals (Figure 7.11). Unfortunately, they are not well ordered enough for X-ray crystallographic analysis.[38]

Taxol, a poison from the yew tree, hyperstabilises tubules and is an extremely useful reagent in microtubule research, especially for the isolation of associated protein components.[32] When low concentrations (5–20 μM) of taxol are added, microtubules become stable and attached components can be washed off under conditions that would normally dissolve the tubulin as well.

All these drugs have been used to interfere with microtubule assembly *in vivo* and thereby gain information about microtubule function. Colcemid (a colchicine analogue) and nocodazole are preferred for experiments because their effects are more readily reversible. Some of the less toxic drugs, especially vinblastine, are used in the treatment of cancer as they are inhibitors of mitosis (see Chapter 11).

7.4.7　GTP hydrolysis and its function[19]

As in the case of actin, the hydrolysable nucleotide apparently has the role of switching the state of tubulin between two main alternatives. When bound

(a) 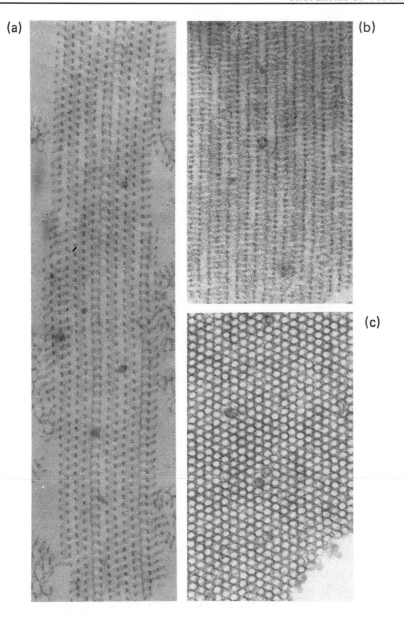 (b)

(c)

Figure 7.11
Helical structures formed from
tubulin protofilaments to which
the drug vinblastine sulphate has
bound. (a) A longitudinal section
through a small crystal formed in
a solution of purified brain
tubulin. (b) & (c) Longitudinal
and cross-sectional views of
crystals induced in living
sea-urchin eggs, illustrating the
close packing together of helices
in a hexagonal arrangement.
× 70 000.

initially, the unhydrolysed molecule puts free protein subunits into a condition
that is favourable for assembly. Then, after hydrolysis during the assembly
process, having only diphosphate in the E-site makes the polymer unstable.
GTP hydrolysis is not essential for any step in the assembly process: non-
hydrolysable GTP analogues also support microtubule assembly. Tubulin differs
from actin in that no assembly occurs if the only nucleotide present is GDP,
but not all dimers need to have GTP on the E-site for *co-assembly with MAPs*.
Thus, the species of bound nucleotide may affect the conformation of the protein
or the distribution of charge on its surface, but other factors can have similar
effects.

Attempts have been made to measure the lag time between addition of
tubulin dimer onto a microtubule end and hydrolysis of the E-site GTP. The

results suggest that if it exists at all it must be fairly short. The loss of cleaved phosphate (P_i) may be slower—see Section 3.2 on actin.

Microtubules resemble actin filaments in having a fast (active, or plus) end and a slow (inactive, or minus) end. **Treadmilling** of subunits from the plus end to the minus end, using energy derived from GTP-hydrolysis, has been postulated for microtubules in exact analogy to the phenomenon in F-actin. There are, however, differences from actin which are discussed in Section 7.6.

7.5　Microtubule polarity[7,8]

Hook formation, described above, has been used by McIntosh and colleagues to determine the polarity of tubules within cells, in a manner analogous to the decoration of actin filaments with myosin fragments. Lysed cells are incubated with a solution of exogenous tubulin, in a buffer that favours aberrant assembly (poor nucleation). The polarity of the pre-existing microtubules can be determined by fixing and sectioning the cells and then observing cross-sectional views by electron microscopy. Hook-decoration has proved to be a more reliable method than decoration with flagellar dynein, which does not always show an obvious direction of curvature.

The method depends on the fact that hooks form in only one of the two possible mirror-image orientations. The correct orientation (Figure 7.10) was established by using the method on flagellar tubules, where the basal ends and distal tips can be determined from the accessory structures (see Section 9.3). The rule is that if you point the thumb of your *right* hand to the *basal* (minus) end of the tubule, your fingers curl in the direction of the hook.

The definite structural polarity of axonemes is also useful in experiments which involve polymerising microtubules from the ends of axonemes and comparing the rates of assembly at the two ends (see below). It has been established that the distal (or plus) end initiates assembly more readily and is much more active.

7.6　Dynamic instability[1,2,11]

Some experiments carried out by Mitchison and Kirschner first revealed that the plateau phase in the assembly curve conceals a highly dynamic activity of the microtubule ends. When they diluted a solution of microtubules and investigated the results by electron microscopy, instead of finding that all of the tubules had partially depolymerised, as expected, they found that some had grown longer. Thus, it is possible in the same solution to have some tubules shrinking and others growing.

The tentative explanation proposed by Mitchison and Kirschner was that if hydrolysis lags behind assembly, a rapidly growing tubule might have a stabilising cap of dimers with GTP bound to them. The end may lose its cap only when hydrolysis happens to catch up with the addition of subunits. Then, if GTP-bearing dimers were released at a much slower rate than GDP-dimers, ends with caps would grow whilst ends that were not capped would continue to lose GDP-dimers. Mitchison and Kirschner predicted that, if the cap theory were correct, uncapped microtubules would shrink to nothing and the sample should gradually consist of fewer and longer microtubules.[11]

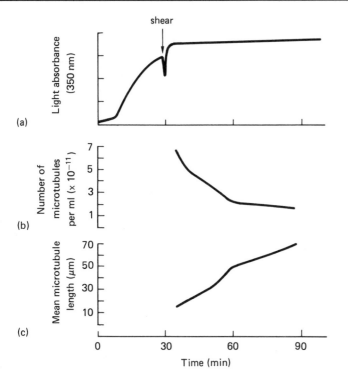

Figure 7.12

Some of the original evidence for dynamic instability of microtubules. (a) The assembly of tubulin as monitored by turbidity, which is known to provide a measure of the total mass of microtubule polymer. After assembly had almost reached a plateau level, the sample was sheared to break the tubules into shorter lengths. This resulted in a large transient drop in turbidity, suggesting that the broken ends were more prone to lose subunits than the original growing ends. After further incubation, the turbidity reached a new plateau. (b) & (c) Results obtained by electron microscopy of samples taken at intervals after shearing. Even after the turbidity became constant, the number of microtubules dropped, presumably because some depolymerised completely. The average length of the others increased as they incorporated the released subunits. End-to-end annealing of microtubules may also play some part in the length redistribution.

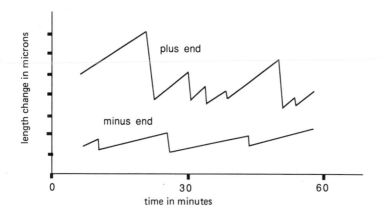

Figure 7.13

Dynamic instability at the two ends of an individual microtubule (based on data obtained by light microscopy[3]). Growing and shrinking phases alternate at each end but the two ends behave independently. Shrinkage is more rapid than growth, but under steady-state conditions (as here) continues for shorter times.

Electron microscopy suggested that microtubules in equilibrium did become longer and fewer, whilst the total amount of polymer remained constant (Figure 7.12).

7.6.1   Observation of individual microtubule behaviour[3,4]

Direct observations of individual microtubules by light microscopy have confirmed the dynamic instability of microtubules. However, it is not clear that the data fit a simple cap theory: most microtubules observed undergo *cycles*

Table 7.1
Association and dissociation rates for pure tubulin dimers[4]

	GTP-tubulin dimers (growing phase)		GDP-containing tubulin? (shrinking phase)	
	plus end	minus end	plus end	minus end
k_{on} ($\mu M^{-1} s^{-1}$)	8.9	4.3	0	0
k_{off} (s^{-1})	44	23	733	915
critical concentration (μM)	5	5	—	—

Figure 7.14

Rates of growth and rapid shrinkage of microtubules as a function of protein concentration. (a) During growing phases, the assembly of pure GTP–tubulin is fairly similar to ATP–actin assembly (Figure 3.6a) except that subunit off-rates are higher, leading to higher critical concentrations at both ends. Also, the two ends of a microtubule differ less from each other than do the two ends of an actin filament. (b) During shrinking phases, the rates of subunit loss from the two ends are similar and independent of the surrounding protein concentration. If the rate represents the off-rate of GDP–tubulin (the on-rate being zero), this is 100–1000× greater than the off-rate of ADP–actin (Figure 3.6b). [Data from Walker, *et al.* (1988). *J. Cell Biol.* **107**, 1437–1448.]

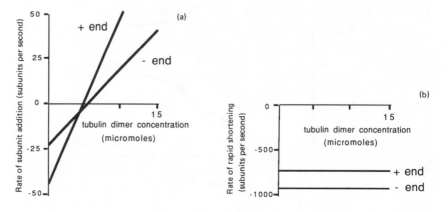

of growing and shrinking, at *both ends* (Figure 7.13). The same kind of behaviour occurs independently at each end. The main difference observed between ends is that the plus end is more active, with both addition and loss of subunits occurring more rapidly. An exchange of labelled subunits at steady state had been measured earlier by several groups and was assumed to result from treadmilling. However, dynamic behaviour is so marked that this is now thought to be mainly responsible for the exchange, at least in the case of pure tubulin microtubules (but see Section 7.6.6).

Walker and colleagues have determined on- and off-rate constants for the active and inactive ends of pure tubulin microtubules (Figure 7.14). This was done by assembling tubulin at a range of concentrations onto axoneme seeds and observing the growing microtubules by video-enhanced light microscopy (see Figure 8.15). The measurements were analysed as described for F-actin growth in Section 3.2. The values in Table 7.1 can be converted back to the measured rates of change in microtubule length by assuming that 13 subunits are needed to change the length by 8 nm.

Since tubulin dimers with GDP on the E-site will not assemble, only the rates for GTP-tubulin could be determined directly. The set of rates on the right was evaluated from measurements made during the shrinking phase; if the cap theory were correct, these rates would apply to GDP-tubulin. It seems unlikely that they are appropriate for individual GDP-tubulin dimers; detailed studies of assembly by Weisenberg and others have shown that, although microtubules do not grow in the presence of GDP-tubulin, there is a critical concentration at which pre-assembled microtubules are stable in its presence.[39]

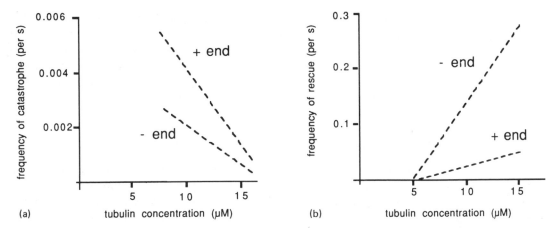

(a)

(b)

Figure 7.15

The probability of switching in a given time from growing to shrinking phases and vice versa is dependent on protein concentration. (a) A catastrophe, defined as the switch from growth to shrinkage, becomes less probable at higher concentrations. (b) Rescue, defined as the switch from shrinkage to growth, becomes more probable. A minus end is *less* likely to begin shortening and *more* likely to be rescued than a plus end.

The very high rates of shortening suggest that some sort of cooperative process may be responsible for the loss of tubulin subunits during rapid disassembly.

As well as these rate constants, Walker *et al.* determined the frequencies at which growing ends converted to the shrinking state (defined by them as a **catastrophe**) and shrinking ends started to grow again (defined as **rescue**). The way these frequencies vary with tubulin concentration is shown in Figure 7.15. The dependence on protein concentration is not as strong as might be expected. If the cap results from a lag between addition of a new subunit to the end of a microtubule and the hydrolysis of its GTP, higher protein concentrations and consequently higher elongation rates should produce longer caps. The likelihood of a catastrophe should rapidly become very small as protein concentration increases.

Walker *et al.* suggested instead that the exchangeable GTP on an end subunit is hydrolysed as soon as another subunit binds to it, but not hydrolysed as long as it is exposed. This would mean there would never be more than a monolayer of subunits with E-site GTP. The measured off-rates for GTP–tubulin are high enough to allow the *lateral* extent of the cap to vary (and possibly vanish completely) by the occasional loss of GTP-containing subunits from the ends, rather than as the result of asynchrony between subunit addition and GTP hydrolysis.

Note that the values in Table 7.1 cannot necessarily be used to predict the precise behaviour of microtubules in cells. All the rates vary with conditions in the medium. Measurements have been made only over a small range of conditions, since the choice of buffer solutions that support microtubule assembly *in vitro* is quite limited. Also, the dimer concentration cannot be raised very much, otherwise it becomes impossible to observe individual microtubules. There may be important new effects at higher concentrations. Finally, the rates will be affected by MAPs present in cells (see Section 7.6.7). Nevertheless, values deduced from observations on cells are similar (see Section 10.2).

7.6.2 Severed microtubules[5]

Experiments in which microtubules were cut with a UV microbeam provide evidence that the presence of dimers containing GDP–tubulin on the ends of microtubules may not necessarily prompt rapid shrinkage. When a tubule is cut, one cut end (corresponding to a plus end) immediately proceeds to shrink rapidly. The other appears quite stable and may even start to elongate. GDP-containing tubulin should have been present at both ends, yet a catastrophe occurred only at the more active end. The fact that a severed minus end behaves in the same way as a minus end under steady-state conditions leads one to wonder whether this end of a protofilament is *ever* capped by subunits whose E-site GTP is unhydrolysed.

7.6.3   The uniform behaviour of minus ends

It can be seen from Figure 7.16 that only at one end is a layer of *β-tubulin* subunits left 'exposed' until another layer of heterodimers binds. At the other end, any newly binding β-tubulin is automatically covered up by its α-tubulin partner, which could mean that the GTP bound to the β-subunit is hydrolysed when that very dimer, rather than the next one, adds to the end. Mandelkow and Mandelkow originally suggested that if this occurred at the minus end it would explain the surprisingly constant properties of minus ends with different histories.[6] The end where β-subunits remain exposed and unhydrolysed is probably the plus end. The polarity of tubulin heterodimers in a microtubule has, however, not yet been ascertained directly.

7.6.4   A possible modification to the simple capping mechanism

As discussed above, it seems unlikely that the dynamic instability of minus ends can be explained in terms of acquisition or loss of a terminal layer of GTP-dimers. Whether an end is growing steadily or shrinking rapidly probably depends instead on the structural state of subunits at the end of the microtubule. It may perhaps depend on whether the end is 'neat' or 'frayed'. In this case, rapid disassembly might be a stochastic process whereby the ends of microtubules are in danger of fraying at any time and then disassembling cooperatively.

The end dimer of each protofilament is probably a key subunit. The start of fraying might be caused by the random failure of a critical number of lateral bonds to these subunits. The relative weakness of lateral bonds, evident from Section 7.4, is probably an important factor in the instability of microtubules. Individual bonds may be broken and remade at frequent intervals, but the central part of a microtubule would remain intact because of the combined forces of long series of individually weak lateral bonds.

The problem is to explain the apparent cooperativity of rapid disassembly from the ends. One possibility is that an end dimer freed from all its lateral bonds might undergo a conformational change which can be propagated in some way along the protofilament, breaking the lateral bonds of each successive dimer in turn. The disruptive conformation may be related to that found in a tubulin ring (Section 7.4.5). Fraying might continue either until a subunit finally failed to transmit the conformational change to its neighbour or until a fresh subunit binding to the end of the protofilament reasserted the lateral bond-forming conformation.

7.6.5   The modulating effect of a GTP cap on the plus end

Since the behaviour at the two ends of a microtubule is qualitatively similar, it is likely to result from a single mechanism. That suggested in Section 7.6.4 is a possibility. However, it is clear that behaviour at the plus end must be

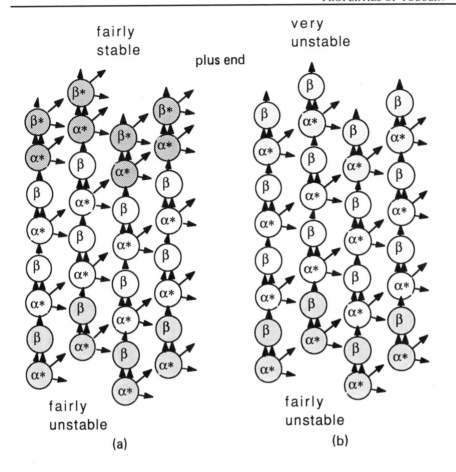

fairly
stable

very
unstable

plus end

fairly
unstable

fairly
unstable

(a)

(b)

Figure 7.16

The fact that assembly is favoured if GTP is bound to the E-site (on β-tubulin) must mean that GTP–tubulin subunits (shown starred) make stronger bonds (arrows) with neighbours in the microtubule lattice than GDP–tubulin subunits. Probably the lateral bonds are most affected, since GDP–tubulin can still form ring polymers (Figure 7.10). For simplicity, GDP–β-tubulin subunits are shown making *no* lateral bonds. It is assumed that GTP–β-tubulin and GTP–α-tubulin subunits make lateral bonds of similar strength.

A possible explanation for the differential behaviour of the two ends of a microtubule is that E-site GTP is hydrolysed immediately at the minus end but awaits the addition of another subunit at the plus end.[6] During growth, most protofilaments would be capped at the plus end by GTP-subunits (diagram (a)). These presumably make good bonds with lateral neighbours and inhibit a switch to the rapid shrinkage phase. If the capping subunits are missing (diagram (b)), either because of individual subunit loss or because the microtubule has been severed, lateral bonding at the plus end is poor and the polymer is unstable. The unvarying presence of α-tubulin, containing non-hydrolysable GTP, at the minus ends of protofilaments may stabilise the polymer to an intermediate extent.

strongly modulated by the nucleotide bound to the terminal β-tubulin. How might this operate?

A likely scheme is illustrated in Figure 7.16. An α-tubulin subunit should always be capable of making good bonds with its neighbours because of the permanent presence of GTP, but a β-tubulin subunit should only make good bonds when it has GTP (or GDP + P_i) bound to it. Three possible end states are shown in Figure 7.16. At the minus end, the final α-tubulin subunits hold the protofilaments together. When the plus end is fully capped (Figure 7.16a), the α-tubulin and β-tubulin subunits of the end dimer *both* contribute to lateral stability. When the cap is totally missing (Figure 7.16b), only monomers second from the end can make good bonds. It is likely that this is a less stable arrangement than that at the minus end.

Thus, a terminal layer of GTP-containing subunits may affect the *probability* of a catastrophe by strengthening the lateral bonds at the ends of the protofilaments. Because of the structural difference between the two ends, the effect of exchangeable GTP-hydrolysis may be important for behaviour only at the plus end, the end that normally grows and shrinks *in vivo*.

7.6.6 A comparison with F-actin

Having microtubules assembled from heterodimers with hydrolysable GTP on only one of the monomers may allow a much greater difference between

GTP–tubulin and GDP–tubulin than between ATP–actin and ADP–actin, without compromising the strength of the whole polymer. Alternate lateral bonds can still keep the protofilaments together. In contrast, the bonds made by every actin monomer are important in keeping an actin filament intact (see Figure 3.1). It must be particularly important to have no weak links in a contractile system where the filaments are involved in producing tensile forces.

Microtubules are thought to provide pushing but not pulling forces. Thus, the presence of poorly-bonding GDP–β-tubulin may only be important at the very tip of a plus end. Having a large difference between the properties of GDP–tubulin and GTP–tubulin, and restricting the effect of this difference on the stability of a microtubule to subunits at its end, allows the microtubule network to be changed very rapidly and be used to perform a pioneering role in the determination of cell structure.

7.6.7 The effect of MAPs on assembly and disassembly rates: treadmilling?[3,31]

Microtubule-associated proteins are covered in the next chapter, but it is necessary to mention here their effects on the properties of tubulin. MAPs are known to reduce greatly the critical concentration required for microtubule assembly (by changing the balance of forces discussed in Section 7.3.1). Most of the published measurements of on- and off-rates in the presence of MAPs, and the conclusions drawn from them, are not reliable since the possible effects of dynamic instability were not taken into account. The net addition to or loss from a microtubule end, as observed by electron microscopy, will depend unpredictably on the time taken over the experiment and give a misleading impression of the actual growth and shrinkage rates.

In their study of individual microtubule ends by dark-field microscopy, Horio and Hotani reported that MAPs have most effect on the shrinking phase, reducing the frequency and extent of shrinkage. However, more recent reports from Salmon's group suggest that the precise effects on growth and shrinkage depend on the species of MAP added. The final picture may well be quite complex. However, it seems clear that MAPs can suppress dynamic instability by altering the on- or off-rates at each end of a microtubule. Thus, a suitable choice of MAPs could produce different critical concentrations of protein at the plus and minus ends and make treadmilling, as originally envisaged for F-actin (see Section 3.2.4), possible also for microtubules.

7.7 Summary

The assembly of tubulin into microtubules depends on a critical balance of forces. At high temperatures, hydrophobic forces drive the subunits together; their highly-charged C-terminals tend to drive them apart. The binding of GTP (or suitable analogues) and magnesium ions is needed to produce a dimer conformation that favours assembly. Dynamic assembly and disassembly, even under steady-state conditions, are a manifestation of the delicate balance of these forces. Subtle variations in intersubunit interactions are responsible for the various polymorphic forms.

Comparison of Table 7.1 with the values given for actin in Table 3.1, especially for the off-rates, shows how labile microtubules are compared with actin filaments. Many of the associated proteins described in the next chapter are able to shift the balance towards polymerisation, even to the point of producing very stable tubules.

7.8 Questions

7.1 Work out the average net rates of growth or shrinkage at the two ends of an average microtubule, for tubulin dimer concentrations of $10 \, \mu M$ and $15 \, \mu M$. Use the association and dissociation rates in Table 7.1. Assume that the average growing and rapid shrinking phases last for the following times:

Dimer concentration	Time spent growing		Time spent shrinking	
	Active end	*Inactive end*	*Active end*	*Inactive end*
$10 \, \mu M$	250 s	400 s	40 s	7 s
$15 \, \mu M$	750 s	1250 s	25 s	3.5 s

Is there a net flux of subunits (possibly treadmilling) along the average microtubule at either monomer concentration? If so, in which direction is it?

7.2 What factors control tubulin polymerisation?

7.3 Describe the various ways in which tubulin dimers associate with one another. Which types of bonding are more conserved?

7.9 Further reading

Key papers and reviews

1 Kirschner, M. W. & Mitchison, T. (1986). Beyond self-assembly: from microtubules to morphogenesis. *Cell* **45**, 329–342.
   (A stimulating review.)

2 Mitchison, T. & Kirschner, M. W. (1987). Some thoughts on the partitioning of tubulin between monomer and polymer under conditions of dynamic instability. *Cell Biophys.* **11**, 35–55.

3 Hotani, H. & Horio, T. (1988). Dynamics of microtubules visualised by darkfield microscopy; treadmilling and dynamic instability. *Cell Motil. & Cytoskel.* **10**, 229–236.

4 Walker, R. A., O'Brien, E. T., Pryer, N. K., Soboeiro, M. F., Voter, W. A., Erickson, H. P. & Salmon, E. D. (1988). Dynamic instability of individual microtubules analyzed by video light microscopy: rate constants and transition frequencies. *J. Cell Biol.* **107**, 1437–1448.
   (Not easy to read but the authors have made it as readable as possible.)

5 Walker, R. A., Inoué, S. & Salmon, E. D. (1989). Asymmetric behaviour of severed microtubule ends after ultraviolet-microbeam irradiation of individual microtubules *in vitro*. *J. Cell Biol.* **108**, 931–937.

6 Mandelkow, E. & Mandelkow, E.-M. (1989). Tubulin, microtubules and oligomers: molecular structure and implications for assembly. In *Cell Movement*, vol. 2, ed. Warner, F. D. & McIntosh, J. R. pp. 23–46. New York: Alan R. Liss.

7 Heidemann, S. R. & McIntosh, J. R. (1980). Visualization of the structural polarity of microtubules. *Nature* **286**, 517–519.

8 McIntosh, J. R. & Euteneuer, U. (1984). Tubulin hooks as probes for microtubule polarity: an analysis of the method and evaluation of data on microtubule polarity in the mitotic spindle. *J. Cell Biol.* **98**, 525–533.

9 Cleveland, D. W. (1989a). Autoregulated control of tubulin synthesis in animal cells. *Curr. Opinion Cell Biol.* **1**, 10–14.

10 Cleveland, D. W. (1989b). Use of DNA transfection to analyze gene regulation and function in animal cells: dissection of the tubulin multigene family. *Cell Motil. & Cytoskel.* **14**, 147–155.

Papers providing more detail

11 Mitchison, T. & Kirschner, M. W. (1984a). Dynamic instability of microtubule growth. *Nature* **312**, 237–242.
    (The original experimental evidence for dynamic instability.)

12 Mitchison, T. & Kirschner, M. W. (1984b). Microtubule assembly nucleated by isolated centrosomes. *Nature* **312**, 232–237.

13 Dustin, P. (1984). *Microtubules*, 2nd edn. New York: Springer-Verlag.

14 Roberts, K. & Hyams, J. S. (1979). *Microtubules*. London & New York: Academic Press.
    (Comprehensive background to older work.)

15 Avila, J., ed. (1989). *Microtubule Proteins*. Boca Raton, FA: CRC Press.

16 McKeithan, T. W. & Rosenbaum, J. L. (1984). The biochemistry of microtubules. In *Cell and Muscle Motility*, vol. 5, ed. Shay, J. W. pp. 255–288. New York: Plenum.

17 Amos, L. A. & Eagles, P. A. M. (1987). Microtubules. In *Fibrous Protein Structure*, ed. Vibert, P. J. & Squire, J. M., pp. 215–246. London & San Diego: Academic Press.

18 Mohri, H. & Hosoya, N. (1988). Two decades since the naming of tubulin—the multi-facets of tubulin. *Zool. Sci.* **5**, 1165–1185.
    (Review of diversity in sequences and post-translational modifications, with extensive references.)

19 Carlier, M.-F. (1988). Role of nucleotide hydrolysis in the polymerization of actin and tubulin. *Cell Biophys.* **12**, 105–117.

20 Correia, J. J. & Williams, R. C. (1983). Mechanisms of assembly and disassembly of microtubules. *Annu. Rev. Biophys. Bioeng.* **12**, 211–235.

21 Timasheff, S. N. & Grisham, L. M. (1980). *In vitro* assembly of cytoplasmic microtubules. *Annu. Rev. Biochem.* **49**, 565–592.

22 Erickson, H. P. & Pantaloni, D. (1981). The role of subunit entropy in cooperative assembly, nucleation of microtubules and other two-dimensional polymers. *Biophys. J.* **34**, 297–309.

23 Beese, L., Stubbs, G. & Cohen, C. (1987). Microtubule structure at 18 Å resolution. *J. Mol. Biol.* **194**, 257–264.

24 Wais-Steider, C., White, N. S., Gilbert, D. S. & Eagles, P. A. M. (1987). X-ray diffraction patterns from microtubules and neurofilaments in axoplasm. *J. Mol. Biol.* **197**, 205–218.

25 Kirschner, K. & Mandelkow, E.-M. (1985). Tubulin domains responsible for assembly of dimers and protofilaments. *EMBO J.* **4**, 2397–2402.

26 Linse, K. & Mandelkow, E.-M. (1988). The GTP-binding peptide of β-tubulin. *J. Biol. Chem.* **263**, 15 205–15 210.

27 Bayley, P. M. & Manser, E. J. (1985). Assembly of microtubules from nucleotide-depleted tubulin. *Nature* **318**, 683–685.

28 Littauer, U. Z., Giveon, D., Thierauf, M., Ginzburg, I. & Ponstingl, H. (1986). Common and distinct tubulin binding sites for microtubule-associated proteins. *Biochem. J.* **83**, 7162–7166.

29 Maccioni, R. B., Serrano, L., Avila, J. & Cann, J. R. (1986). Characterization and structural aspects of the enhanced assembly of tubulin after removal of its carboxyl terminal domain. *Eur. J. Biochem.* **156**, 375–381.

30 Breitling, F. & Little, M. (1986). Carboxy-terminal regions on the surface of tubulin and microtubules. *J. Mol. Biol.* **189**, 367–370.

31 Farrell, K. W., Jordan, M. A., Miller, H. P. & Wilson, L. (1987). Phase dynamics at microtubule ends: the coexistence of microtubule length changes and treadmilling. *J. Cell Biol.* **104**, 1035–1046.

32 Schiff, P. B., Fant, J. & Horwitz, S. B. (1979). Promotion of microtubule assembly *in vitro* by taxol. *Nature* **277**, 665–667.

33 Sullivan, K. F. (1988). Structure and utilization of tubulin isotypes. *Annu. Rev. Cell Biol.* **4**, 687–716.

34 Cleveland, D. W. & Sullivan, K. F. (1985). Molecular biology and genetics of tubulin. *Annu. Rev. Biochem.* **54**, 331–366.

35 Raff, E. C. (1984). Genetics of microtubule systems. *J. Cell Biol.* **99**, 1–10.

36 Burton, P. R. & Himes, R. H. (1978). Electron microscope studies of pH effects on assembly of tubulin free of associated proteins. *J. Cell Biol.* **77**, 120–133.

37 Frieden, C. (1985). Actin and tubulin polymerization: the use of kinetic methods to determine mechanism. *Annu. Rev. Biophys. Biophys. Chem.* **14**, 189–210.

38 Amos, L. A., Jubb, J. S., Henderson, R. & Vigers, G. (1984). Arrangement of protofilaments in two forms of tubulin crystal induced by vinblastine. *J. Mol. Biol.* **178**, 711–729.

39 Weisenberg, R. C. (1980). Role of co-operative interactions, microtubule-associated proteins and guanosine triphosphate in microtubule assembly: a model. *J. Mol. Biol.* **139**, 660–677.

40 Weisenberg, R. C. (1981). A microtubule assembly model. In *Microtubules and Microtubule Inhibitors*, ed., De Brabander, M. & De Mey, J., pp. 161–174. Amsterdam: Elsevier/N.-Holland.

41 Caplow, M., Ruhlen, R., Shanks, J., Walker, R. A. & Salmon, E. D. (1989). The stabilization of microtubules by tubulin.GDP.P$_i$ subunits. *Biochemistry* **28**, 8136–8141.

8 PROTEINS ASSOCIATED WITH MICROTUBULES

There are probably as many proteins involved in controlling the assembly, disassembly and interactions of microtubules in cells as there are for actin. But fewer have been studied in great detail and it is possible that some major categories remain to be discovered. In addition to the structural components described below, a number of minor enzymatic components are known to copurify with tubulin. These include a transphosphorylase that can convert bound GDP back to GTP, an enzyme that specifically removes tyrosine from the C-terminus of α-tubulin monomers, and another that can replace this terminal residue.

8.1 MAPs[1-4]

Proteins that copurify with tubulin through several cycles of assembly and disassembly, that promote microtubule assembly and that appear to bind with fairly constant stoichiometries have been given the general name of **microtubule-associated proteins (MAPs)**. Many seem to correspond to long wispy structures attached to the outer surfaces of microtubules which do not show up very clearly in electron micrographs (see Figure 8.1a–c). They vary in size and sequence between species and, more particularly, between different tissues. The use of monoclonal antibodies has revealed great specificity in MAPs: in the same neuron, for example, some MAPs in the axon differ from those in the dendrites. On the other hand, many different MAPs appear to have some common antigenic epitopes: the available protein sequences suggest that some tubulin-binding domains are quite highly conserved. Also, many varieties of MAPs may arise from alternative splicing of the same messenger RNA, to give alternative MAPs with the same tubulin-binding regions.

8.1.1 The effect of MAPs on microtubule assembly

Most variation in microtubule stability appears to depend on MAP characteristics. For example, fish and other organisms which are not homeothermic

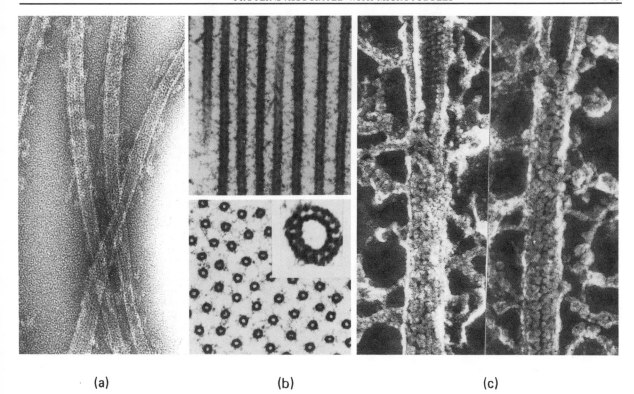

(a) (b) (c)

have microtubules which resemble flagellar tubules in being cold-stable, although their tubulin shows no obvious difference from that of mammals.

The MAPs present in mammalian brain have been best characterised, though even their precise functions are still not clear. It is not a 1:1 relationship: each MAP molecule is thought to control several tubulin dimers, and may not bind identically to each. There is evidence that some stretches bind to the acidic C-termini of tubulin monomers (see Section 7.3.1) and neutralise their repulsive force. The effect of GTP and GDP (Section 7.4.7) may also be charge-related, though this involves a different domain of the tubulin monomer (see Figure 7.6). In both cases, part of the assembly-promoting effect may simply be due to a change in the net charge at tubulin–tubulin interfaces, leading to improved bonding (Section 7.6). But it is likely that a bound MAP domain, or GTP rather than GDP, also affects the relative conformation of the separate domains within the tubulin molecule.

It has been suggested that the main effect of MAPs can be measured as a reduction in the off-rates for tubulin dimers. But most of the rates reported were calculated without taking into account the effects of dynamic instability (Section 7.6), so these conclusions may not be totally correct. The participation of tubulin–MAP oligomers in assembly might be expected to have some effect on on-rates also. Tubulin assembled in the presence of MAPs has an absolute *need* for only *substoichiometric* levels of exchangeable GTP. Weisenberg has pointed out that this may be accounted for by assuming that MAPs induce the formation of tubulin oligomers, even under depolymerising conditions, and that it is not necessary for all the tubulin dimers in these oligomers to have GTP bound to them for the whole complex to add on to a microtubule end.[41]

Figure 8.1

Electron micrographs of reassembled brain microtubules composed of tubulin and brain MAPs. The latter are evident as fine filamentous projections from the surfaces of the tubules. (a) In negative stain, the MAP projections provide very little contrast since they tend to bind heavy-metal salt rather than excluding it. Here the specimens are suspended over a hole to maximise the contrast.[44] × 250 000. (b) Positively-stained specimens in thin section, seen in longitudinal (above) and end-on views. × 100 000; inset × 500 000. (c) Freeze-dried microtubules shadowed with platinum: the projections show up well. Notice also the views of tubulin subunits where microtubules have broken to reveal their interiors; the images compare well with the model (Figure 7.3), reconstructed from negatively-stained specimens. × 400 000. [Reproduced with permission from Amos, L. A. (1977). *J. Cell Biol.* **72**, 642–654; Kim *et al.* (1979). *J. Cell Biol.* **80**, 266–276 and Heuser, J. E. & Kirschner, M. W. (1980). *J. Cell Biol.* **86**, 212–234.]

Table 8.1
Proteins that bind to microtubules (MTs)

Name	Polypeptides (kD)	Sources	Molecule
Motor proteins			
Axonemal dyneins	415–480	Cilia and flagella: *Chlamydomonas*, *Tetrahymena*	Bouquet: 3-headed
Outer arms		Sea-urchin sperm	2-headed
Inner arms	>400	All above	2-headed
Cytoplasmic dyneins (MAP 1C)		Mammalian brain	2-headed
Brain HMWP		Mammalian brain	?
Squid HMW		Squid axon, optic nerve	2-headed
Nematode		*C. elegans*	?
Sea urchin		Egg	2-headed?
Amoeba HMWP	440	*Reticulomixa*	?
Kinesins	134, 75	Sea-urchin egg	60 nm rod + 2 small heads at N-terminal end, a 'mop' at the other
	116, 65	Squid nerve, mammalian brain, *Drosophila*	
	110, ?	*Acanthamoeba*	
	105, ?	*Dictyostelium*	
	130	*Aspergillus*	
bimC gene product[1]	80	*Saccharomyces*	Motor domains at C-termini
KAR3 gene product[1]	80	*Drosophila*	
ca[nd+] gene product[2]	100, 75	Mammalian brain, sea-urchin egg	25 nm crosslinks between microtubules
Dynamin			
Vesikin	292	Squid axoplasm	Binds to vesicles
Bundling proteins			
Axolinin	255	Cortical layer of squid axon	105 nm rod
Brain spectrin	220, 240		Up to 190 nm rod
Dynamin	100	Sea-urchin egg	Bind at both ends even if ATP present
Cytoplasmic dynein	400	Brain	

Crosslinks to other structures			
Synapsin I	76, 78	Binds to vesicles, NFs, actin	Rod
Thermolabile assembly-promoting proteins (? = unreliable molecular weights, estimated from mobility on SDS gels)			
MAP 1A	350?	Vertebrate brain	20 nm projection from MT
MAP 1B (= MAP 5, MAP 1.2, MAP 1.X)	256	Vertebrate brain, esp. embryonic	Repetitive sequence—stretched along MT?
Chartins	69, 72, 80?		
Thermostable, assembly-promoting proteins			
MAP 2(A + B, v. similar)	200	Vertebrate neurons (dendrites)	Up to 100 nm projection from MT
MAP 2C	70	Short alternatively spliced product of MAP 2A gene	Shorter projection
210 kD HeLa MAP	210?	Mammalian cells	Probably similar to MAP 2
MAP 4	240?	Vertebrate, non-neuronal	
205 kD *Drosophila* MAP	205?	*Drosophila*	
Tau (group)	35–40	Vertebrate neurons (axonal)	Short (up to 20 nm) projection from MT
Cold stabilisation proteins			
STOPs	60–80	Mammalian brain	?
Less well characterised assembly-promoting proteins			
125 kD MAP	125	Vertebrate brain	?
Sea-urchin egg MAPs	76, 250	Sea-urchin egg	?
Buttonin	75	Sea-urchin egg	?
Aster-forming protein	51	Sea-urchin egg	Spherical
Trypanosome MAP	300	Trypanosome cortex	?
Dimer/oligomer-sequestering proteins			Repetitive sequence probably stretched along MT
33 kD brain protein	33	Porcine brain	?

[1] See Vale, R. D. & Goldstein, L. S. B. (1990). *Cell* **60**, 883–885.
[2] See Sawin, K. & Mitchison, T. (1990). *Nature* **345**, 22–23.

8.1.2 Variation in size and thermostability

The enormously varied sizes of MAP polypeptides (see Table 8.1) do not seem to be related to their polymer-stabilising properties. Instead, the domains that project away from the microtubule (see Section 8.2) seem to be highly variable in size.

In addition to occurring in a variety of sizes, brain MAPs also sort into two classes, depending on their thermostabilities. One group retains its tubulin-binding and assembly-promoting properties even after boiling, the other does not. What this distinction represents in terms of structure or function remains a matter for conjecture.

8.1.3 Tau proteins

These are a group of thermostable polypeptides that run on SDS gels as though their molecular masses are in the 45–80 kD range, though known sequences show a true range of 36–45 kD and suggest that tau molecules are extended and mainly hydrophilic.[38] Sequences for a set of human tau proteins[40] suggest that they arise by alternative splicing at several points in a common message. Other species have strongly homologous proteins, especially in the C-terminal region.

By electron microscopy, isolated tau molecules shadowed with platinum appear as rods up to 90 nm long.[36] The molecules can form paracrystals, which appear polar and stain in transverse bands. The axial periodicity of the banding ranges from 22 nm to 68 nm.[37] Curled-up paracrystals exhibited a gradual change in longitudinal spacing across their width. This observation suggests that the molecule may be fairly elastic. This flexibility may be related to the ability to retain activity after heating; the molecule may be able to vary its conformation quite freely, without becoming fixed in a 'wrong' arrangement, and only assume a definite structure when bound to a microtubule lattice.

When purified tau is attached to microtubules, projections up to 35 nm (though mostly less than 20 nm) long have been observed, which presumably leaves a domain up to 55 nm in length attached to the microtubule surface.[36] Hirokawa and colleagues found that the arrangement of projections seemed to fit a 12-dimer (96 nm) or 6-dimer (48 nm) axial periodicity along each protofilament (see Section 8.2). Known sequences have three or four repeats of a motif containing an excess of positively-charged amino acids (Figure 8.2). It is believed to represent one type of tubulin-binding site, possibly associating with the acidic C-terminus of a tubulin monomer. There may be other types of binding, if each tau molecule lies alongside a number of tubulin dimers.

Tau proteins have multiple sites which can be phosphorylated. The ability of the phosphorylated forms to bind to tubulin and stimulate assembly into microtubules is reduced compared with unphosphorylated molecules. Antibody staining suggests that tau is found mainly in axons (see also Section 10.5). Unlike tubulin, which pre-exists as a soluble protein pool, tau is apparently synthesised in the cell body just before neurites grow out, so it may regulate the quantity of microtubules exported into an axon[42] (see the similar story for tektins in Section 9.6). Microinjected tau increases the total amount of polymerised tubulin and decreases the rate of microtubule disassembly.

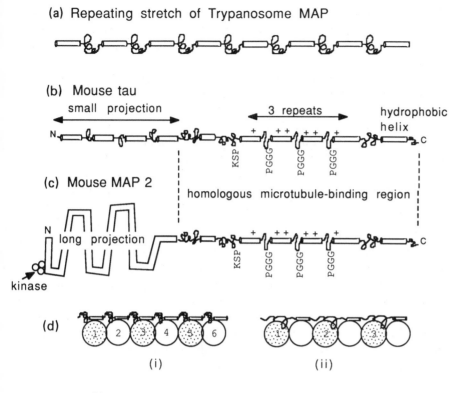

(a) Repeating stretch of Trypanosome MAP

(b) Mouse tau

small projection   3 repeats   hydrophobic helix

N — C

KSP   PGGG   PGGG   PGGG

(c) Mouse MAP 2

homologous microtubule-binding region

N   long projection   C

kinase

KSP   PGGG   PGGG   PGGG

(d)

(i)   (ii)

Figure 8.2

Structures predicted from the sequences of MAPs using standard computer programs [Staden, R. (1982). *Nucl. Acid Res.* **10**, 2951–2961.] Rods represent α-helix; no attempt has been made to distinguish between possible β turns and random coiling. (Current methods cannot predict polypeptide conformations in detail but do give a general idea of the types of structure present.) The highly repetitive trypanosome MAP sequence[46] (a) seems well suited for multiple interactions with successive tubulin subunits (d), most probably monomers (i) or dimers (ii) along a protofilament. (b) & (c) The C-terminal sections of tau and MAP 2 are closely related to each other and include three or four repeats of a tubulin-binding sequence that includes the PGGG motif (proline and three glycines—predicted to form a sharp turn in the polypeptide) and a predominance of positively-charged residues.[38,39] The small number of repeats suggests a fairly complex interaction with microtubules (see Figure 8.6). The KSP motif (lysine, serine, proline) may indicate a phosphorylation site that controls the binding affinity (see text).

8.1.4 STOPs[51]

A class of MAPs that has been found associated with cold-stable tubules has been named **STOPs (stable tubule-only proteins)**. One that has been purified from rat brain, a nominally 64 kD protein referred to as **switch protein**, is thought to confer stability against both cold and drugs when in the dephosphorylated state, but allow disassembly when phosphorylated. Thus, its properties suggest that it may be a form of tau-like protein with extreme stabilisation properties.

8.1.5 MAP 2

These high-molecular-weight polypeptides have a strong homology with tau proteins. **MAP 2** can be resolved on SDS gels into two bands, **MAP 2A** and **MAP 2B**, both of which are thermostable. The DNA sequence has been obtained for a mouse-brain MAP 2 (see Figure 8.2).[39] Although its mass had been estimated from its mobility on SDS gels as 280–300 kD, the sequence data give a value of 199 kD—still much greater in size than tau. The chief difference is a longer (up to 90 nm) flexible projection (see Figure 8.1). This corresponds to a major N-terminal domain, which does not by itself bind to microtubules. MAP 2B seems to be antigenically very similar to MAP 2A, though not identical.

Isolated MAP 2 appears in shadowed electron-microscopic preparations as a long molecule of variable length, up to a maximum of 185 nm,[45] which could mean that nearly 100 nm of the molecule contacts the microtubule surface. There is evidence for periodic long projections from microtubules with a 12-dimer (96 nm) axial repeat (Section 8.2). Exposure of bound MAP 2A to proteolytic enzymes releases a major portion, leaving a 28–39 kD basic peptide now known to be the C-terminal portion (with 67 per cent homology to tau) and three similar copies of the basic microtubule-binding motif (Figure 8.2). This fragment binds strongly to the microtubule and has the assembly-promoting property.

The MAP 2 molecule can also be multiply phosphorylated. A specific protein kinase is associated with the projection domain of some molecules; 39 kD and 54 kD polypeptides that copurify with MAP 2A are the catalytic and regulatory components. Phosphorylated MAP 2A is less able to bind and promote microtubule assembly: this is presumably due to modification of sites on the assembly-promoting domain, possibly including the **KSP** (lysine, serine, proline, in the one-letter code) motif that MAP 2 shares with tau (see Figure 8.2).

The projection domain of MAP 2 has most of the phosphorylation sites; the sequence is qualitatively similar to that of neurofilament (NF) protein projections (Section 2.3.3), including irregular repeats of the KSP motif. Phosphorylation of the MAP 2A projection domain may serve to regulate interactions with other cellular components, as proposed for the equivalent modification of NF heavy-chain projections. MAP 2A has been localised in neuronal cell bodies and dendrites, but not in axons (see also Section 10.6.8).

MAP 2C is a lower-molecular-weight version of MAP 2, produced by splicing the MAP 2A gene in a different manner.[50] It is more widely distributed than MAP 2A and occurs in embryonic brain where the larger polypeptides are not found. Like MAPs 2A and 2B, it differs from tau in being mainly a component of dendrites rather than axons.

8.1.6 MAP 1A

This high-molecular-weight component seems to share considerable sequence homology with MAP 2 and to stimulate tubulin assembly to a similar extent. Its sequence is not yet known. MAP 1A is reported to produce 20 nm long projections, arranged on the microtubule wall in the same manner as the longer projections of MAP 2A. The striking difference from MAP 2A is that MAP 1A loses its activity when heated. Two low-molecular-weight polypeptides, of 28 kD and 34 kD, copurify with MAP 1A. They are not needed for the binding of MAP 1A to tubulin but seem to be essential for its assembly-promoting activity. The 34 kD polypeptide has been shown to be associated with a 120 kD fragment of the heavy chain that also binds to microtubules.

Using antibodies to MAP 1A, Vallee's group have labelled all parts of neuronal cytoplasm and numerous other cell types. Experiments by other groups showed staining of neurons only. There may be subspecies of MAP 1A that are ubiquitous and others that are restricted.

8.1.7 Other neuronal MAPs[1,3]

SDS gels of microtubule preparations from adult vertebrate brain often show minor bands below the major MAP 1A component. They have, perhaps unfortunately, been referred to as **MAP 1B** and **MAP 1C**. The latter differs from the other MAPs in being moderately resistant to proteases and has recently been identified as **cytoplasmic dynein** (see Section 8.6). MAP 1B is also known as **MAP 5** and is the major MAP in developing embryonic neurons. It lacks the repeated motif present in tau and MAP 2.

MAP 3 is present in brain tissue in smaller amounts than MAP 1 or MAP 2 and appears to be a smaller polypeptide. It is found in both glial cells and neurons; in the latter case, apparently only in NF-rich axons. Some even smaller MAPs, named **chartins**, are also thermolabile.

8.1.8   Non-neuronal MAPs

When microtubules are prepared from animal cells other than those from brain, the major MAP bands usually appear well below the positions of MAP 1 and MAP 2A, commonly in positions corresponding to 200–220 kD. The latter may, therefore, be of a similar size to MAP 2A but lack the structural peculiarity that makes MAP 2A run anomalously on SDS gels. There are also polypeptides in non-neuronal cells that share similarities with tau proteins.

Very little is known about MAPs in cells other than those from vertebrates, but high-molecular-weight polypeptides have been found associated with microtubules in many cases, even in protozoa. The sequence is known for a segment of a 300 kD MAP[46] from **trypanosomes** (the organisms that cause sleeping sickness). It contains eight almost perfect repeats and its structure is predicted to be alternating regions of α-helix and random coil (Figure 8.2). This structure would give the sort of elasticity apparently found in tau and may be a common feature of MAPs.

8.1.9   MAP 2-like proteins that bind to other structures

Several proteins have been described as being similar to MAP 2, either antigenically or in showing a tendency to bind to microtubules, though they are not primarily associated with microtubules. One example is a heat-stable IF-associated (nominally 300 kD) polypeptide named **plectin**.[4] Similarly, **vesikin**, a protein that copurifies with axonal vesicles (Section 8.6.5), binds to microtubules and is heat-stable like MAP 2. **Synapsin I** may also be related; as well as binding to vesicles it associates with microtubules when phosphorylated, or with actin filaments when not.[54]

It has been shown that brain spectrin (**fodrin**) can bundle microtubules[48] and there is some tentative immunological evidence of a relationship between MAP 2 and spectrin, as well as one between MAP 1A and **ankyrin** (Section 4.5). These properties provide a glimpse of what may be a complex and subtle web of interactions between different components of the cytoskeleton.

8.2   MAP structure

As mentioned above, many MAPs are thought to consist of a domain that projects out into the medium, as shown in Figure 8.3, and one that binds directly to the microtubule surface. The binding domain almost certainly binds to a number of different tubulin dimers. These are probably restricted to one or very few protofilaments, since MAPs remain quite strongly bound to tubulin when microtubules depolymerise into protofilament rings and coils.

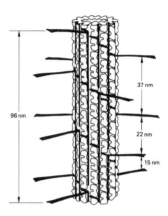

Figure 8.3

MAPs appear to form a fairly regular arrangement of projections from the surface of a microtubule.[44] There tends to be a saturation level of one MAP molecule to about 12 tubulin dimers, which implies an axial spacing of 96 nm for projections from each protofilament. The staggering of molecules on adjacent protofilaments to produce a symmetrical arrangement gives the spacings indicated for molecules seen projecting from the sides of a microtubule (see also the top right of Figure 8.4). In the model shown here, projections are physically separated from one another by tubulin-binding domains that run along the protofilaments, contacting each of the 12 dimers.[14] Small lateral excursions organise the relative displacements of MAP molecules on adjacent protofilaments. This scheme is probably oversimplified for MAP 2 and tau, since current evidence suggests that each of these molecules binds to a much smaller number of tubulin dimers;[38,39] it may be relevant for more repetitive MAPs. Figure 8.4 shows an alternative way of achieving the same superlattice.

8.2.1 Arrangement of MAPs on microtubules[44]

Both tau protein and high-molecular-weight MAPs are able to bind to microtubules so as to give projections at regular intervals, although they sometimes appear to bind in an irregular way. The preferred axial spacing tends to be one MAP per 12 dimers (as illustrated in Figures 8.3 & 8.4), like the overall repeat distance of accessory proteins on flagellar outer doublets (see Section 9.3).

MAPs spaced at 12-dimer intervals along each protofilament could form the helical superlattice marked out by the black circles in Figure 8.4, on an A-tubule-like dimer lattice (defined in Section 7.1). Projections from adjacent protofilaments would produce the pattern illustrated in Figure 8.3. Measurements of spacings observed between projections in electron micrographs of purified microtubules have given values that fit this pattern better than any corresponding to other possible superlattices. However, many reassembled microtubules are aberrant, having 14 protofilaments rather than 13. Since it is impossible to fit a truly symmetrical superlattice onto microtubules possessing an extra protofilament, aberrant arrangements will always occur *in vitro*.

Hirokawa and colleagues have measured the spacings between projections on bundles of microtubules decorated with tau proteins. The histograms contained a set of peaks similar to those obtained for microtubules in cells, namely a mixture of spacings predicted from 12-dimer and 6-dimer lattices. It

Figure 8.4

To set up a '12-dimer' superlattice of MAP molecules, each needs to influence three points on the superlattice. The diagram shows a possible way of stretching out a 45 nm long domain (65 nm tau molecule minus a 20 nm long projecting domain)[36] to make contact with neighbouring molecules at three such points. An arrangement of this type might allow a random number of additional molecules to bind to tubulin at intermediate sites (such as the lighter molecule in the bottom left complex) to produce an unsaturated '6-dimer' lattice.[43]

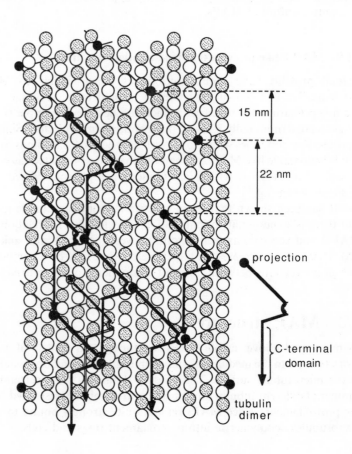

is possible that additional molecules occasionally bind in between those positioned on a 12-dimer lattice (as in Figure 8.4). There is another possible explanation for some 6-dimer intervals. When two microtubules, each with its own projections, lie side by side, both sets will appear as crossbridges. It is known whether these are paired or staggered axially, but a half-staggered arrangement might account for apparent 6-dimer intervals.

8.2.2   Projections *in vivo*[43]

Measurements of projections from microtubules in fixed cells should be most reliable, since few natural microtubules appear to deviate from the standard 13-protofilament structure. Jensen and Smaill have carried out a careful analysis of spacings in images of thin sections. Histograms clearly show that the spacings are not random but show peaks at spacings predicted from the 12-dimer superlattice. Some spacings fit in with an unsaturated 6-dimer axial repeat, as for tau.

8.2.3   Control of MAP spacing

It is not known how the long spacing (96 nm) is controlled but it is probably measured out in some way by the associated protein itself. Why does it matter? For efficient control of microtubule assembly, it is presumably advantageous for the cell to employ a much smaller number of MAP molecules than tubulin dimers. Each MAP molecule should interact with several dimers in an exclusive manner. The most natural way is by the formation of a symmetrical superlattice, whereby the MAP molecules interact with each other to enforce a well-spaced arrangement. MAP 2A is long enough to measure out the full 96 nm axial repeat directly along each protofilament (as in Figure 8.3), as well as producing the observed long projection, but tau is not. The similarity in sequence between the C-terminal domains of tau and MAP 2 and the fact that MAP 2 and tau can compete *in vitro* for the same binding sites on microtubules suggest that the pattern of binding may be similar in both cases.

Figure 8.4 shows a possible scheme of lateral interactions that might enable a 55 nm tau molecule domain to measure out 96 nm axial spacings. The stretches of repeated sequence (Figure 8.2) are strongly implicated as the main tubulin-binding sites. The C-terminal portion probably also associates with the microtubule, since it does not appear to be proteolysed when the N-terminal projection of MAP 2 is removed by trypsin. There may be other weak tubulin-binding sites on the other side of the repeating stretch.

8.2.4   Possible functions of different MAP projections

Specific localisations of tau and MAP 2 in different parts of neurons (Sections 8.1 & 10.5) suggest that the differences between them are important. The obvious difference lies in the projecting domain. MAP 2 may carry out in dendrites the lateral spacing and crosslinking roles which have been suggested for neuro-filament extensions in axons (see Section 2.3.3). Embryonic brain uses a much smaller version of MAP 2 (MAP 2C), lacking most of the projection domain; a reduction in crosslinking may be important in allowing for the plasticity of developing neurons.[50]

Do different protruding MAP domains also provide tags that specify what types of vesicles are to be transported along particular routes? There is some evidence that different bundles of microtubules lying side by side in the same axon provide distinct tracks for transport of components to separate branches of the axon. MAP projections may perhaps provide the distinguishing characteristic.

8.3 Specialised associated proteins

8.3.1 Bundling proteins

Many types of stable microtubule bundles exist (see Section 10.1) but only in a few cases have the proteins responsible been identified. The outer doublet microtubules of **axonemes** are held together by elastic straps composed of a protein named **nexin**. Little else is known about it. A 255 kD polypeptide that forms rod-shaped crosslinks between microtubules has been purified from axons and given the name **axolinin**.[52] It is apparently restricted to the cortical layer of cytoplasm next to the axonal membrane, where fixed crosslinks are unlikely to interfere with fast axonal transport.

8.3.2 Assembly-initiating components

Special sites in cells are foci for microtubule assembly and are often referred to collectively as **microtubule organising centres** (**MTOCs**). Little is known about them, but their constituents are now being studied by a number of groups. The centrosomes and kinetochores (Sections 9.6 & 11.4) of animal cells have been studied most.

A number of proteins have been located antigenically in the region of centrosomes. In most cases their functions are unknown. Some may be involved in centriole assembly, which is a complex process. Others may attach there simply to ensure that they are shared during cell division. But some must be involved in initiating, organising and controlling microtubule assembly.

Sakai and colleagues have purified an interesting candidate, a 51 kD polypeptide from isolated sea-urchin mitotic apparatuses (spindles and associated components) and have shown that it has aster-forming activity. It appears to be a component of the **pericentriolar material** but antibodies to the protein also stain the spindle microtubules.[47]

8.3.3 Assembly-inhibiting proteins

Tubulin dimer-sequestering proteins, analogous in function to actin monomer-sequestering proteins such as profilin (Section 4.2.7), may account for large pools of unpolymerised tubulin that are found in some cells, especially oocytes and eggs. A 33 kD protein with these properties has been isolated from brain.[49] It appears to have little effect on already-assembled microtubules but inhibits assembly of pure tubulin in a 1:1 stoichiometry. If MAPs are also present, the 33 kD protein can inhibit the assembly of 2–3 times as much tubulin; this suggests that it may 'cap' tubulin–MAP oligomers, or hold them in a non-assembling conformation.

8.4   Axonemal dyneins[6-8]

These giant protein molecules were named by Gibbons, who first isolated them from flagella a quarter of a century ago. In recent years, the long-predicted cytoplasmic equivalents have finally been identified and shown to be extremely similar to the flagellar dyneins. Dyneins are emerging as a universal family of microtubule-associated motors, just as myosins appear to be universal agents of actin-associated contractility.

The binding of flagellar dynein provides another electron-microscopic way of determining tubule polarity in thin sections, in addition to the hook method described in Section 7.4. It is also characteristic of flagellar outer-arm dynein that it decorates microtubules, of any type, in a polarised fashion at 24 nm intervals along protofilaments (see Figures 9.2 & 9.12). What evidence there is suggests that this definite periodicity may not be observed for inner-arm flagellar dynein (Section 9.3) or for cytoplasmic dynein (Section 8.6).

8.4.1   Polypeptide components

Dynein, as isolated from flagellar or ciliary axonemes, consists of six or seven distinct large polypeptides and various smaller chains, as enumerated in Table 8.2. The heavy chains run on SDS gels with mobilities similar to the high-molecular-weight MAPs described above, but are estimated from other types of measurement to have molecular masses of more than 400 kD. Such values are deduced from estimates of the polypeptide stoichiometry in intact molecules, together with their sedimentation coefficients or estimates of their mass from measurements of inelastic electron scattering in a scanning transmission electron microscope (STEM—Appendix A.2.5).

8.4.2   Molecular structure

Isolated particles contain one, two or three large polypeptides, depending on which species they come from (see Table 8.2). Most information available is for the outer dynein arms, which can be more easily purified than the inner arms. In the electron microscope each subunit polypeptide appears as a globular particle with a tail; each head is much larger (9–12 nm diameter sphere) than a single myosin head (15 nm × 4.5 nm rod—see Section 4.3). Johnson and colleagues first showed that dynein molecules can resemble small bunches of flowers; those consisting of three heavy chains have three globular heads, molecules with two heavy chains have only two heads (Figure 8.5).[17,18]

In negative stain, the heads have stained centres; some images obtained by Toyoshima show both the bouquet structure of the whole molecule and staining in the centres of heads with roughly circular profiles.[19] This central region may represent the ATP-binding site.

Electron-microscope images from Goodenough and Heuser's group have revealed details of the molecules *in situ* (Figure 9.11).[18] In their original nomenclature, the basic dynein unit is a globular head supported on a 'stem' and sporting a less conspicuous 'stalk' on the opposite side from the stem (see Figure 8.7). The stalks are consistently 13–16 nm long, but the stems vary in length (22–38 nm) and appearance, depending on the precise molecule. The most variable features are small globular subunits that decorate the stems and

Table 8.2
Some dynein components

Source	Molecules (kD)	ATPase	Polypeptide chains (kD)	Molecular structure
Sea-urchin sperm tail outer-arm components				
α-Chain particle		(+)	475, 122(=IC1)	15 S, 1-headed
β-Chain particle		+		12 S, 1-headed
IC2 + IC3 particle		−	90, 76	9–10 S
Total: outer arm	1250	+	475, 450, IC1, IC2, IC3, 4 LCs	20 S, 2-headed
Sea-urchin egg dynein fractions				
Cross-reacts with ciliary dynein	?	+	475, ?	20 S
Non-cross-reacting	?	+	HMr-3	20 S
Tetrahymena cilia outer-arm components				
Total outer arm (resistant to dissociation)	1950	+	400 × 3, 100, 85, 70, 4 LCs	22 S, 3-headed
Tetrahymena cilia inner-arm components (?)				
2 distinguishable particles	510	+	?	14 S, 1-headed
Chlamydomonas flagella outer-arm components				
α-Chain particle		+	480, ?	1-headed
β-Chain particle		+	440, ?	1-headed
α/β Particle	1250	+	480, 440, 86, 73, 6 LCs	18 S, 2-headed
γ-Chain particle	480	+	415, 4 LCs	12 S, 1-headed
Cytoplasmic dynein				
Brain dynein (MAP 1C)	1270	+	410 × 2, 74 × 3, 59, 57, 55, 53	20 S, 2-headed

'LC' stands for light chains—14–24 kD

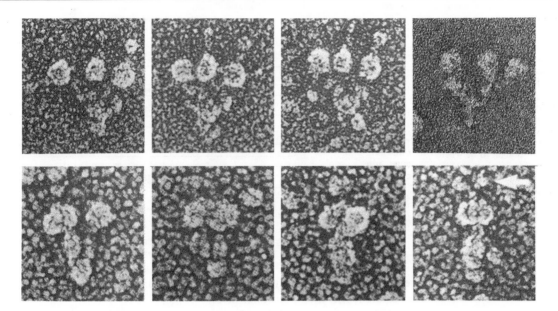

are thought to correspond to some of the intermediate and light chains that copurify with the dynein heavy chains (see Table 8.2). The stems plus associated globular subunits seem to condense down to form 'feet' on the A-tubule surface.

The interactions between flagellar dynein and outer doublet microtubules are discussed in Section 9.4. Briefly, it is assumed that the feet of a bouquet bind in a fixed (though salt-labile) manner to one microtubule. The protruding stalks, which interact with the second microtubule but are not released by ATP, may serve a temporary crosslinking function. During ATP-dependent sliding, some region of the surface of the globular head may interact directly with the surface of a microtubule.

8.4.3   Domains of the heavy chains[6–8]

Each heavy chain is now believed to possess an ATPase activity, its characteristics varying slightly from one to another. A peculiarity of axonemal dynein is a property referred to by Gibbons as **latent ATPase**. If the molecules are isolated under mild conditions their intrinsic ATPase is low compared with activity associated with beating axonemes, but it increases markedly after harsher treatments. This uncoupling of ATPase activity from motility may represent a conformational change or the loss of a regulatory polypeptide chain.

Dynein is characteristically more sensitive than myosin to vanadate ions, which inhibit its ATPase activity, probably by acting as a phosphate analogue. A more surprising feature is that, in the presence of vanadate and MgATP, each heavy chain is cleaved into two large pieces (at site V1 in Figure 8.6) by UV-light irradiation and the ATPase activity is destroyed. The same reaction has been observed for all known dyneins and has become an important method of assaying the involvement of dyneins in cellular activity (Section 10.4). Proteolytic enzymes produce a different pattern of fragments. The heavy chains that have been studied tend to produce a large (200–300 kD) central fragment and smaller terminal fragments (Figure 8.6). A C-terminal domain is most easily

Figure 8.5

Electron microscopy of isolated axonemal dynein molecules. The upper row shows three-headed 'bouquets' from protist axonemes: the first three are platinum-shadowed outer arms from *Chlamydomonas* flagella.[17] The fourth is a negatively-stained *Tetrahymena* outer arm.[19] The lower row shows two-headed outer arms from sea-urchin sperm tails: the first molecule is in the bouquet configuration with its heads splayed apart, the last clearly has its heads fused together to form a 'toadstool'. × 300 000. [Reproduced with permission from Goodenough, U. W. and Heuser, J. E. (1989), in *Cell Movement*, vol. 1, ed. Warner *et al.*, pp. 121–140. New York: Alan R. Liss; and Toyoshima, Y. Y. (1987). *J. Cell Biol.* **105**, 887–895.]

Figure 8.6

Proteolytic sites in dynein heavy-chain polypeptides.[8,6] Gaps between boxes show places where trypsin cleaves readily and may represent flexible links between structural domains. Arrows show sites of photolytic cleavage in the presence of ATP and vanadate ions. The cut at V1 in each chain (heavy arrows) is more easily made and is thought to be where P_i normally binds (when not displaced by vanadate). The V2 sites probably bind to the nucleotide base.

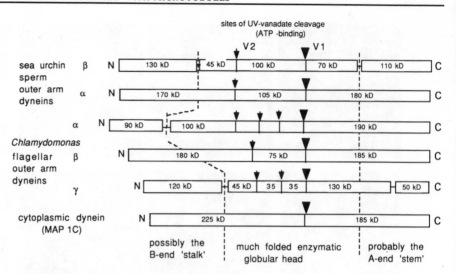

removed *in vitro* from purified dynein heavy chains, leaving the ATPase activity associated with the N-terminal piece.

Gibbons has argued that the easily-lost C-terminal domain may represent the stem that attaches dynein in an ATP-insensitive manner to the A-tubule (see Section 8.5.2) because, after a light trypsin treatment of whole axonemes, much of the ATPase activity can be extracted with ATP. This would leave the N-terminal domain as the main candidate for the stalk at the B-end of the molecule. Since the stalk seems rather constant in length it is probably the central globular domain that varies in size between different heads (Figure 8.6). Goodenough and Heuser have pointed out that one of the heads in a flagellar dynein molecule always appears more elongated than the other or others. The distance between V1 and the C-terminus is remarkably constant; the differences between the stems to different heads (Figure 8.5) are presumably due to the associated intermediate chains.

8.5 Cytoplasmic motors [5–12]

In recent years, cytoplasmic extracts have been searched for proteins which bind to microtubules in the absence of ATP. Extracts from squid axoplasm, various mammalian tissues and sea-urchin eggs have all been shown to contain at least two soluble proteins that bind to microtubules, pellet with them under centrifugation, and can then be released into solution when ATP is added. In each case, the two main classes of molecule obtained have been separated either by gel filtration, where molecules elute according to their dimensions, or by centrifugation through a sucrose gradient, which differentiates molecules according to their densities.

There is some evidence that at least some of these soluble molecules can crosslink microtubules in the same manner as axonemal dyneins and they may therefore be involved in relative sliding between different microtubules. Other evidence suggests that cytoplasmic dyneins may produce transport of membranous vesicles along microtubules. A molecule with two types of microtubule-binding sites might serve both functions, especially if there should turn out to be sites on vesicles that resemble the surface of a microtubule.

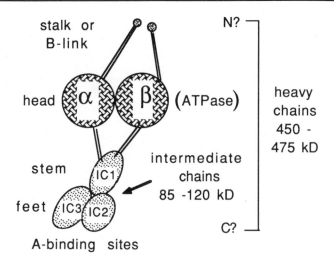

stalk or
B-link

head   α   β   (ATPase)

heavy
chains
450 -
475 kD

stem   IC1   intermediate
chains
85 -120 kD

feet   IC3   IC2

A-binding sites

N?

C?

Figure 8.7

Diagram of the proposed structure of sea-urchin outer-arm dynein, which is a heterodimer of α and β heavy chains.[8,17] Each heavy chain forms one globular head; its C- and N-terminal segments probably protrude from the head to form a stem and a stalk. Intermediate chains (IC1–IC3) are thought to associate with the combined stems. The two-headed molecule breaks down into the α-chain by itself, a β-chain–IC1 complex and an IC2–IC3 complex. The locations of four light chains are unknown.

8.5.1   Brain dynein[22,23]

One molecule purified in this way was found to correspond to a minor protein band previously observed in the high-molecular-weight region of SDS gels of standard microtubule preparations (Section 7.3.2) from mammalian brain. It is a protease-resistant polypeptide originally referred to as **MAP 1C** (see Section 8.1.6) but now clearly unrelated to MAP 1A, MAP 1B or any of the MAP 2 bands. If nucleotide is absent during taxol-induced microtubule assembly, the amount of MAP 1C that pellets with microtubules increases to be comparable with that of MAP 1A.

Its properties in common with flagellar dynein include its heavy chain of more than 400 kD; a comparable ATPase activity, which is readily inhibited by vanadate, a nucleotide analogue; UV-light-induced cleavage in the presence of vanadate (see Figure 8.6); inhibition by agents that oxidise sulphydryls (e.g. N-ethylmaleimide, NEM); and an equivalent direction for transport of microtubules (Figure 8.16). Other dynein-like molecules mentioned below have most of these properties. There are also some differences from axonemal dynein, including absence of 'latent' ATPase.

Cytoplasmic dynein appears to be very similar in structure to two-headed flagellar molecules (see Figure 8.8). Polypeptides of around 75 kD copurify with the heavy chains and some lighter chains are thought to contribute to the molecule (see Table 8.2). The medium and light chains are assumed to be associated with the base of the bouquet formed by the two heavy chains (Figure 8.9). Both ends of the molecule are able to bind to microtubules (Figure 8.10). The association of the stem domain with microtubules is very much more labile than the attachment between flagellar dynein and A-tubules. The heads of cytoplasmic dynein possess stalks that appear identical to the stalks protruding from flagellar dynein heads and also seem to bind to microtubules in a non-ATP-dependent manner. Thus brain dynein apparently has the capability of crosslinking microtubules into bundles.

Similar molecules have since been isolated from various types of non-neuronal vertebrate cells, including chick fibroblasts, as well as from the wide range of species mentioned below.

Figure 8.8

Cytoplasmic dynein isolated from pig brain and seen by negative-stain electron microscopy.[23] Particles are in the same orientation as the flagellar dyneins in Figure 8.5. Those in the upper row have fused heads; two are attached by their fused stalks to microtubules. The molecules in the lower row are in the bouquet configuration. × 300 000.

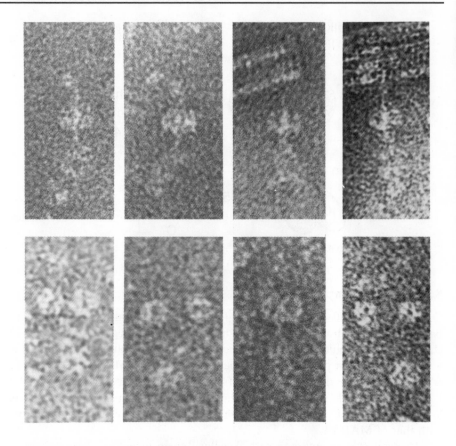

Figure 8.9

Diagram of the proposed structure of a cytoplasmic dynein molecule, which is essentially similar to a sea-urchin outer arm.[22] Whether the two heads are identical or different is not known.

Possible arrangement of subunits in cytoplasmic dynein

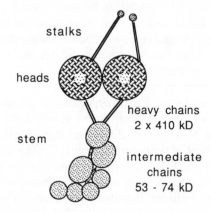

8.5.2 Squid dynein and vesikin[24]

Cytoplasmic dynein with properties equivalent to those of MAP 1C has been isolated from squid giant axon, the source of cytoplasm that originally showed *in vitro* particle transport along microtubules, and also from squid optic lobes, which provide better quantities of material for biochemical fractionation.

Gilbert and Sloboda have succeeded in purifying a form of dynein that binds to both microtubules and vesicles and even translocates the vesicles along the microtubules (at $2 \mu m/s$). Their preparations contain two high-molecular-weight polypeptides. One must be dynein, since it can be photocleaved in the presence of vanadate and MgATP. The other is thermostable and runs in a similar position to MAP 2 on gels. Gilbert and Sloboda have called this polypeptide **vesikin**.

Vesikin is found associated with vesicles after they have been purified on sucrose gradients and chromatographed on columns. It seems likely that vesikin forms the link between dynein and vesicles that is missing from other cytoplasmic dynein preparations. It binds the ATP analogue 8-azidoATP, probably because it has phosphorylation sites. In the absence of ATP it appears to bind to microtubules, rather than to vesicles. It is MAP-like in that it stimulates microtubule assembly; it also shows some antigenic cross-reactivity with antibodies to brain MAPs, as well as being thermostable. Its full role remains to be determined (see also Section 10.4.1).

8.5.3 Sea-urchin egg dynein[16,17]

Soluble dynein was in fact identified first in sea-urchin eggs, where it is abundant and where antibodies to the protein were found to label mitotic spindles. It was argued, however, that cytoplasmic dynein in eggs might just be stored protein waiting for the formation of cilia in the developing embryo. More recently a subpopulation of molecules has been identified, with slightly smaller heavy chains and antigenic characteristics which distinguish them from the rest. These molecules are found bound to microtubules, whereas most egg dynein is soluble; the free molecules are probably inactive until incorporated into cilia. Pratt and colleagues have found a further subfraction apparently stably associated with vesicles. This work suggests that there may be more than one type of cytoplasmic dynein, with different specific functions.

8.5.4 Amoeba dynein[25]

Reticulomyxa is a giant amoeba that produces an extensive network of axon-like extensions. Each strand of the network contains microtubule bundles that support bidirectional organelle movements at rates of up to $25 \mu m/s$. The movements continue unchanged, even after removal of the membrane by detergent treatment. Schliwa and colleagues have isolated and characterised a 20–22 S molecule from the cytoplasm. This protein, referred to simply as _Reticulomyxa_ **high-molecular-weight protein (HMWP)** produces rapid _bidirectional_ movement of latex beads along _Reticulomyxa_ microtubules _in vitro_, at an average speed of $3.6 \mu m/s$. Attempts to produce microtubule-gliding over glass have been unsuccessful (perhaps because of competition between molecules pushing both ways).

The ATPase activity is similar to that of flagellar dynein and is also sensitive to vanadate-dependent, UV-light-induced cleavage. It remains to be seen whether the preparation contains two distinct species of molecule which act in different directions along the microtubule, or whether a single species of dynein-like molecule can be switched between two modes of operation. There

Figure 8.10

Bundling of purified brain microtubules by purified brain dynein.[23] Bundles occur even in the presence of ATP, just as flagellar dynein arms crosslink doublet microtubules under activating conditions (Figure 9.11). The white striations on the microtubules are thought to be the stems of cytoplasmic dynein molecules. The heads appear superimposed on an adjacent microtubule. × 200 000.

is some evidence from experiments on lysed cells that phosphorylation may change particle movements from being bidirectional to retrograde only.

8.5.5 Nematode dynein[26,7]

McIntosh's group have isolated a dynein-like molecule from eggs of the nematode *Caenorhabditis elegans*. Unlike those of echinoderms and molluscs, embryos of nematodes are not ciliated, so dynein will presumably not be made in these eggs for incorporation into cilia. The *C. elegans* protein was originally reported to move in the opposite direction from flagellar and brain dyneins, but an improvement in the method of assaying direction (using whole de-membranated sperm) has shown it to be a retrograde motor like most other dyneins.

8.6 Newly-identified motor molecules

8.6.1 Kinesin[9–12]

A completely different class of cytoplasmic microtubule-associated motor has properties that distinguish it from dynein. The **kinesin** molecule consists of a pair of main chains of 100–140 kD and a pair of medium chains of 65–75 kD. Kinesin has a significantly lower intrinsic ATPase than either flagellar or cytoplasmic dynein, though in the presence of microtubules it is comparable to that of dynein. It is not affected by low levels of vanadate (though levels above 100 μM inhibit its ATPase activity, as in the case of myosin) and cannot be cleaved by UV light.

Other properties (including apparent molecular weights; see Table 8.1) vary among proteins isolated from different species. For example, **AMPPNP** produces rigor when added to squid or mammalian preparations but not when added to extracts from sea-urchin eggs or from *Dictyostelium*; in fact, this ATP analogue even causes detachment of *Dictyostelium* kinesin from microtubules. NEM treatment has no effect on kinesin from squid axoplasm but does seem to inhibit sea-urchin egg kinesin. Moreover, antigenic cross-reactivity has been found amongst squid, *Drosophila*, mammalian and sea-urchin kinesins, but not yet between kinesin purified from *Dictyostelium* and higher species.

Recently, some genes responsible for mitotic defects in a yeast, in a fungus and in *Drosophila* have been sequenced and found to include coding closely homologous to the sequence of the enzymatic domain of kinesin (see below). Other parts of the sequences seem unrelated. This suggests that there is probably a superfamily of kinesins, analogous to the myosin superfamily (Section 4.3), with different specific roles in cells (see also Section 11.6.1).

8.6.2 The structure of kinesin[29–35]

Kinesin, as isolated from vertebrate cells and sea-urchin eggs, is a rod-shaped molecule with some resemblance to two-headed myosins, though the rod domain is shorter (60–70 nm, rather than 150–60 nm) and there is some sort of structure at the tail end of the rod. Electron-microscope images of rapidly-frozen specimens

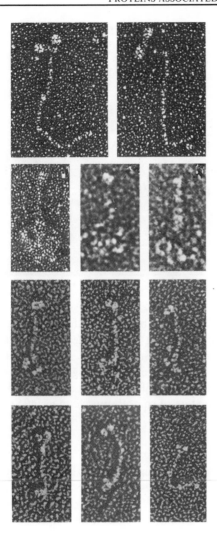

Figure 8.11
EM images of shadowed kinesin molecules compared with two-headed myosin:[34] the upper two molecules are non-muscle myosin; the next three images are of pig brain kinesin molecules, selected to show large structures at the tail end of the rod;[29] the lower six images are of chicken fibroblast kinesin [provided by Drs M. Sheetz & J. Heuser]. In some cases, the tail end structure is no larger than the small pair of heads at the top end. × 220 000.

show a pair of heads which are significantly smaller than myosin heads (Figure 8.11). In negative stain, the paired heads appear as a small fork (Figure 8.12).

The structure at the tail end of the rod often seems to be made up of a larger number of strands, each longer than the heads (see Figures 8.11–8.13). The medium chains (60–75 kD) have been located at this end. Some variability in the appearance of this second structure seems to arise because the strands may lie in a neat bunch, be aggregated into a ball, be splayed out like a mop, or even be missing. Their function may be to hold on to objects that are to be moved along microtubules.

Structure predictions based on the complete sequence of *Drosophila* kinesin (Figure 4.6) confirm that the central part of the molecule is probably an α-helical coiled coil (see Chapter 2). The non-helical length of sequence at the N-terminus fits nicely with the 45 kD globular domains that can be obtained by proteolytic cleavage and still interact enzymatically with microtubules. The C-terminal stretch is predicted to form a smaller globular domain.

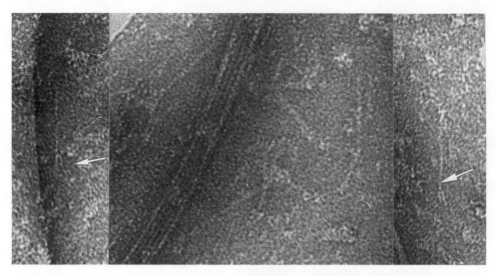

Figure 8.12

EM images of pig brain kinesin in negative stain.[29] Neither the thin rods nor the structures at each end provide much contrast against the stain. Arrows indicate the small forks that represent the two heads in such images. Structures at the tail end are variable, as in Figure 8.11.
× 400 000.

The rod seems capable of bending in the middle, at a point where the sequence suggests a weakening of the coiled-coil structure. Hirokawa and colleagues have found that a majority of the molecules in some kinesin preparations are actually bent double, as diagrammed in Figure 8.13. It is not yet clear whether this folded conformation is active or whether it is equivalent to the folded, inactive state of smooth-muscle myosin (Sections 4.3 & 5.5).

8.6.3   Dynamin[57]

Vallee and co-workers have identified a 100 kD polypeptide that runs ahead of bona fide mammalian kinesin on SDS gels and shows no antigenic cross-reactivity with it. When purified, the 100 kD protein crosslinks microtubules into hexagonally-packed bundles. It has a low MgATPase activity that is not stimulated by microtubules alone. However, in the presence of soluble co-factors, notably a 75 kD polypeptide, ATP-splitting is significantly activated by microtubules and causes bundled microtubules to slide apart. Further work will show whether or not this protein belongs to the kinesin family.

Figure 8.13

A model for the structure of kinesin.[29–35] The 45 kD N-terminal domain of kinesin forms a smaller globular head than the 95 kD head of myosin. The linear map of the heavy chain is compared with that of myosin II in Figure 4.6. The coiled-coil α-helical rod is shorter in kinesin than in myosin II. It is capable of bending near its centre. The rod ends in a small C-terminal knob, to which the 65 kD intermediate chains (ICs) are probably bound. Kinesin ICs may assume a variety of conformations and may thus be responsible for the varying appearance of the tail end structure, which ranges from a large knob to a feathery tail. If the ICs are missing, the C-terminal knobs of the heavy chains may look very similar to the heads (Figure 8.11).

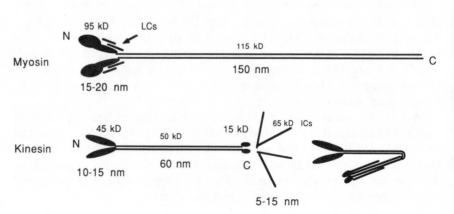

8.7 Kinetic behaviour of dynein and kinesin[11,28,31]

The ATPase of axonemal dynein resembles that of myosin except that it is more sensitive than myosin to vanadate. In axonemes an analogous condition to muscle rigor exists: if the ATP level is reduced to micromolar (from the usual millimolar) level, sperm axonemes go stiff, whereas ATP plus an inhibitor of ATPase produces 'relaxation' in the form of a smoothing out of the bending wave. In sections of rigor axonemes, dense crossbridges seem to reach right across from the A-tubule to the B-tubule of the adjacent doublet.

The cycle outlined for myosin in the kinetic diagram (Figure 5.9) has been shown by Johnson and colleagues to apply equally well to dynein. In the case of both these enzymes, it is thought that the rate-limiting step in the cycle of ATP hydrolysis is the dissociation of the products (step 9 in Figure 5.9). The interaction with polymer catalyses this process (step 5) and thereby accelerates the turnover of ATP. After releasing ADP and P_i, the enzyme rapidly binds and hydrolyses a fresh molecule of ATP, then reattaches to the polymer. If the basic mechanism is the same, perhaps the enormous size of a dynein head (250–350 kD) as compared with a myosin head (95 kD) or a kinesin head (45 kD) results from the insertion of additional machinery to put the motor into reverse gear.

Studies of kinesin kinetics are still in the preliminary stages. Observations by Scholey and colleagues on the rigor state suggest that it is likely that the kinesin crossbridge cycle follows essentially the same route round the scheme of states in Figure 5.9 as myosin or dynein. However, ATP hydrolysis by kinesin is unusually strongly coupled to motility. Most probably the differences from myosin and dynein observed *in vitro* are due to a strong cooperativity between the two heads of a kinesin molecule; Schnapp and colleagues have found that AMPPNP (which probably mimics ADP + P_i) remains strongly bound to one of the two heads if free nucleotide is missing from the medium. When ATP is added, it binds to the second head and AMPPNP is released from the first head. These results suggest a scheme along the lines shown in Figure 8.14. Such behaviour would ensure that one head is usually bound to the microtubule

Figure 8.14

Scheme for cooperative conformational changes of kinesin heads during the crossbridge cycle, basically the same as for myosin (Figures 5.9 & 5.10), except that step 5 may be inhibited for one head until ATP binds to its partner.

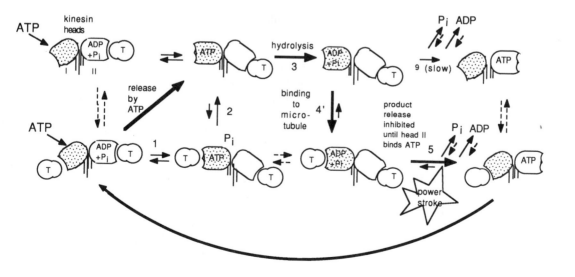

surface while the other is going through the detached parts of the cycle. Dynein may be just as strongly cooperative *in situ* (see Section 9.2).

An interesting observation by Scholey and colleagues is that some monoclonal antibodies which bind specifically to the head domain of sea-urchin egg kinesin *uncouple* its ATPase from motility. Microtubules are inhibited from sliding even though the tubulin-activated ATPase is stimulated 2–4-fold by having these antibodies bound to the heads. The coupling between ATPase and motility is also lost after proteolysis of the molecule, though Gelfand and colleagues found that the 45 kD head domain retains a high ATPase activity in the presence of microtubules.

8.8 *In vitro* assays of microtubule-associated motility[7,10,13–15]

Kinesin, flagellar dyneins and cytoplasmic dynein all bind quite readily to clean glass slides and, in the presence of ATP, cause microtubules to slide over the surface. Single flagellar dynein heads have been shown to be active, like individual myosin heads (Section 4.3), but it is not yet known whether individual kinesin heads produce motility. The attachment to glass is thought to be a non-specific interaction with negative charges on its surface. Any molecules that stick by parts of their surface that are not directly involved in motile activity are able to interact with suitably orientated microtubules. Kinesin has also been found to bind to negatively-charged polystyrene beads and to transport them along microtubules. All these movements can be observed directly by video-enhanced light microscopy (Figure 8.15).

Using microtubules of known polarity, it has been shown that most types of molecule produce movement only in one direction. If repolymerised microtubules are seeded from flagellar axonemes with their dynein arms removed, the difference in extent of assembly from the two ends (see Section 7.6) reveals their polarity; if purified centrosomes are used to initiate the assembly, growth is entirely in the plus direction, producing an aster-like structure. Axonemes can even be used without brain tubulin extensions, since fraying also distinguishes the distal ends. Kinesin can be described as moving particles towards the active end of the microtubule or as moving the tubule with its inactive end leading (Figure 8.16). By analogy with axonal transport (see Section 10.4), this is often referred to as the **anterograde** or **orthograde** direction. Flagellar and brain dyneins push a microtubule so that its active end leads, and would be expected to produce **retrograde** transport of particles. Single dynein heads are capable of producing movement.

Dynamin apparently does not produce microtubule sliding over glass. However, bundles induced by the addition of dynamin and co-factors to purified microtubules have been seen to slide apart into individual polymers at speeds between 0.2 and 1 μm/s. Neither the relative polarities of microtubules in the bundles nor their sliding direction are known at present.

8.8.1 Kinesin steps[27]

The motion of polystyrene beads being transported along stationary microtubules has been analysed in detail by Gelles and colleagues. After smoothing, curves

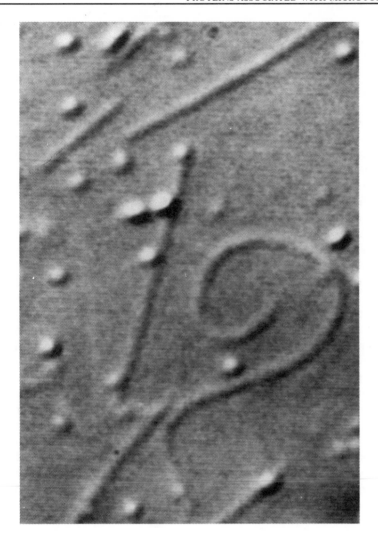

Figure 8.15
Video-enhanced light micrograph (using DIC optics) of individual microtubules in isolated axoplasm gliding over a glass surface. Notice the difference in actual magnification from Figures 7.1a & 8.1a: microtubule diameters appear expanded ($\times 10$, approx.) in the light-microscope image, owing to the lower resolving power. $\times 20\,000$. [Micrograph by Dr Dieter Weiss.]

plotting the distance from a fixed point of a slowly-moving bead against time showed evidence of steps at intervals of 4 nm. The movement of such beads appeared to have only a very restricted lateral component, which could be accounted for simply as rocking about the point of attachment. These observations suggest that kinesin moves along a microtubule, binding to successive tubulin monomers along a protofilament (see Section 7.1) without becoming fully detached between steps. It remains to be shown that two heads are essential for this behaviour.

8.8.2   Rotation caused by 14 S flagellar dynein[14]

14 S *Tetrahymena* dynein is unique, so far, in causing sliding microtubules to *rotate* during the motility assay. Vale and Toyoshima measured the revolution of curved pieces of axoneme by dark-field light microscopy and found that they rotated 1–3 times per micrometre. This is too frequent for the heads to be following the slow twist of the protofilaments often observed for reassembled

microtubules. And if this were the cause, one would expect kinesin and two-headed dyneins to produce rotation. There is no other physically meaningful helix of an appropriate pitch in the microtubule lattice (Sections 7.1 & 8.2) that 14 S dynein might follow, so the rotation is more likely to result from a stochastic process, as described next.

The interaction of 14 S dynein with the microtubule lattice must produce a force with a lateral component; this torque may perhaps play a role in the functioning of inner-arm dynein, in helping to generate bending waves. The long pitch of the *in vitro* rotation suggests that the lateral force component is not quite enough to shift the dynein head round to the next protofilament in a single step; so, on rebinding to the nearest monomer, the head mostly ends up on the same protofilament. There may be, however, a 10 per cent probability of making the quantum leap in the lateral direction, leading to rotation. The observed variation in the pitch of the rotation would not be surprising under such a scheme.

8.9   Summary

The best-known tubulin-associated proteins, at present, fall into two main classes. The first main group are the MAPs, whose probable roles lie in regulating the polymeric state and forming sideways projections for interaction with other structures. In addition, there must be specialised proteins with assembly-initiating functions at one end (MTOCs) and possibly capping proteins at the other, but little is known about the molecules involved.

The second main category are molecules that produce movement (summarised in Figure 8.16). Microtubule-associated and actin-associated motility differ in that the former exhibits bidirectional sliding. In most tissues, different molecules seem to be responsible for sliding towards the two ends of microtubules. For movement towards the centre of the cell there are cytoplasmic dyneins with similar properties to flagellar dyneins. Kinesin slides in the reverse direction—that is, towards the plus end of the polymer. In this respect, and in its structure, it

Figure 8.16

Microtubule-associated motors and the motility they produce *in vitro*.[5-10] The anterograde motor, kinesin, moves particles towards the plus end of a microtubule stuck to glass (lower left). Above, if kinesin molecules are fixed to the glass, they cause microtubules in solution to slide with their minus ends leading. One-, two- or three-headed dyneins (on the right) move microtubules with their plus ends leading, showing that they are retrograde motors. They have not been found to move beads, but in axoneme preparations the dynein arms carry the doublet microtubule to which they are fixed towards the minus end of the adjacent B-tubule (bottom right). (See also Table 10.1.) Dynamin apparently also produces sliding between microtubules, but their relative polarity is not known at present.[57]

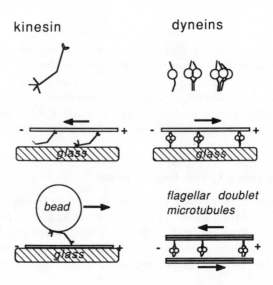

shares some similarities with myosin. Its kinetic behaviour *in vitro* shows some interesting differences from myosin and dynein. Other motors, possibly related to dynein but not at present well characterised, may work bidirectionally.

8.10   Questions

8.1 What are the probable functions of MAPs? What might be the functions of their being arranged on a symmetrical superlattice on the microtubule? Suggest reasons for their restricted distributions within brain tissue.

8.2 How is the pattern of polypeptide bands in a whole flagellar dynein extract interpreted in terms of molecular structures?

8.3 Compare the properties and probable roles of brain dynein and kinesin.

8.11   Further reading

Reviews: MAPs

1 Olmsted, J. B. (1986). Microtubule-associated proteins. *Annu. Rev. Cell Biol.* **2**, 421–457.
   (Comprehensive up to 1986.)

2 Vallee, R. B., Bloom, G. S. & Theurkauf, W. E. (1984). Microtubule-associated proteins: subunits of the cytomatrix. *J. Cell Biol.* **99**, 38s–44s.
   (Less comprehensive, more readable.)

3 Matus, A. (1988). Microtubule-associated proteins: their potential role in determining neuronal morphology. *Annu. Rev. Neurosci.* **11**, 29–44.

– Matus, A. (1990). Microtubule-associated proteins. *Curr. Opinion Cell Biol.* **2**, 10–14.

4 Wiche, G. (1985). High-molecular-weight microtubule-associated proteins: a ubiquitous family of cytoskeletal connecting links. *Trends Biochem. Sci.* **10**, 67–70.

Reviews: motors

5 Gelfand, V. I. (1989). Cytoplasmic microtubular motors. *Curr. Opinion Cell Biol.* **1**, 63–66.

– Vale, R. D. (1990). Microtubule-based motor proteins. *Curr. Opinion Cell Biol.* **2**, 15–22.

6 McIntosh, J. R. & Porter, M. E. (1989). Enzymes for microtubule-dependent motility. *J. Biol. Chem.* **264**, 6001–6004.

7 McIntosh, J. R., Satir, P., Gibbons, I. R. & Warner, F. D. (1988). Force production and microtubule-coupled cell movement. *Cell Motil. & Cytoskel.* **11**, 182–186.
   (Conference summarised by organising committee. See also the two-volume book of the conference, 'Cell Movement': vol. 1, ed. Warner, F. D., Satir, P. & Gibbons, I.; vol. 2, ed. Warner, F. D. & McIntosh, J. R. New York: Alan R. Liss.)

8 Gibbons, I. R. (1988). Dynein ATPases as microtubule motors. *J. Biol. Chem.* **263**, 15 837–15 840.
   (A compact review of dynein biochemistry.)

9 Schliwa, M. (1989). Head and tail. *Cell* **56**, 719–720.
(A minireview of kinesin discoveries.)

10 Vale, R. D. (1987). Intracellular transport using microtubule-based motors. *Annu. Rev. Cell Biol.* **3**, 347–378.

11 Porter, M. E. & Johnson, K. A. (1989). Dynein structure and function. *Annu. Rev. Cell Biol.* **5**, 119–152.

12 Vale, R. D., Scholey, J. M. & Sheetz, M. P. (1986). Biological roles for a new microtubule-based motor. *Trends Biochem. Sci.* **11**, 464–468.

Original papers for further reference

13 Allen, R. D., Weiss, D. G., Hayden, J. H., Brown, D. T., Fujiwake, H. & Simpson, M. (1985). Gliding movement of and bidirectional transport along single native microtubules from squid axoplasm: evidence for an active role of microtubules in cytoplasmic transport. *J. Cell Biol.* **100**, 1736–1752.

14 Vale, R. D. & Toyoshima, Y. Y. (1988). Rotation and translocation of microtubules *in vitro* induced by dyneins from *Tetrahymena* cilia. *Cell* **52**, 459–469.

15 Sale, W. S. & Fox, L. A. (1988). Isolated β-heavy chain subunit of dynein translocates microtubules *in vitro*. *J. Cell Biol.* **107**, 1793–1797.

16 Porter, M. E., Grissom, P. M., Scholey, J. M., Salmon, E. D. & McIntosh, J. R. (1988). Dynein isoforms in sea urchin eggs. *J. Biol. Chem.* **263**, 6759–6771.

17 Goodenough, U. W., Gebhart, B., Mermall, V., Mitchell, D. R. & Heuser, J. E. (1987). High-pressure liquid chromatography fractionation of *Chlamydomonas* dynein extracts and characterization of inner-arm dynein subunits. *J. Mol. Biol.* **194**, 481–494.

18 Goodenough, U. W. & Heuser, J. E. (1989). Structure of the soluble and *in situ* ciliary dyneins visualized by quick-freeze deep-etch microscopy. In *Cell Movement*, vol. 1, ed. Warner, F. D., Satir, P. & Gibbons, I. pp. 121–140. New York: Alan R. Liss.

19 Toyoshima, Y. Y. (1987). Chymotryptic digestion of *Tetrahymena* ciliary dynein. I: Decomposition of three-headed 22S dynein to one- and two-headed particles. *J. Cell Biol.* **105**, 887–895.

20 Haimo, L. T. & Fenton, R. D. (1988). Interaction of *Chlamydomonas* dynein with tubulin. *Cell Motil. & Cytoskel.* **9**, 129–139.

21 Porter, M. E. & Johnson, K. A. (1983). Characterization of the ATP-sensitive binding of *Tetrahymena* 30S dynein to bovine brain microtubules. *J. Biol. Chem.* **258**, 6575–6581.

22 Pascal, B. M., Shpetner, H. S. & Vallee, R. B. (1987). MAP 1C is a microtubule-activated ATPase which translocates microtubules *in vitro* and has dynein-like properties. *J. Cell Biol.* **105**, 1273–1282.

23 Amos, L. A. (1989). Brain dynein crossbridges microtubules into bundles. *J. Cell Sci.* **93**, 19–28.

24 Do, C. V., Sears, E. B., Gilbert, S. P. & Sloboda, R. D. (1988). Vesikin, a vesicle associated ATPase from squid axoplasm and optic lobe, has characteristics in common with vertebrate brain MAP 1 and MAP 2. *Cell Motil. & Cytoskel.* **10**, 246–254.

25 Euteneuer, U., Koonce, M. P., Pfister, K. K. & Schliwa, M. (1988). An ATPase with properties expected for the organelle motor of the giant amoeba, *Reticulomyxa*. *Nature* **332**, 176–178.

26 Lye, R. J., Porter, M. E., Scholey, J. M. & McIntosh, J. R. (1987). Identification of a microtubule-based cytoplasmic motor in the nematode *C. elegans. Cell* **51**, 309–318.

27 Gelles, J., Schnapp, B. J. & Sheetz, M. P. (1988). Tracking kinesin-driven movements with nanometre-scale precision. *Nature* **331**, 450–453.

28 Cohn, S. A., Ingold, A. L. & Scholey, J. M. (1987). Correlation between the ATPase and microtubule translocating activities of sea urchin egg kinesin. *Nature* **328**, 160–163.

29 Amos, L. A. (1987). Kinesin from pig brain studied by electron microscopy. *J. Cell Sci.* **87**, 105–111.

30 Hisanaga, S., Murofushi, H., Okuhara, K., Sato, R., Sakai, H. & Hirokawa, N. (1989). The molecular structure of adrenal medulla kinesin. *Cell Motil. & Cytoskel.* **12**, 264–272.

31 Kuznetsov, S. A. & Gelfand, V. I. (1986). Bovine brain kinesin is a microtubule-activated ATPase. *Proc. Natl. Acad. Sci., USA* **83**, 8530–8534.

32 Yang, J. T., Laymon, R. A. & Goldstein, L. S. B. (1989). A three-domain structure of kinesin heavy chain revealed by DNA sequence and microtubule binding analyses. *Cell* **56**, 879–889.

33 Hirokawa, N., Pfister, K. K., Yorifuji, H., Wagner, M. C., Brady, S. T. & Bloom, G. S. (1989). Submolecular domains of bovine brain kinesin identified by electron microscopy and monoclonal antibody decoration. *Cell* **56**, 867–878.

34 Scholey, J. M., Heuser, J., Yang, J. T. & Goldstein, L. S. B. (1989). Identification of globular mechanochemical heads of kinesin. *Nature* **338**, 355–357.

35 Kuznetsov, S. A., Vaisberg, Y. A., Rothwell, S. W., Murphy, D. B. & Gelfand, V. I. (1989). Isolation of a 45 kDa fragment from the kinesin heavy chain with enhanced ATPase and microtubule-binding activities. *J. Biol. Chem.* **264**, 589–595.

36 Hirokawa, N., Shiomura, Y. & Okabe, S. (1988). Tau proteins: the molecular structure and mode of binding on microtubules. *J. Cell Biol.* **107**, 1449–1459.

37 Hagestedt, T., Lichtenberg, B., Wille, H., Mandelkow, E.-M. & Mandelkow, E. (1989). Tau protein becomes long and stiff upon phosphorylation: correlation between paracrystalline structure and degree of phosphorylation. *J. Cell Biol.* **109**, 1643–1652.

38 Lee, G. L., Cowan, N. & Kirschner, M. (1988). The primary structure and heterogeneity of tau protein from mouse brain. *Science* **239**, 285–288.

39 Lewis, S. A., Wang, D. & Cowan, N. J. (1988). Microtubule-associated protein MAP 2 shares a microtubule binding motif with tau protein. *Science* **242**, 936–939.

40 Goedert, M., Spillantini, M. G., Potier, M. C., Ulrich, J. & Crowther, R. A. (1989). Cloning and sequencing of the cDNA encoding an isoform of microtubule-associated protein tau containing four tandem repeats: differential expression of tau protein mRNAs in human brain. *EMBO J.* **8**, 393–399.

41 Weisenberg, R. C. (1980). Role of co-operative interactions, microtubule-associated proteins and guanosine triphosphate in microtubule assembly: a model. *J. Mol. Biol.* **139**, 660–677.

42 Drubin, D. G. & Kirschner, M. W. (1986). Tau protein function in living cells. *J. Cell Biol.* **103**, 2739–2746.

43 Jensen, C. G. & Smaill, B. H. (1987). Analysis of the spatial organization of microtubule-associated proteins. *J. Cell Biol.* **103**, 559–569.

44 Amos, L. A. (1977). The arrangement of high molecular weight microtubule-associated proteins. *J. Cell Biol.* **72**, 642–654.

45 Voter, W. A. & Erickson, H. P. (1982). Electron microscopy of MAP 2 (microtubule-associated protein 2). *J. Ultrastruct. Res.* **80**, 374–382.

46 Schneider, A., Hemphill, A., Wyler, T. & Seebeck, T. (1988). Large microtubule-associated protein of *T. brucei* has tandemly repeated, near-identical sequences. *Science* **241**, 459–462.

47 Ohta, K., Toriyama, M., Endo, S. & Sakai, H. (1988). Localization of mitotic-apparatus-associated 51 kD protein in unfertilized and fertilized sea urchin eggs. *Cell Motil. & Cytoskel.* **10**, 496–505.

48 Ishikawa, M., Murofushi, H. & Sakai, H. (1983). Bundling of microtubules *in vitro* by fodrin. *J. Biochem. (Japan)* **94**, 1209–1217.

49 Ishikawa, M., Murofushi, H., Nishida, E. & Sakai, H. (1984). 33 K protein—an inhibitory factor of tubulin polymerization in porcine brain. *J. Biochem. (Japan)* **96**, 959–969.

50 Papanrikopoulou, A., Doll, T., Tucker, R. P., Garner, C. C. and Matus, A. (1989). Embryonic MAP 2 lacks the crosslinking sidearm sequences and dendritic targeting signal of adult MAP 2. *Nature* **340**, 650–652.

51 Margolis, R. L. & Rauch, C. T. (1981). Characterization of rat brain crude extract microtubule assembly: correlation of cold stability with the phosphorylation state of a microtubule-associated 64 K protein. *Biochemistry* **20**, 4451–4458.

52 Masumoto, G., Tsukita, S. & Arai, T. (1989). Organization of the axonal cytoskeleton: differentiation of the microtubule and actin filament arrays. In *Cell Movement*, vol. 2, ed. Warner, F. D. & McIntosh, J. R., pp. 335–356. New York: Alan R. Liss.
      (Includes axolinin molecule.)

53 Pratt, M. M. (1986). Stable complexes of axoplasmic vesicles and microtubules: protein composition and ATPase activity. *J. Cell Biol.* **103**, 957–968.

54 Harada, Y. & Yanagida, T. (1988). Direct observations of molecular motility by light microscopy. *Cell Motil. & Cytoskel.* **10**, 71–76.

55 Kim, H., Binder, L. I. & Rosenbaum, J. L. (1979). The periodic association of MAP 2 with brain microtubules *in vitro*. *J. Cell Biol.* **80**, 266–276.

56 Staden, R. (1982). An interactive graphics program for comparing and aligning nucleic acid and amino acid sequences. *Nucl. Acid. Res.* **10**, 2951–2961.

Late additions

57 Shpetner, H. S. & Vallee, R. B. (1989). Identification of dynamin, a novel mechanochemical enzyme that mediates interactions between microtubules. *Cell* **59**, 421–432.

58 Lee, G. (1990). Tau protein: an update on structure and function. *Cell Motil. & Cytoskel.* **15**, 199–203.

59 Vallee, R. B. (1990). Molecular characterization of high molecular weight microtubule-associated proteins: some answers, many questions. *Cell Motil. & Cytoskel.* **15**, 204–209.

60 Vale, R. D. & Goldstein, L. S. B. (1990). One motor, many tails: an expanding repertoire of force-generating enzymes. *Cell* **60**, 883–885.

THE AXONEMES OF 9
CILIA AND FLAGELLA

An **axoneme** is the specialised bundle of microtubules found within the membrane of a eukaryotic flagellum or cilium, organelles which are distinguished from one another really only on the basis of details of their modes of beating. Axonemes and **basal bodies**, which are closely related in structure, are found throughout the animal kingdom, in algae, protists, fungi and primitive plants such as ferns. The bundles are highly ordered and very stable in comparison with microtubules in general. There are many other motile organelles consisting of bundles of stable microtubules in different arrangements, found in numerous invertebrate species. There are also variations on the usual '9 + 2' arrangement (shown for a sperm tail in Figure 1.13) throughout the animal kingdom. However, the '9 + 2' axoneme seems to have been selected during evolution as the optimum, geared for maximum efficiency. It is to microtubules what a striated muscle is to actin: a powerful machine in which efficiency is maximised, at the expense of flexibility in the redeployment of the components.

9.1 The '9 + 2' arrangement[2,3]

Although the lengths of axonemes vary from a few microns up to 1 or 2 mm, their outer diameters are consistently around 210 nm. This is because the internal structural arrangement is so highly conserved. The basic arrangement of doublet and singlet microtubules and the accessory structures attached to them was determined by careful electron-microscopic studies, mainly of thin-sectioned intact material (illustrated by Figure 9.1), throughout the 1960s and 70s. Studies of disrupted specimens, either negatively stained (as in Figures 9.2 & 9.3) or shadowed (as in Figure 9.11) have added detailed information about individual molecular components of axonemes. Figure 9.4 summarises the arrangement in the complete structure. Note the asymmetry which allows one to know whether the structure is being viewed from the basal end or the distal tip. As described in Section 7.5, the faster-growing (plus) ends of axonemal microtubules are at the tip.

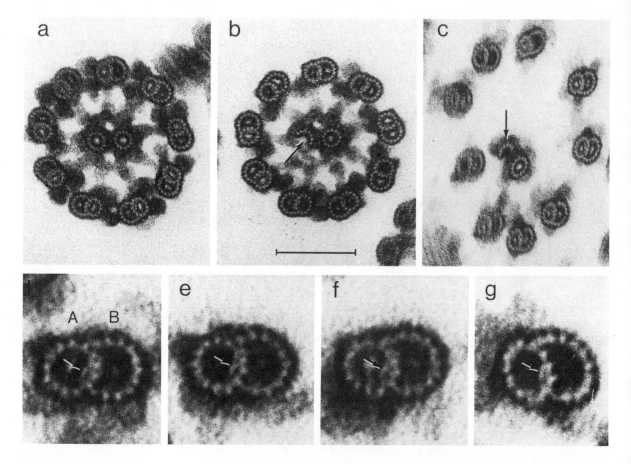

Figure 9.1

Cross-section through sea-urchin sperm-tail components. Treatment of the specimen with tannic acid before staining has produced a negative-staining effect around individual protofilaments of the microtubules. (a) An intact isolated axoneme. (b) & (c) Axonemes with some components extracted: outer arms and part of one central microtubule are missing from (b); both sets of arms and most of the central complex are missing from (c). (d)–(g) Individual doublet tubules. Arrows indicate extra material near the inner junction of A- and B-tubules. See Figures 9.4 & 9.6 for an interpretation of the structure. (a)–(c) × 100 000; (d)–(g) × 400 000. [Reproduced with permission from Linck, R. W. *et al.* (1981). *J. Cell Biol.* **89**, 309–322.]

9.1.1   Doublet microtubules[17,18]

Nine doublet microtubules lie symmetrically around the axoneme axis. Each doublet appears to be made up of one complete closed 13-protofilament microtubule, which is referred to as the **A-tubule**, and an incomplete 10-protofilament tubule, referred to as the **B-tubule** (Figure 9.4b). In cross-section it is usually described as C-shaped. The free edges of the B-tubule wall make T-junctions with the outer surface of the A-tubule. The section of A-tubule between two B-tubule edges is referred to as the **common wall**. The doublets are skewed so that the long axis of each lies at an angle of about 80°, rather than 90°, to a radial line through its centre. Thus, the B-tubules are slightly further from the centre than are the A-tubules. Attached to the surface of the A-tubules are two rows of **dynein arms** which, under some conditions, can be seen making close contact with the B-tubule of the next doublet. There is also some evidence for a further type of link between adjacent doublets. This appears as fine strands of connecting material, thought to provide a permanent elastic link. The protein believed to be involved has been given the name **nexin**.

9.1.2   The central structure[17,19,26]

The central component along the axis of a standard axoneme consists of a complex based on two singlet microtubules (C_1 and C_2), usually referred to as

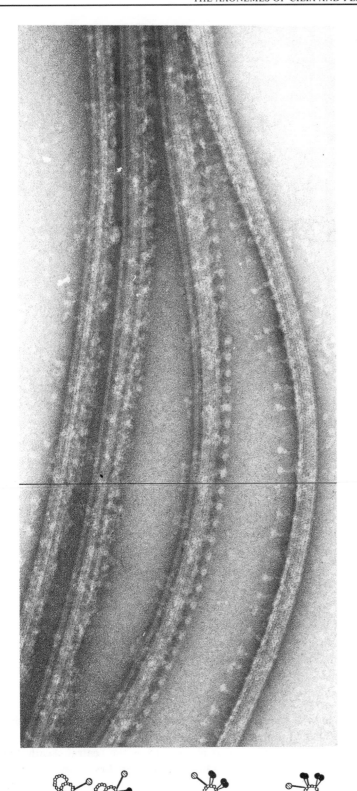

Figure 9.2

A group of four doublet microtubules viewed in negative stain. The group is part of a fraying isolated sea-urchin axoneme. The diagram below shows how the doublet tubules are probably lying on the substrate. Since the electron micrograph is a projected image, arms and spokes are seen most clearly when they are not superposed on the doublet microtubules. × 200 000. [Micrograph taken by Dr Richard Linck.]

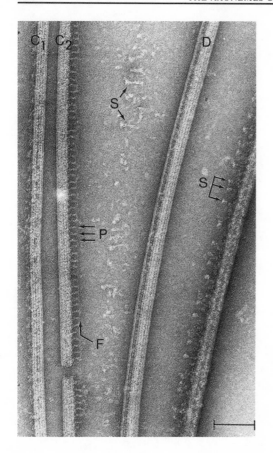

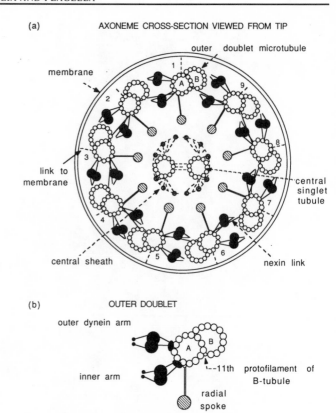

Figure 9.3

Microtubules from an axoneme in a more disintegrated state than in Figure 9.2. On the right are doublet tubules (D). S indicates attached or detached radial spokes. The pair of singlet tubules on the left came from the central complex. The ladder of projections (P) attached to C_2 is part of the central sheath complex. Sometimes the ends appear connected to form a filament (F). Scale bar = 100 nm. [Reproduced with permission from Linck, R. W. *et al.* (1981). *J. Cell Biol.* **89**, 309–322.]

Figure 9.4

(a) Diagram of the '9 + 2' axoneme structure seen in cross-section, as viewed from the distal tip (plus end).[24] Some structures, such as the central sheath and the link to the membrane, have not yet been fully characterised. A doublet tubule and dynein arms are shown in more detail in (b).

the **central pair**. A complex ladder-like structure (see Figure 9.3) links the two central singlets and projects out so that the complete central complex is roughly cylindrical. In older reviews it is referred to as the **central sheath**. Small rods with globular ends, known as **radial links** or **spokes** (shown in Figure 9.2), project inwards from the A-tubules and interact with the surface of this central complex.

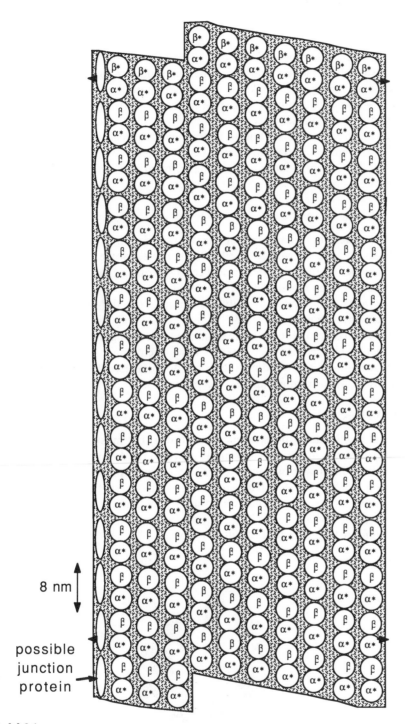

8 nm

possible
junction
protein

Model 9.1

Add a B-tubule to your microtubule model. Leave white strips along each side and fold them
back to glue on to Model 7.1: see Figure 9.4b for the relative positions of the two edges.

9.1.3 Numbering of the outer doublets

The doublet tubules can be numbered consistently, relative to the plane defined by the central pair of microtubules; according to a convention defined by Afzelius, doublet number 1 lies along a radial line which is perpendicular to the central tubule plane. Numbering then continues in the direction indicated by the arms, that is, anti-clockwise in the view from the tip (Figure 9.4a). The **effective stroke** of a cilium or flagellum is often approximately planar and has been shown to be in the direction defined by a line drawn from doublet 1 to the gap between doublets 5 and 6. In some cilia, instead of the usual arms on doublet 5, there are apparently permanent bridges between doublets 5 and 6. These **5/6-bridges** introduce an additional element of asymmetry, related to the stroke direction.

9.1.4 Structural variations

In certain invertebrates and teleost fish, motile axonemes lack the central pair of tubules and are referred to as **'9 + 0'** axonemes. Some invertebrate species have an elaborate radially symmetrical central structure instead; in other cases, there seems to be no replacement structure. The number of outer doublet tubules can also vary without loss of motility; for example, '6 + 0', '12 + 0', and '14 + 0' axonemes have been documented.

It is very common for the basic structure to be strengthened with additional components. The sperm tails of many species, including mammals, have a ring of **accessory fibres** surrounding the outer doublets (see Figures 9.5 & 1.13). These probably have important elastic properties.

9.1.5 ATPase activity[1]

Chemical fractionation of axonemes has shown that most, but not all, of the ATPase activity is in the dynein arms attached to the A-tubule of each doublet. The inner and outer arms consist of different dynein polypeptides, as detailed in Section 8.5 & Table 8.2. Dialysis in a low-ionic-strength buffer extracts all of the arms and leaves axonemes immotile. Washing in 0.5 M KCl extracts just the outer arms and halves the beat frequency of axonemes. The outer arms can be rebound, with restoration of full motility.

Figure 9.5

Cross-sectional view of the axoneme in the tail of a squid sperm. It is essentially similar to Figure 9.1, except for the large accessory fibres attached to the outside of each doublet microtubule. Similar accessory fibres are found in the sperm of very many species. × 50 000. [Reproduced with permission from Linck, R. W. *et al.* (1981). *J. Cell Biol.* **89**, 309–322.]

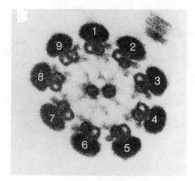

9.2 Molecular arrangements in doublet microtubules

The arrangements of tubulin dimers in A- and B-tubules are thought to differ, as described in Section 7.1. The A-tubule is a 13-protofilament cylinder, with what seems to be a helically-symmetrical, axially-staggered arrangement of dimers; the B-tubule has only 10 tubulin protofilaments, does not close on itself, and has an unstaggered dimer arrangement (see Figure 9.6). In some cross-sectional views, what looks like a thinner 11th protofilament occurs at the inner junction (see Figure 9.1). This is probably not tubulin and may be a special component required to hold this junction together. If axonemes are solubilised in urea or detergent, material at this junction appears to be the least soluble part of the structure. The major components of the insoluble residue are the tektin proteins[20] (see Section 9.6.2), though it is not yet clear that they actually form the 11th B-tubule protofilament. The latter may be a protein somewhat more soluble than tektins, and thus absent from the final residue, while tektins may form a more integral part of the A-tubule wall.

9.2.1 Longitudinal spacing of accessory structures

The spacings of outer arms and radial spokes, as shown in Figure 9.6, was established some years ago from electron micrographs of thin-sectioned axonemes and from negatively-stained components (see Figures 9.2 & 9.3). The complexity in the arrangement of the inner arms was unsuspected until more recently.

Clear images of dynein arms *in situ* were obtained by Heuser and co-workers using the rapid-freezing technique, followed by deep etching to remove ice from the surface structures.[19] The arrangement and shapes of the arms was best seen in regions of axoneme where the doublet tubules were well spaced from one

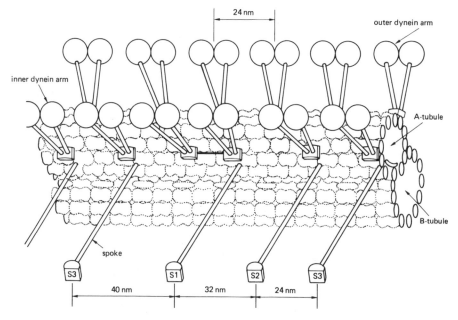

Figure 9.6

Arrangement of the molecular components in a flagellar doublet microtubule. Outer dynein arms, whether two-headed (as shown here) or three-headed, are spaced at regular intervals of 24 nm. Inner arms have the same average spacing but are bunched to match the unequal spacings of the radial spokes.[29] Spoke S1 is associated with the equivalent of 2 arms, the second being so distorted by the bunching that it appears to be only single-headed in many micrographs. [Redrawn from Amos, W. B. *et al.* (1986). *J. Cell Sci. Suppl.* **5**, 55–68.]

another. In such regions the connections to the B-subfibre were fine rod-like projections, while the globular heads appeared to have collapsed towards the A-tubule.

Tetrahymena and *Chlamydomonas* axonemes each revealed outer arms with three globular heads (see Table 8.2), apparently grouped together to give a mushroom-shaped aggregate.[27] As expected from earlier work, the bunches of heads were spaced at 24 nm intervals, three times the tubulin dimer repeat. Sea-urchin sperm tails showed two-headed arms, also at axial intervals of 24 nm.

The spacing of the inner arms and radial spokes is more complex, as shown in Figure 9.6, but apparently the same in different species. Inner arms appear to be bunched in pairs and quadruplets; each bunch is associated with one of the radial spokes, as shown. The relationship between the groups of inner arms and the various particles isolated from *Chlamydomonas* flagella (listed in Table 8.2) is unknown. The overall axial repeat of twelve times the dimer repeat (i.e. nominally 96 nm) is the same as the repeat distance postulated for high-molecular-weight cytoplasmic MAPs (see Section 8.2). There are three radial spokes per repeat. Curiously, the 24 nm/32 nm/40 nm spacings between them bear some resemblance to the spacings (22 nm/37 nm/37 nm) of projections arising from the MAP helical superlattice. Perhaps the arrangement of spokes (and inner dynein arms) is related in some way to a similar helical superlattice.

9.3   Genetic analysis of flagellar components [5-8]

Two-dimensional gel electrophoresis of axonemes shows that more than two hundred different polypeptides are present. Analysis of mutant flagella lacking particular structures, such as inner or outer dynein arms or radial spokes, have shown which polypeptides may help form such accessory structures. Perhaps some of the proteins are essential only in determining the correct self-assembly of the structure (a problem of complexity similar to or even greater than that of assembling a bacteriophage). At present it is not possible to guess how many are important for motility.

A possible route to understanding their interaction in the axoneme is an elegant complementation test devised by Lewin. Haploid *Chlamydomonas* cells occur as one of two types, known as *plus* and *minus* (as opposed to female and male). A *plus* and a *minus* cell can be mated (fused) to produce a temporary dikaryon. Normally in such a dikaryon the flagella of the *plus* partner are active and those of the *minus* partner are turned off. Sometimes, when a *plus* mutant with paralysed flagella is mated to a wild-type *minus*, motility is restored to the paralysed flagellum of the *plus* partner.

Luck and colleagues have performed this type of experiment using a radioactive mutant and a cold wild type. When the flagella are isolated and run on a 2–D gel, the one spot out of hundreds which is *not* radioactive corresponds to the polypeptide donated to the flagellum of the mutant by the wild-type partner, and this polypeptide must be the one responsible for the restoration of function. This method is being used to reveal complex assembly sequences and functional interdependence between the flagellar polypeptides. For example, some defects in *spoke* polypeptides can be made good by compensatory mutations in the *dyneins*—a totally unexpected result (see Section 9.4.3).

Mutants with defective basal bodies have recently revealed a chromosome (or linkage group) reserved for genes influencing flagellar biogenesis.[10] Luck's group, using *in situ* hybridisation techniques, have tentatively localised sequences belonging to this chromosome in the basal bodies of *Chlamydomonas*, rather than in the nucleus. A small chromosome might form the amorphous matrix packed in the lumen of each basal body and determine basal body length in much the same way that the genome of a virus determines the size of the capsid. Even more interesting is the possibility of such a chromosome having some relationship to the 'entity' whose limited supply controls the number of centrioles in a cell (Section 11.2).

9.4 The motile mechanism

Progress in understanding how the motion of the '9 + 2' axoneme is controlled is slow because of the complexity of the system. Active sliding between adjacent doublets is generally accepted, but the production of bending waves in a flagellum is less easy to explain in detail. Each step in the sequence of events probably depends on completion of the previous structural change and also on a programme of enzymatic events. In particular, there is evidence for phosphorylation of numerous axonemal proteins.[37] Many of these modifications may be involved in controlling motility, but few details are known at present.

9.4.1 The sliding-filament model for flagellar motility

Relative sliding between doublet microtubules was originally predicted by Satir from electron-microscopic observations of the tubules at the tips of cilia.[3] Some doublets were found to extend further than their neighbours, as illustrated in Figure 9.7a.

Experiments by Summers and Gibbons later provided the first direct demonstration of active shearing at the surface of microtubules.[11] Using demembranated cilia from *Tetrahymena*, they slightly digested the axonemes with trypsin to loosen the bundle. When Mg^{2+} and ATP were applied, dark-field light microscopy showed that the axonemes elongated, turning into thinner

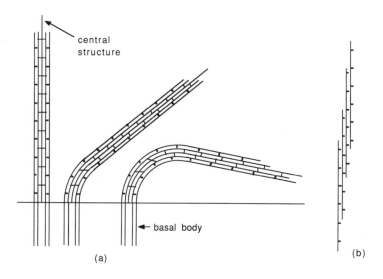

(a)

(b)

Figure 9.7

(a) Sliding of axonemal microtubules induced by dynein arms (schematically represented by the shorter crossbridges) normally produces bending *in vivo*,[3] because the extent of sliding is limited by crosslinking components, including the radial spokes (represented by longer crosslinks—see also Figure 9.8).[14] (b) Mild digestion with trypsin loosens the crosslinks and allows extensive sliding *in vitro*, without bending.[11]

structures up to seven times longer than the original axonemes (Figure 9.7b). Subsequent EM work showed that the thin structures were separate doublets which had slid relative to each other like the sections in a fireman's ladder.

When trypsinised axoneme preparations are fixed in the process of sliding, the polarities of the doublet microtubules can be determined by EM (from the radial spoke spacing, for example—Section 9.2) and compared with the direction of sliding. This has been done for normal axonemes, for axonemes whose outer arms have been chemically extracted, and also for axonemes isolated from mutant *Chlamydomonas* cells that lack outer dynein arms. The direction of movement is the same in all cases: inner or outer dynein arms attached to a doublet have always been found to push an adjacent doublet towards the tip of the axoneme (i.e. to transport their own doublet towards the minus end: the *retrograde* direction of Section 10.4).

9.4.2 Roles of inner and outer arms[4,6-8]

The cilia of mutant *Chlamydomonas* cells lacking outer arms are capable of motility, which appears normal except that the beat frequency is reduced to between one-half and one-third. Differential extraction of sea-urchin sperm outer arms gives a similar result. This shows that outer arms are not essential for normal bend propagation; they simply increase the speed of sliding. In contrast, the only known *Chlamydomonas* mutant that lacks inner arms is immotile. After isolation and treatment with trypsin, the axonemes exhibit feeble interdoublet sliding.[34] One must conclude that the inner arms probably govern the sliding in some way.

9.4.3 Roles of spokes and nexin links[6-8,14]

It is not known whether the radial spoke heads have an ATPase activity, but this seems possible. The central pair appears to *rotate* in some paralysed mutants.[35] Perhaps the normal interaction between radial spokes and the central structure has an azimuthal force component (as postulated for the force between single-headed *Tetrahymena* inner-arm dynein and microtubules—Section 8.9.1). The mutant may lack some structure that normally anchors the central pair and prevents free rotation. High-voltage electron microscopy of normal sperm axonemes fixed in motion suggests there is some degree of rotation of the central pair over actively bending stretches, but the original orientation seems to be restored on the return stroke.

Figure 9.8

Spoke heads appear to interact in a ratchet-like manner with features projecting every 16 nm from the central sheath. (a) The unequal spacings between successive spokes mean that the spokes cannot all contact projections simultaneously in a straight orientation. (b) Spokes may, however, tilt during sliding to remain in contact with a projection for a short distance. (c) & (d) In regions of bending, whole groups of spokes are seen tilted and, presumably, are strongly attached to the sheath. It is postulated that the tilt produced by interdoublet sliding reaches some maximum angle (>33°?) before the spoke heads detach. In straight regions, the maximum tilt angle seems to be much smaller (see (b)), suggesting that the interaction is weaker where there is no bending. [Redrawn from Warner, F. D. & Satir, P. (1974). *J. Cell Biol.* **63**, 35–63.]

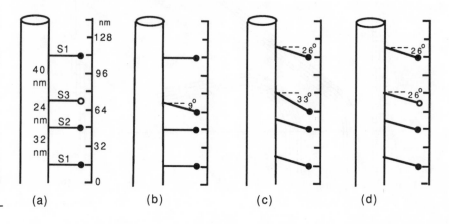

It has been suggested that the spokes and central sheath complex in normal '9 + 2' axonemes form a transverse ratchet-like connection between groups of doublets.[14] EM images show changes in successive spoke angles over regions of bending, as diagrammed in Figure 9.8. A spokeless mutant of *Chlamydomonas* shows normal sliding in the Summers and Gibbons experiment but is unable to produce bending waves. Surprisingly a further mutation, this time involving the outer dynein arms, suppresses the effect and produces cells that can swim; the axonemes still lack radial spokes and the central pair complex. Also, as mentioned above, there are simple types of native axoneme which lack any central structure yet are still able to produce propagated waves.

All of this suggests a regulatory role for the central structure, which may switch on the activity of different sets of arms (which are otherwise switched *off* by some other mechanism) in succession, according to some programme that determines the pattern of beating. In this respect it is interesting that Goodenough and Heuser have found the inner arms—which dominate in bending motility—to be grouped so closely around the bases of the radial spokes (Section 9.2.1).

Very little is known about the so-called **nexin links** between doublet tubules. There is EM evidence for thin, apparently elastic, filaments in addition to the dynein arms in the gap between doublets. Stephens has identified a possible polypeptide band. Permanent links, as these are thought to be, would help to hold the bundle together and also limit the net sliding between doublets.

9.4.4 Simple models of flagellar motility

The only two absolutely essential components of an axoneme seem to be doublet microtubules and dynein arms. Flagella lacking many of the other structures are still able to beat. Although relative sliding between adjacent doublet tubules is probably the primary motile mechanism, how this is converted into periodic beating may depend as much on the mechanical properties of the doublet tubules as on cooperative interactions between the dynein arms. Much theoretical work has focused on the propagation of bending waves by elastic rods, and it is possible to devise models that account for the observed waveforms.

In recent years, attempts have been made by Brokaw and others to simulate sliding-filament behaviour by computation.[15] Of necessity, the models are highly simplified; they involve deformable, elastic, rod-shaped crossbridges, each working independently. One point to emerge from such work is that for tight coupling between ATPase activity and motility, independent crossbridges would need to interact systematically with each binding site along the B-tubule (either every 4 nm, if tubulin monomers are all equivalent, or every 8 nm for dimers). However, it seems likely that dynein heads cooperate closely *in situ* (see Figure 9.11), perhaps in a similar way to that proposed for kinesin (Figure 8.14), and may take turns in interacting with tubulin. In addition, there may be close co-operation between different dynein arms.

9.4.5 Coiling: a possible analogy with bacterial flagella[5]

The elastic properties of microtubules may include a 'flip-flop' feature, whereby the doublet tubules have energy minima for one or more curved conformations. Isolated individual doublet tubules are often seen in a coiled configuration.

Figure 9.9

EM image of bacterial flagella in negative stain. They are superficially similar to singlet microtubules but are not in fact closely related. Bacterial flagella are composed of flagellin, a protein that is quite distinct from tubulin. The outer diameter is smaller than a microtubule (20 nm rather than 25–30 nm) and the central channel is much narrower. × 150 000. [Micrograph taken by Dr John Finch.]

Figure 9.10

Representation of the subunits in a bacterial flagellum (not to scale). The subunits are thought to drop into two slightly different stable conformational states, represented here by filled and empty ellipses. All subunits in the same longitudinal protofilament are in the same state. The proportion and distribution of states amongst different protofilaments is variable, however, and depends on environmental conditions. The strain produced by the structural differences between protofilaments is taken up by an overall twisting of the flagellum. [Redrawn from Calladine, C. R. (1978). *J. Mol. Biol.* **118**, 457–479.]

Costello originally suggested that this coiling might be part of the beating mechanism; Brokaw has since proposed that curvature of the microtubules is in some way responsible for propagating the signal that activates successive dynein arms.[16]

There is a possible analogy here with bacterial flagella, although they are not closely related in structure (Figure 9.9) or mechanism. The primary motility in bacterial flagella is due to rotation of the baseplate within the cell body. The flagellum extending into the external medium is gently twisted like an elongated corkscrew, so its rotation produces the hydrodynamic force. It consists of a single 'tubule', made of a protein called **flagellin** which is not obviously related to tubulin and which assembles into a different helical lattice, based on eleven nearly longitudinal protofilaments. According to a theory devised by Calladine, each protofilament has a bistable structure, assuming either an extended or a contracted state.[5,33] Depending on the number of contracted protofilaments, the whole flagellum can be in one of eleven different states. In two of the states the flagellum is straight, in the others it is twisted to varying degrees into helices of either hand. A diagrammatic example is shown in Figure 9.10.

Normal flagella assume a particular helical waveform under normal conditions, but others predicted by the theory have been observed at higher and lower pH values. Asakura's group have also found mutant flagella in a number of the other predicted states. Such a bistable version of elasticity means that under stress (mechanical overload) the bacterial flagellum can assume different forms.

The contribution of such a property to eukaryotic flagellar motility is not yet clear. The combined forces of active and inactive crosslinks may 'click' the doublet tubules through a series of straight and curved states, rather than merely bending them by overcoming a steadily increasing elastic resistance. The subunit conformational change may even be propagated longitudinally along the protofilaments and contribute to the coordination of events in the axoneme.

9.5   Interactions between dynein and microtubules[27-9]

The interaction of dynein arms with microtubules is usually discussed in terms of ATP-sensitive B-tubule binding sites at one end and an ATP-insensitive A-tubule binding site at the other. This seems to be an oversimplified view, since electron micrographs show outer arms binding to B-tubules in the presence of ATP. Specimens rapidly frozen *in vivo* consistently show some rows of arms with thin projections extending from the heads to contact the B-tubule, as in Figure 9.11 and the upper part of Figure 9.12a. Similar types of contact have also been seen in thin sections through pellets of purified flagellar outer arms combined with reassembled brain microtubules. They are thought to represent the fused pairs or triplets of ultra-fine stalks seen sprouting directly from each globular head in images of isolated molecules (Section 8.5). Isolated *inner* arms possess similar stalks and are presumed to bind to microtubules in similar ways.

It has been suggested that the role of the fine connection is to resist sliding movements and thereby help to promote flagellar bending; active sliding presumably takes place only between a minority of doublet pairs at any given

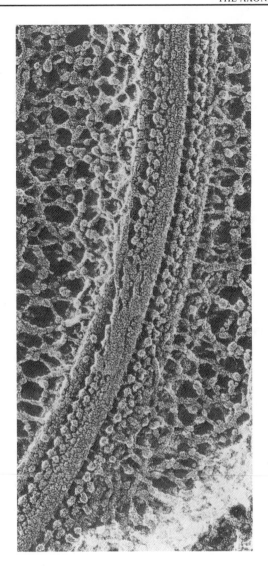

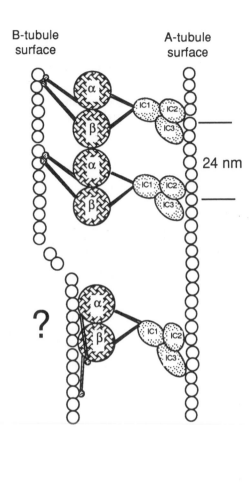

B-tubule
surface

A-tubule
surface

24 nm

?

Figure 9.11

View of an isolated, demembranated axoneme that was reactivated with ATP. Two doublet tubules stand out above the amorphous background (which is due to salt and other components of the frozen medium). The double row of globular particles on each tubule represents dynein outer arms; the smaller blobs closest to the tubule surface are the 'feet'; the larger blobs are the fused heads. Fine stalks bridge the heads to a neighbouring tubule. × 140 000. [Reproduced with permission from Goodenough, U. W. & Heuser, J. E. (1989) in *Cell Movement*, vol. 1, ed. Warner, F. D. *et al.*, pp. 121–140. New York: Alan R. Liss.]

Figure 9.12

The possible interactions between dynein molecules and doublet microtubules in an axoneme. Interactions via the fine stalks (upper two molecules) have been described in detail by Goodenough and Heuser (see Figure 9.11).[4,19] The heads of each arm interact so closely *in situ* that they appear to form a single 'toadstool'; the stalks fuse to form a composite 'B-link'. The same basic arrangement has been seen for axonemes in rigor and in the presence of ATP, except for a change in the conformation of the heads relative to the feet; heads relaxed with ATP are tilted closer to the feet. The globular heads have not been observed interacting directly with the surface of the B-tubule, as in the hypothetical configuration shown below. This could be because only a minority of dynein arms are actively pushing at any moment.

moment, so it is reasonable that there should be some mechanism to prevent sliding elsewhere. Highly elastic nexin links would probably not be adequate for this purpose. Whilst the active interaction is taking place, presumably between the B-tubule surface and the globular heads that carry the ATPase activity (lower part of Figure 9.12), either the fine projections may be displaced from contact with the B-tubule or the sliding force may simply overcome the weaker attachment force. Thus, cooperative interplay between two modes of binding of dynein heads to B-microtubules may play a major role in the mechanism for producing the bending motions of axonemes. Obviously there are additional control mechanisms in the case of '9 + 2' axonemes (Figure 9.8).

9.6   Assembly of axonemes and basal bodies [5,6,9,21,22]

The large number of constituent polypeptides provides an indication that assembly of an axoneme must be a complex process. A basal body appears to be essential for initiation of the assembly and to provide a template for the 9-fold cylindrical arrangement.

9.6.1   Assembly of doublet and triplet microtubules

The formation of complex microtubules requires additional kinds of bonding between tubulin protofilaments as well as the normal side-to-side bonds found in the lattice of a singlet microtubule. As described in Section 7.4, pure tubulin alone is almost able to form doublet tubules. An incomplete microtubule—that is, a curved sheet of protofilaments—can attach along one edge to the outer surface of another complete or incomplete microtubule.

When viewed from the distal end, the sheet curves clockwise, which means that the interaction between the complete microtubule and the curved sheet is equivalent to the outer junction of a doublet. This junction could form in doublet and triplet microtubules without help from accessory proteins, except for an initiating substance (at the base, perhaps) that must trigger assembly onto the outside of a *particular* protofilament of the A-tubule.

The inner junction, on the other hand, probably needs some kind of accessory protein glue all along its length. There is electron-microscopic evidence for additional material at this junction, as described in Section 9.1. **Tektin** filaments are thought to derive from this region of the doublet wall, either forming the 11th B-tubule protofilament or being an integral part of an adjacent A-tubule protofilament. There are almost certainly some other non-tubulin components in this region, which become solubilised during fractionation of doublet microtubules.

9.6.2   Control of assembly in the axial direction [9,24]

The active ends (Sections 7.5 & 7.6) of all the microtubules in an axoneme are at the distal tip and addition of tubulin subunits to the doublet microtubules has been shown to take place here *in vivo*. Diffusion of tubulin dimer along the axoneme from its base may be sufficiently fast to account for the rate of growth. Assembly of the B-tubule lags somewhat behind that of the A-tubule. Forward extensions to growing A-tubules, known as **distal filaments**, have also been observed by electron microscopy. Possibly these are tektin filaments, or tektin–tubulin complexes, assembling ahead of the bulk of the A-tubule tubulin.

Sea-urchin embryos begin synthesis of one of the tektin polypeptides at the time of ciliogenesis, whereas most other axonemal components, including tubulin and dynein, are present in large pools even before fertilisation. Tektin A is made in limited quantities which are almost completely used up during ciliogenesis.[31] If the cilia are experimentally removed, the cell makes fresh tektin A to incorporate into new cilia. It is postulated, therefore, that the lengths of cilia and flagella are determined by the number of tektin polypeptides produced (similar to the way that tau production may control the production of axonal microtubules—see Section 8.1.3).

The entire axoneme seems to be stabilised by the tektin filaments and various accessory structures bound to them. It is possible to solubilise most of the tubulin from isolated ciliary axonemes but still leave a recognisable 9-fold ring structure.[30] Thus it seems probable that assembled tubulin needs to be stabilised by these structures and that any assembly continuing after the supply of tektin has run out will produce unstable polymer. Tektins are also likely to define an axial periodicity that helps to determine the positioning of many of the projecting accessory structures. A MAP 2-like superlattice may also be involved (see Sections 9.2.1 & 8.2.1).

There is a cap-shaped structure at the tip of the central pair throughout flagellar growth.[24] It has been found to inhibit the assembly of tubulin onto the distal ends of central-pair microtubules in isolated axonemes but this does not necessarily mean this end is blocked *in vivo* (see the case of actin filaments growing at ends apparently embedded in membranes—Section 5.3). On the other hand, the proximal ends of these two microtubules are free, ending close to where the outer doublets grow from the triplet tubules of the basal body. It is possible that assembly occurs here, at the less active ends of the central microtubules, and that the central structure slides up the centre of the axoneme during flagellar elongation.

9.6.3 Basal-body structure formation[21-3]

Basal bodies and centrioles (Figure 9.13) consist of a cylinder of nine *triplet* microtubules; a second incomplete tubule, the C-tubule, is attached to each B-tubule in much the same way as the B-tubule binds to the A-tubule. There are no dynein arms; possibly one function of the C-tubules is to reserve space for dynein arms in the axoneme that may grow out from the plus end of the basal body (Figure 9.14).

The assembly of a basal body involves the creation of the complex 9-fold symmetry from scratch and probably requires an even larger variety of highly specific proteins than the assembly of an axoneme onto it. Control of the very specific cylinder length must also involve a somewhat different process, perhaps using a very small number of copies (e.g. 1 per triplet microtubule) of some component—a piece of nucleic acid of the right length is one possibility.

Basal-body assembly has been studied in most detail in *Chlamydomonas*. Each of the cell's two flagella has a single basal body. Shortly before cell division (see Figure 11.9) the flagella are resorbed and a new basal body appears next to each pre-existing one, so there are four. Each new basal body seems to be

Figure 9.13

Basal bodies and centrioles consist of a 9-fold arrangement of triplet microtubules.[22] There are no dynein arms between triplets but a series of permanent crosslinks holds the bundle together. A molecular cartwheel fills the minus end of the cylinder; it is probably involved in initiating the assembly of the structure. When one of the cylinders is acting as a basal body, axonemal doublets assemble onto the plus ends of the triplet microtubules. In centrosomes (Section 11.2), the cylinders—now called centrioles—are always found in pairs orientated at right angles to one another. Dense clouds of satellite material associated with the outer surfaces of the cylinders are thought to be responsible for the initiation of cytoplasmic microtubules.

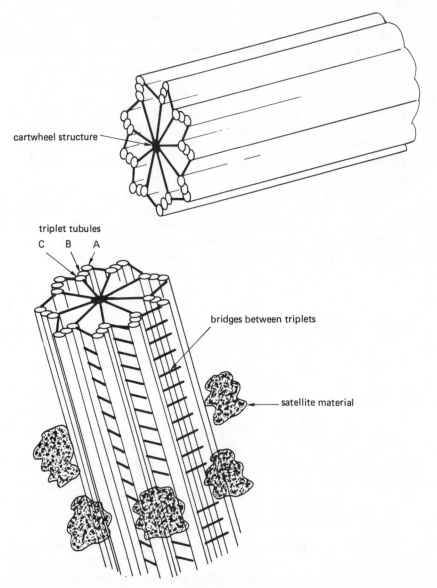

cartwheel structure

triplet tubules

C   B   A

bridges between triplets

satellite material

nucleated by at least two fibres which protrude from the side of an old one. In this type of cell, basal-body production appears to be controlled quite rigidly by the pre-existing structure.

In other cells, particularly sea-urchin eggs, there is evidence for *de novo* assembly, not carried out in the vicinity of pre-existing centrioles. The evidence does not rule out the existence of smaller nucleating structures, made on and then released from the existing centrioles.

Organelles called **probasal bodies** are assembled initially and are complex structures in themselves. They each consist of a short (100–50 nm) cylinder of triplet microtubules. A ring of A-tubules appears first, then B-tubules and C-tubules add on in quick succession. In the centre, cartwheel structures form in several layers and gradually become more pronounced. In some species, apparently, a cartwheel appears *before* the cylinder of triplet tubules. When the

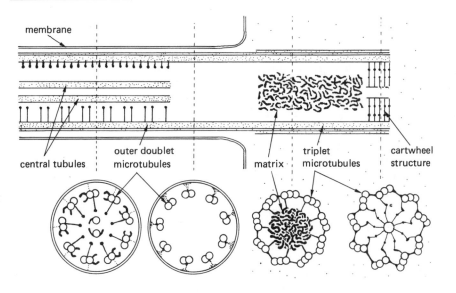

membrane

central tubules | outer doublet microtubules | matrix | triplet microtubules | cartwheel structure

Figure 9.14

The various structural levels found for an axoneme growing out from a basal body.[21] The basal body apparently assembles onto the initiating cartwheel structure and extends to a fixed length;[22] the amorphous matrix within may play a role in determining its size.[10] Axonemal outer doublet microtubules, decorated with dynein arms and radial spokes, assemble directly onto the basal body triplets but central singlet microtubules never make contact with the basal body. They may be initiated at the distal tip and grow proximally as the axoneme extends.[24] The radial spokes presumably play an important role in guiding assembly of the axonemal central structure. Some sort of interaction with the plasma membrane produces the closely-fitting protrusion of membrane around the axoneme, though the basal body remains embedded in the cytoplasm.

probasal body is complete, the triplet microtubules extend, A-tubules leading the way, and the full basal body is formed. The lengths of both the probasal body and of the full basal body are strictly controlled by some means. The cartwheel structure does not extend beyond the length of the probasal body; further up, the centre is filled by an amorphous matrix.

9.7   Summary

A flagellar or ciliary axoneme is a superbly evolved structure for producing rapid propulsion through fluid media. Whereas in a muscle filaments are connected together in a three-dimensional arrangement such that relative sliding between filaments leads to a contraction in length, lateral connections in an axoneme are organised to convert microtubule sliding into propagated bending waves. The assembly of axonemes and basal bodies seems to be very precisely controlled, in a manner more reminiscent of complex viruses than of other organelles in the cytoplasm of multicellular organisms.

9.8   Questions

9.1 In what simple way might the shearing interaction between dynein arms and microtubules in an axoneme be converted to bending?

9.2 How may the siting of axially-repeating accessory structures in an axoneme be specified, and what might determine its final length? What are the additional problems involved in assembling a basal body?

9.9   Further reading

Reviews

1 Gibbons, I. R. (1981). Cilia and flagella of eukaryotes. In *J. Cell Biol.* **91** (*Discovery in Cell Biology*), 107–124.
(*An excellent but rather detailed review of '9 + 2' flagellum work, with an accent on the history of the subject.*)

2 Warner, F. D. (1972). Macromolecular organization of eukaryotic cilia and flagella. *Adv. Cell & Mol. Biol.* **2**, 193–235.

  (A fairly old review of the overall structure of axonemes: still relevant, except for fine details of some molecular structure.)

3 Satir, P. (1974). How cilia move. *Sci. Am.* **231**(4), 44–63.

4 Goodenough, U. W. & Heuser, J. E. (1985). Outer and inner arms of cilia and flagella. *Cell* **41**, 341–342.

  (A good short review.)

5 Amos, W. B. & Duckett, J. G., eds (1982). *Prokaryotic and Eukaryotic Flagella.* Society for Experimental Biology, Symp. No. 35.

6 Dutcher, S. K. & Lux, F. G. (1989). Genetic interactions affecting flagella and basal bodies in *Chlamydomonas. Cell Motil. & Cytoskel.* **14**, 104–117.

7 Huang, B. (1986). *Chlamydomonas reinhardtii*: a model system for the genetic analysis of flagellar structure and motility. *Int. Rev. Cytol.* **99**, 181–216.

8 Luck, D. J. L. (1984). Genetic and biochemical dissection of the eucaryotic flagellum. *J. Cell Biol.* **98**, 789–794.

9 Lefebvre, P. A. & Rosenbaum, J. L. (1986). Regulation of the synthesis and assembly of ciliary and flagellar proteins during regeneration. *Annu. Rev. Cell Biol.* **2**, 517–546.

10 Goodenough, U. W. (1989). Basal body chromosomes? *Cell* **59**, 1–3.

Original papers

For brief scanning; the micrographs are certainly worth looking at

11 Summers, K. E. & Gibbons, I. R. (1971). ATP-induced sliding of tubules in trypsin-treated flagella of sea urchin sperm. *Proc. Natl. Acad. Sci., USA* **68**, 3092–3096.

12 Warner, F. D. & Mitchell, D. R. (1978). Structural conformation of ciliary dynein arms and the generation of sliding forces in *Tetrahymena* cilia. *J. Cell Biol.* **76**, 261–277.

13 Sale, W. S. & Satir, P. (1977). Direction of active sliding of microtubules in *Tetrahymena* cilia. *Proc. Natl. Acad. Sci., USA* **74**, 2045–2049.

14 Warner, F. D. & Satir, P. (1974). The structural basis of ciliary bend formation. *J. Cell Biol.* **63**, 35–63.

15 Brokaw, C. J. & Johnson, K. A. (1989). Dynein-induced microtubule sliding and force generation. In *Cell Movement*, vol. 1, ed. Warner, F. D., Satir, P. & Gibbons, I., pp. 191–198. New York: Alan R. Liss.

16 Brokaw, C. J. (1989). Operation and regulation of the flagellar oscillator. In *Cell Movement*, vol. 1, ed. Warner, F. D., Satir, P. & Gibbons, I., pp. 267–279. New York: Alan R. Liss.

17 Warner, F. D. (1970). New observations on flagellar fine structure. *J. Cell Biol.* **47**, 159–182.

  (An alternative to reference 2 above.)

18 Hopkins, J. M. (1970). Subsidiary components of the flagella of *Chlamydomonas reinhardtii. J. Cell Sci.* **7**, 823–839.

19 Goodenough, U. V. & Heuser, J. E. (1985). Substructure of inner dynein arms, radial spokes, and the central pair/projection complex of cilia and flagella. *J. Cell Biol.* **100**, 2008–2018.

20 Linck, R. W., Amos, L. A. & Amos, W. B. (1985). Localization of tektin filaments in microtubules of sea urchin sperm flagella by immunoelectron microscopy. *J. Cell Biol.* **100**, 126–135.

21 Cavalier-Smith, T. (1974). Basal body and flagellar development during the vegetative cell cycle and the sexual cycle of *Chlamydomonas reinhardtii*. *J. Cell Sci.* **16**, 529–556.

22 Gould, R. R. (1975). The basal bodies of *Chlamydomonas reinhardtii*: formation from probasal bodies, isolation and partial characterization. *J. Cell Biol.* **65**, 65–74.

23 Miki-Noumura, T. (1977). Studies on the *de novo* formation of centrioles: aster formation in the activated eggs of sea urchin. *J. Cell Sci.* **24**, 203–216.

24 Dentler, W. L. & Rosenbaum, J. L. (1977). Flagellar elongation and shortening in *Chlamydomonas*. III: Structures attached to the tips of flagellar microtubules and their relationship to the directionality of flagellar microtubule assembly. *J. Cell Biol.* **74**, 747–759.

25 Linck, R. W. & Langevin, G. L. (1981). Reassembly of flagellar B ($\alpha\beta$) tubulin into singlet microtubules: consequences for cytoplasmic microtubule structure and assembly. *J. Cell Biol.* **89**, 323–337.

26 Linck, R. W., Olson, G. E. & Langevin, G. L. (1981). Arrangement of tubulin subunits and microtubule-associated proteins in the central pair microtubule apparatus of squid (*Loligo pealei*) sperm flagella. *J. Cell Biol.* **89**, 309–322.

27 Goodenough, U. W. & Heuser, J. E. (1984). Structural comparison of purified proteins with *in situ* dynein arms. *J. Mol. Biol.* **180**, 1083–1118.

28 Sale, W. S., Goodenough, U. W. & Heuser, J. E. (1985). The substructure of isolated and *in situ* dynein arms of sea urchin sperm flagella. *J. Cell Biol.* **101**, 1400–1412.

29 Goodenough, U. W. & Heuser, J. E. (1989). Structure of the soluble and *in situ* ciliary dyneins visualized by quick-freeze, deep-etch microscopy. In *Cell Movement*, vol. 1, ed. Warner, F. D., Satir, P. & Gibbons, I., pp. 121–140. New York: Alan R. Liss.

30 Stephens, R. E., Oleszko-Sluts, S. & Linck, R. W. (1989). Retention of ciliary ninefold structure after removal of microtubules. *J. Cell Sci.* **92**, 392–402.

31 Stephens, R. E. (1989). Quantal tektin synthesis and ciliary length in sea urchin embryos. *J. Cell Sci.* **92**, 403–413.

32 Amos, W. B., Amos, L. A. & Linck, R. W. (1986). Studies of tektin filaments from flagellar microtubules by immunoelectron microscopy. *J. Cell Sci. Suppl.* **5**, 55–68.

33 Calladine, C. R. (1978). Change of waveform in bacterial flagella: the role of mechanics at the molecular level. *J. Mol. Biol.* **118**, 457–479.

34 Okagaki, T. & Kamiya, R. (1986). Microtubule sliding in mutant *Chlamydomonas* axonemes devoid of outer or inner dynein arms. *J. Cell Biol.* **103**, 1895–1902.

35 Kamiya, R. (1982). Extrusion and rotation of the central pair microtubules in detergent-treated *Chlamydomonas* flagella. *Cell Motil.*, Suppl. 1, 169–173.

36 Linck, R. W. (1989). *Tektins and Microtubules*. Advances in Cell Biology, vol. 3, pp. 35–63, ed. Miller, K. R. Greenwich, Connecticut: JAI Press.

37 Tash, J. S. (1989). Protein phosphorylation: the second messenger signal transducer of flagellar motility. *Cell Motil & Cytoskel.* **14**, 332–339.

10 MICROTUBULES IN CELLS

Chapter 7 describes some of the remarkable properties of tubulin. Tubulin itself is essentially the same in all microtubules; cytoplasmic microtubules can be decorated with flagellar dynein and cytoplasmic motors interact with flagellar microtubules. The arrangement of tubulin monomers has been found to be similar in all microtubules where this has been investigated. Doublet and triplet microtubules only occur in flagella, basal bodies and centrioles, but tubulin from other sources is capable of forming at least one of the lateral junctions required to form complex microtubules (see Section 7.4.2).

Microtubules in cells exhibit a range of behaviours, some being as labile as those studied *in vitro*, or even more so, others apparently as stable as those in axonemes. Their properties and behaviour are strongly modulated by their associated proteins.

10.1   Microtubule networks

10.1.1   Typical animal cells[1-3,19,20,50-2]

In animal cells such as fibroblasts, a network of microtubules generally originates from a single region close to the nucleus known as the **cell centre** or **centrosome** (see Figure 10.1). In most such cells this includes a pair of **centrioles**, each equivalent in structure to a basal body (see Section 9.6). The centrioles appear in the electron microscope to be embedded in a cloud of darkly-staining amorphous **peri-centriolar material**. The latter is believed to be an important component for microtubule initiation. The inactive ends of many microtubules are embedded in this material.

Microtubules mostly appear to be arranged with a definite polarity, usually the same as the polarity of axonemal tubules with respect to the cell centre. From this region, the tubules spread outwards towards the periphery. In a stationary cell they generally extend equally to all regions of cytoplasm. In a *moving* fibroblast they are orientated preferentially towards the 'leading edge'

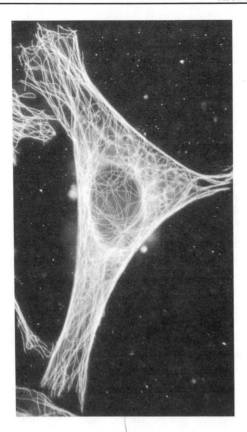

Figure 10.1

3T3 cell stained for microtubules with anti-tubulin antibodies and fluorescently-labelled second antibodies. Notice the central point from which many of the microtubules appear to radiate; presumably this is the centrosome. [Micrograph by Dr John Kilmartin.]

(see Section 6.2), with the cell centre usually on the leading side of the nucleus; in a nerve cell body, a smaller initiation site lies opposite each neurite, into which many microtubules extend. Thus, microtubules often (though not always) appear to exert some sort of directional control over regions of the cell that advance by means of actin-associated motile forces.

Before cell division, the number of centrioles doubles and they separate to form two centrosomes. Then, during division, microtubules are arranged in a symmetrical spindle with a centrosome at each pole. Every microtubule is orientated with its inactive end towards one of the poles. Details of this arrangement are given in Chapter 11.

There are, however, other kinds of arrangements, even in animal cells. In epithelial cells, such as MDCK cells for example, the arrangement is less obviously ordered, though there is a general tendency for them to be oriented at right angles to the baso-lateral and apical surfaces (see Figure 1.14). In nucleated red blood cells, such as occur in non-mammalian vertebrates, and in unactivated platelet cells, the overall circular shape is maintained by a ring of bundled microtubules (see Figure 6.2).

10.1.2   Plant cells[8]

The central region of a plant cell is usually occupied by a water-filled vacuole (Figure 1.15) and the cytoskeleton is arranged around it. Microtubules generally lie near to and parallel with the cell wall. Immunofluorescent staining of microtubules in carefully fixed plant cells has revealed that they lie in a close-packed helical arrangement. Components of the cellulose-containing cell wall are usually lined up in parallel with the underlying microtubule array.

10.1.3   Protists[21,53-6]

Distinctive bundles of microtubules are found in many unicellular organisms (e.g. Figure 1.12). The bundles are often arranged to form organelles which may be very much more elaborate than the ubiquitous '9 + 2' axonemes described in Chapter 9. The highly-differentiated cytoplasm, apparently organised by bundles of microtubules and various other fibres, seems to enable some of these single cells to have behaviours as complex as whole multicellular organisms.

The motile axostyle described in Section 10.3.2 is an example of a microtubule-based organelle used, like an axoneme, to produce motility. Many protists also have remarkable microtubular organelles involved in feeding. Solid food enters such organisms through an invagination of the cell surface, which pinches off food vacuoles at its base. The flagellate *Peranema* has two rods in the cytoplasm next to the mouth, which appear to be used like chopsticks to seize prey and pull it into the interior of the cell (Figure 10.2). The oral regions of *Nassula* and other ciliates have a funnel-shaped palisade or basket of rods, each consisting of a bundle of microtubules (Figure 10.3). There is evidence that the membrane around food particles is drawn into the cell by the action of microtubule motors (see Section 10.6).

Centrohelidians have fine protrusions known as **axopods**, that behave in a remarkable manner. The axopod membrane is held out from the cell body by a crossbridged paracrystalline bundle of many microtubules. When triggered

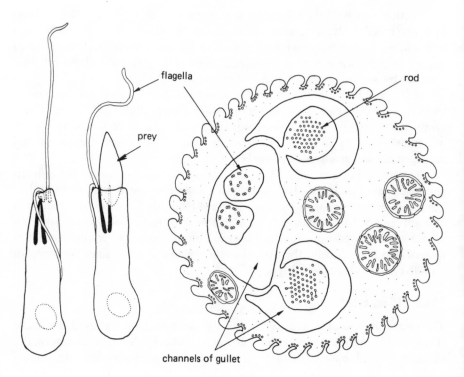

Figure 10.2

Microtubule-containing organelles in the protist *Peranema*.[21] The organism's two flagella emerge from its gullet. In the cytoplasm around the gullet are two rods (shown black). A prey organism is shown being pulled into the gullet. The tips of the rods move further apart and the food is drawn down past them to make phagocytosis possible. The cross-section on the right shows the circular profiles of microtubules bundled together to form the two rods. There is also a pallisade of microtubules underlying the pellicle.

flagella

prey

rod

channels of gullet

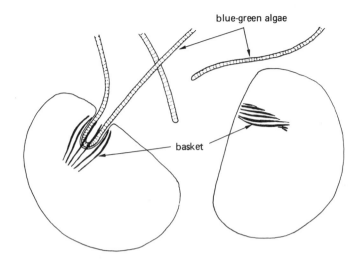

blue-green algae

basket

Figure 10.3

Two individuals of the ciliate *Nassula*. The oral apparatus of this protist resembles a basket consisting of many bundles of microtubules. *Nassula* feeds on algal filaments, which are dragged through the basket and enveloped in membrane to form very extended digestive vacuoles. [Based on studies by Dr John Tucker.[55]]

by contact with prey organisms, an axopod will shorten rapidly, while the microtubules break down. The contraction, at a rate of greater than 100 μm/s, is much too fast to be explained by known rates of rapid disassembly from the microtubule ends (a catastrophe, as defined in Section 7.6). Presumably the microtubules are induced to fall apart all along their lengths.

10.2 Dynamics of microtubules in cells[4,22-5]

As described in Chapter 7, pure tubulin microtubules are readily depolymerised by low temperatures, high pressures or the presence of anti-tubule drugs. Microtubules in living cells exhibit this property in differing degrees. Variations in the properties of microtubules depend on the cell type and the stage of the cell cycle. There may even be several co-existing categories of microtubules, apparently assembling and disassembling under independent control mechanisms. This is especially true in the case of the more complex cell types of protozoa. Apart from forming different arrangements in various regions of the cell, microtubule variation may be apparent in the electron microscope as decoration by distinct accessory-protein structures. Obviously, a cell must have a variety of controls over the assembly of microtubular systems. We are only beginning to understand what these might be.

10.2.1 Severed microtubules[25]

The most labile microtubules are those assembled during mitosis: their behaviour is discussed in Chapter 11. Microtubules in animal cells show dynamic behaviour in interphase, but to a lesser degree. Native microtubules have been severed by means of a sharply-focused UV laser beam and their behaviour monitored by video-enhanced microscopy. The ends show dynamic behaviour essentially similar to that of purified microtubules (Section 7.6). Some active ends grow, whilst others may shrink. The proximal ends of the detached pieces of microtubule are consistently inactive, just as they are *in vitro*, and the pieces take a long time to disappear.

10.2.2 The injection of labelled tubulin[22-4]

Injected tubulin can be labelled, either with a fluorescent molecule such as **rhodamine** or with some reagent such as **biotin** which can be detected with antibodies after the cells have been fixed. In the first case it is possible to observe assembly of fluorescent tubulin directly in real time, using low-light-level video microscopy; in the second case, the sites where fresh assembly has taken place can be studied in detail by immunofluorescence or electron microscopy.

Pulse-labelling of cultured fibroblast-type cells with microinjected fluorescent tubulin[22] showed that new tubulin dimers are incorporated into the interphase microtubule array at only two sites: the ends of pre-existing tubules and the centrosome. Similar experiments with a different tubulin label[23] showed that microtubules are continually growing out from the centrosome at 3.7 μm/min. The average half-life of a tubule was only 10 min, although some lasted much longer and possibly had a modified (detyrosylated) tubulin.

Such results confirm that microtubules in cells are subject to dynamic instability at their ends, with disassembly occurring faster than regrowth. Conversion from one mode to another is frequent, as *in vitro*. Additional features emerge from detailed measurements made by Schultze and Kirschner. Growth rates varied over a wide range (1–10 μm/min, i.e. 30–300 dimers/s). The variation was shown not to be due to stopping and restarting; a single phase of elongation appeared to be carried out at a consistent rate. However, a given microtubule would show different growth rates during different elongation phases.

Shrinkage rates apparently depended on the immediate history of the microtubule end. Microtubules that had remained stable for 40 s or more before starting to depolymerise did so at rates of 1–15 μm/min (30–450 dimers/s); those which had just previously been growing tended to shrink faster, at up to 25 μm/min (750 dimers/s). These observations suggest that factors in the cytoplasm interfere locally with the dynamic behaviour of microtubule ends. The very high growth rates that are possible may mean that MAPs assist in the incorporation of tubulin oligomers. The greater stability of older microtubule ends may reflect a slow redistribution of MAPs over the microtubule surface, until a more stabilising arrangement is achieved. Chemical modifications to assembled tubulin and MAPs are also thought to play a part in microtubule maturation.

10.3 Functions

10.3.1 Structural supports

Until fairly recently, the main function suggested for microtubules in cells was as structural elements, supporting the shapes of cells. This idea arose from the fact that microtubules are never seen strongly bent but seem fairly rigid and are generally longitudinally positioned in elongated cells. Anti-tubule drugs (or chilling) cause the retraction of elongated portions of cells such as fibroblasts and neurons. However, more recent ideas on microtubule function (Section 10.5) suggest that retraction may be due to loss of adhesion, resulting from a failure to transport vital substances along the microtubules, rather than to the absence of microtubules as structural supports (see also Sections 6.3 & 6.4).

In some cases, it has been proposed that a microtubule array may act as a *scaffold*; for example, longitudinal microtubules are laid down in certain types of developing muscle cells before the myofilaments assemble. The tubules vanish as the sarcomeric structure appears. Thus, the microtubules may play a role in maintaining the elongated shape of the cells while actin and myosin filaments are assembled in position. Delivery of components along microtubule tracks for assembly in different parts of the cell could perhaps be analogous to using the scaffolding around a building to winch up buckets full of bricks.

10.3.2 Sliding of microtubules[5–7,17]

The flagellar axoneme discussed in Chapter 9 was historically the first microtubular system known to be actively involved in cell motility and is still the system about which most detail is known. Here, microtubules clearly form the main structural elements but are also an essential component of the interaction that causes motility. Experiments carried out *in vitro* on purified components (Section 8.8) show that dynein arms are the only other protein molecules that are absolutely needed for microtubule sliding.

Sliding of structures along cytoplasmic microtubules has also been well established, although the motors responsible in particular cases *in vivo* may not be so certain. A direct demonstration of relative sliding of cytoplasmic microtubules equivalent to that seen for trypsinised axonemes (telescoping) has been provided by the experiments by Schliwa and colleagues on bundles from

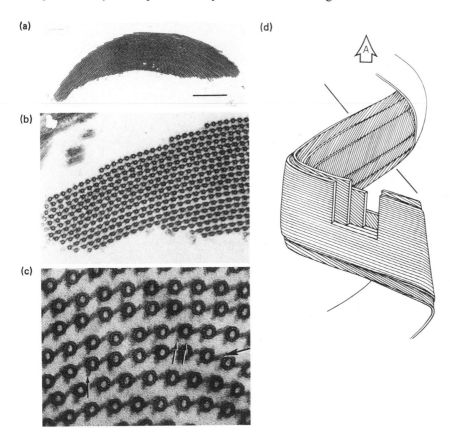

Figure 10.4

(a–c) EM images of a thin section through the axostyle of *Saccinobaculus*, at three different magnifications. The specimen has been treated with tannic acid to achieve a negatively-stained effect around the microtubules, as in Figures 9.1 and 9.5. The microtubules are linked into rows with one type of crosslink. Another type projects towards neighbouring microtubules in the next row and is more variable in orientation. (d) The postulated arrangement of microtubules into sheets, whose edges run at an angle to the microtubule axes. The angle may increase progressively from one sheet to the next (shown exaggerated for the four sheets here) and probably changes as bending waves pass along the organelle. The arrow shows the direction in which the wave travels. [Reproduced with permission from Woodrum, D. T. & Linck, R. W. (1980). *J. Cell Biol.* **87**, 404–414.]

Table 10.1 **Transport rates associated with microtubules (MTs)**	*Type of movement*	*Velocity (µm/s)*	*Transported components*
	Organelles		
	In animal cells	up to 2.5	Endosomes and lysosomes
	Pigment-granule movement	up to 1.0	Granules (bidirectional)
	Axonal fast transport		
	Anterograde	2.3–4.6	Small vesicles
	Mitochondria	0.6–1.2	Mitochondria
	Retrograde	2.3	Large vesicles
	Axonal slow transport (all anterograde)		
	Slow component b	0.02–0.09	Actin, clathrin, spectrin, 'soluble' enzymes, 'carrier proteins'
	Slow component a	0.002–0.01	MTs, NFs, spectrin
	Isolated organelles		
	From squid axon	1.6–7.0	Organelles (bidirectional)
	Endoplasmic reticulum	0.5 + 0.5	ER membrane networks (anterograde)
	Sliding of trypsinised axonemes		
	From sea-urchin sperm	15	MT doublets (retrograde)
	Sliding induced by purified motor molecules		
	Flagellar dyneins (all retrograde)		
	Tetrahymena 22 S	7–10	MTs on glass
	Tetrahymena 14 S	3–7	MTs (rotating) on glass
	MAP 1C	1.0–1.6	MTs on glass (retrograde)
	Reticulomyxa HMWP	3.6	Beads on MTs (bidirectional)
	Nematode motor	1–2	MTs on glass (retrograde)
	Kinesin	0.4–0.8	Beads on MTs, MTs on glass (anterograde)
	Particle movement during MT gelation		
		0.003–0.02	MT asters, amorphous particles
	Chromosome separation		
	In vivo	0.003–0.16	Chromosomes (average speeds)
		0.5–2.0	Individual chromosomes (max. speeds)
	Microtubule ends—growth and shrinkage phases		
	In vivo		
	Elongation	0.015–0.3	
	Shortening	0.015–0.25	
	In vitro—pure tubulin (see Chapter 7)		
	Rapid shrinking	0.4–0.6	
	(Elongation depends on tubulin concentration—see Chapter 7)		
	Chromosomes on the active ends of microtubules *in vitro*		
	(These speeds may correspond to *net* shrinkage (rapid shrinkage—growth) or may be slowed by the stabilising influence of the kinetochore)		
	Shrinking (1 µM tubulin)	up to 0.3	
	Shrinking (8 µM tubulin)	up to 0.03	
	20 µM tubulin	0	

the microtubule-containing networks of a branched amoeba called *Reticulomyxa*; the motor responsible for the sliding may be the same dynein-like molecule that is thought to cause bidirectional transport of particles along the microtubules (see Section 8.6.3).

Various microtubular organelles in protists appear to be designed to produce motility through intertubule sliding. Figure 10.4 shows a striking example: the **axostyle** of the flagellate *Saccinobaculus ambloaxostylus* is a long twisted ribbon that propagates undulating bends. In EM sections it is seen to consist of layers of microtubules. The layers are apparently maintained by fixed crosslinks and are thought to slide relative to one another due to the action of a second type of crossbridge.[53] The crossbridges between layers are probably dynein-like.

Relative sliding between microtubules almost certainly occurs during chromosome separation, as discussed in Section 11.6.1, and similar movements are probably part of the mechanism of slow axoplasmic transport (Section 10.5). The motors involved in these movements are as yet unidentified.

10.3.3 Movement of particles along microtubules[5-7,30-3]

The most detailed observations have been made on axonal microtubules. Transport along axons and dendrites (see Table 10.1) ranges from the fast movement (50–500 mm/day) of vesicles, which can be observed directly by light microscopy, to the slow progress (less than 10 mm/day) of cytoskeletal proteins. Recent work has illuminated several aspects of these processes, with implications for transport in less specialised cells.

10.4   Fast axonal transport[9-12,28]

Particle movement in a nerve axon ensures that neurosecretory vesicles produced in the cell body make net progress towards the termini while larger vesicles, travelling in the opposite direction, apparently return waste products for lysosomal degradation. Mitochondria may travel fast in either direction.

Studies of **axoplasm** (i.e. the cytoplasm from nerve cells), particularly from the giant axons of squid, have shown that fast axonal transport occurs along microtubule tracks. In axoplasm extruded into a suitable buffer, movement has been recorded by video-enhanced microscopy (see Figure 8.15). There is bidirectional movement of particles along filaments and the filaments themselves glide over the surfaces of the glass slide and coverslip. Electron microscopy has been used to demonstrate that the 'filaments' are actually microtubules and the 'particles' are axoplasmic vesicles.

Further evidence for sliding between microtubules and membranes has been found by combining vesicles purified from brain with either brain or flagellar microtubules; in both cases the vesicles moved actively over the tubules when supplied with ATP. All this movement could be stopped in a rigor-like state by adding AMPPNP, a non-hydrolysable analogue of ATP.

10.4.1 Fast transporting motors[5-7,30-3]

As described in Sections 8.6 and 8.7, all the molecules purified from axoplasm or brain that definitely have properties of microtubule-associated motors are *soluble* proteins. Simply washing axoplasmic vesicles with high-salt solutions is sufficient to remove their motility. However, the data do not rule out dynein-like membrane-bound proteins in systems other than axoplasmic vesicles.

Both cytoplasmic dynein and kinesin have characteristics suitable for fast particle-transport motors (see Table 10.1). Either molecule attached to a glass substrate will move microtubules at rates that qualify as 'fast'. As summarised in Figure 8.16, the directions in which the purified motors move would allow kinesin to transport vesicles away from a neuronal cell body (fast anterograde transport) and dynein to return material to the cell body (fast retrograde transport). Kinesin-induced motility (0.3–0.6 μm/s) is not quite as fast *in vitro* as normal anterograde transport (1.4–2.0 μm/s) but this may be because the conditions in the medium after purification have not yet been optimised.

Though neither purified kinesin nor purified dynein will transport salt-washed vesicles along microtubules, it has been known since kinesin was first discovered that vesicles in crude cytoplasmic extracts will move *in vitro* on purified microtubules, and there is evidence that anterograde and retrograde movements under these conditions are due to kinesin and dynein respectively. Anterograde movements of vesicles stop after addition of the non-hydrolysable analogue AMPPNP, which specifically causes kinesin to bind tightly to microtubules. Retrograde movement stops after UV irradiation in the presence of vanadate (see Section 8.4.3). In the latter case, movement can be restored by adding purified cytoplasmic dynein to the mixture.

This evidence suggested that there must be additional factors in the cytoplasmic extract that are needed, either to form a physical link between the motor proteins and sites on the vesicle surface, or to modify enzymatically one or the other to make possible a direct association. Gilbert and Sloboda have recently purified a complex consisting of squid cytoplasmic dynein and a MAP 2-like factor named **vesikin**. When this was added to salt-washed vesicles, they were found to move along purified microtubules. Thus, for retrograde transport at least, a motile system can be reconstituted from purified components. Whether kinesin uses the same linking protein or a different one remains to be determined.

Finally, immunofluorescence experiments using antibodies to kinesin or to vesikin have shown staining of small bodies within fixed cells. These observations help to confirm that the two known motor proteins associate with vesicles *in vivo*.

10.4.2   Control of transport direction[30–3,42–4]

The ability of individual vesicles to travel in either direction along microtubules raises the question of how the direction of travel is controlled *in vivo*. There is evidence to suggest that individual vesicles may move in either direction. If a nerve is constricted at a particular point, particles accumulate on either side of the blockage but many start to move in the opposite direction from their original movement (Figure 10.5). The characteristics of the motors involved, including their sensitivity to inhibitory reagents, suggest that, in each case, the reversal in direction is due to the alternative motor taking over from the one that had been operating.

On the other hand, after organelles undergoing retrograde motion in crude extracts were immobilised by vanadate-mediated photocleavage of their motors, there was no detectable increase in the number of organelles moving in the

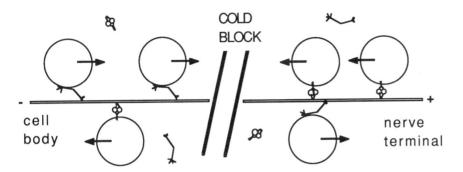

Figure 10.5

Fast transport can be blocked by cooling a limited section of an exposed giant axon.[28] Vesicles accumulate at either end of the block, unable to continue in their original direction of travel. Many begin to move in the reverse direction, revealing that alternative motors are available though not used normally.

anterograde direction. If both soluble motors are present in the cytoplasm at all times, it is clear that there must be some mechanism associated with the organelle which biases its transport towards a particular direction. The nature of this mechanism is completely unknown at present.

10.5 Slow axoplasmic transport[9–13]

The movement of cytoskeletal components has also been followed, using radio-labelled proteins. The rate of 0.2–8.0 mm/d is faster than can be explained by diffusion. Also the differential speeds of various components exclude cytoplasmic-streaming models; relative to a stationary axonal membrane, microtubules and neurofilaments (NFs—see Section 2.3.3) move at 0.01 μm/s, actin and many other proteins at 0.1 μm/s (Figure 10.6). Instead, it is thought that transport is effected by molecular motors, as yet unidentified, which generate force between specific components of the axoplasm. Actin filament transport is dependent on microtubules and may involve linkage of the actin to tubulin by some special motor.

Slow transport is unidirectional and away from the cell body, where most components are synthesised. An intriguing finding is that the same groups of components continue to move at the same rates after axon growth has ceased: the cytoskeleton is evidently replaced continuously thoughout the life of the

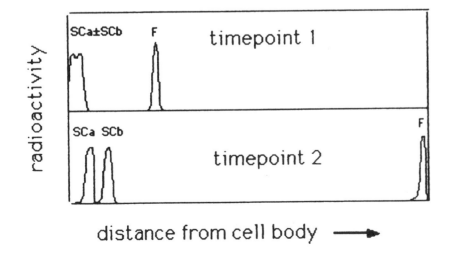

Figure 10.6

Different rates of axonal transport can be demonstrated by following radioactively-labelled proteins along an axon.[10–12] Fast vesicular transport rapidly outstrips the cytoskeletal components. These eventually separate out to give two or more peaks of label. The peaks remain detectable during their progress, suggesting that the proteins are moving in an aggregated/ polymeric form.

cell. In some cases, such as the NF proteins, it has been shown that molecules are degraded at the nerve terminus; possibly this is true for all the cytoskeletal components.

10.5.1   Slow component a (SCa)

The slowest group of actively transported proteins consists of neurofilament and microtubule proteins, together with some brain spectrin (usually classed as an actin-associated protein—Section 4.4; also known to interact with microtubules—Section 8.1). Lasek and colleagues, after careful observation of the moving label, concluded that tubulin and neurofilament proteins tend to move *en bloc* and suggested that these proteins are transported in the form of the same interlinked mass of polymers as that which forms the cytoskeleton. This is known as the **structural hypothesis**.[10] A pulse of radioactively-labelled protein travelling as separate monomers or small oligomers would become more and more spread out as it flowed down the axon. In fact, though, a substantial fraction of the label remains within sharp boundaries over long distances and times.

The main alternative model involves cytoskeletal proteins being assembled in their final positions; that is, either at the nerve terminal or alongside existing polymers within the axon, as this grows thicker. To explain the observed transport of protein *en bloc*, it is assumed that tubulin and NF proteins are transported in some unknown polymeric or aggregated form that is able to lose a proportion of subunits during transport, for assembly into stationary cytoskeletal structures. Finally, at the terminal, the unknown structures are completely disassembled and the subunits reused. This model has found some favour recently, because of evidence that significant amounts of labelled protein do stop moving.[9] However, the evidence can also be incorporated easily into a slightly modified version of the structural hypothesis, as should become clear in the following sections. There is no *necessity* to postulate that tubulin and NF protein are transported in polymeric forms other than microtubules and 10 nm filaments.

10.5.2   Axonal microtubules[18]

Microtubules are not continuous along axons, which can extend to metres in length. A millimetre or two seems to be about the maximum length for a microtubule *in vivo*; the average is a tenth of this length. Also, the active ends lie away from the cell body. These facts eliminate a 'chimneysweep's brush' model, in which fresh subunits are added to microtubule ends in the cell body and the polymers pushed into the axon by their own growth. Instead, it must be supposed that significant lengths of microtubule are pre-assembled onto individual microtubule-organising centres in the cell body and then transported into the axon (see Figure 10.7a). Lengths of NF could also be prefabricated in the cell body.

Electron microscopy has shown that the distal (plus) ends of the microtubules lie near to the surface membrane and may have the appearance of being capped. However, the plus ends of microtubules in growing axons exhibit dynamic behaviour similar to that of other cytoplasmic microtubules.[27] This finding requires some modification of the original structural hypothesis, in which a

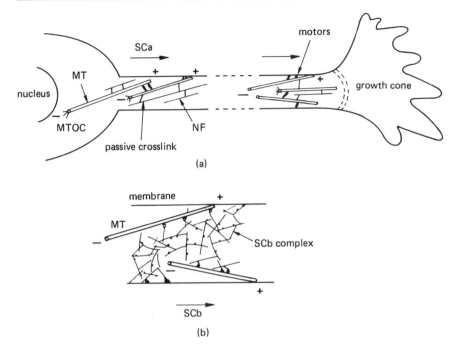

Figure 10.7

Modified version of the structural hypothesis of axoplasmic transport.[10] Microtubules (MT) initially prefabricated in the cell body (on some form of microtubule organising centre, MTOC) are transported actively into the axon and along its length to the terminus (a growth cone in the case of a growing axon, or a synapse for a mature axon). Neurofilaments (NF) appear to be co-transported with microtubules, along with a few minor components; the mass of polymers is known as slow component a (SCa). The plus ends of the tubules are dynamically unstable;[27] their shrinkage and regrowth leads to some exchange of tubulin subunits along the route. Some NF subunits and even whole filaments also become detached from the mass and are left behind.[9] Microtubules are disassembled and NFs are degraded at the terminus.

Slow component b (SCb) complexes consist of a larger number of proteins, including actin and clathrin, and are transported somewhat faster;[10-12] they may move using the microtubules as tracks or the axonal membrane as a substrate.

microtubule, once assembled in the cell body, was proposed to be transported, unaltered, along the axon. Dynamic instability of the distal ends will lead to some exchange of subunits amongst moving polymers and any that may have dropped out of the transport process. There is even recent evidence for some dynamic exchange of NF subunits. It is also probable that some whole filaments dissociate from the mass of moving polymers and remain stationary. This would fit with the observation that axons gradually thicken during growth and that the total number of NFs seen in cross-section increases with axon age.

10.5.3   Cold-stable microtubules[15]

Cold-stable segments of microtubule have been detected all along the lengths of axons and are thought to be stabilised by STOPs (see Section 8.1.4). The segments may represent a subset of microtubules, selectively stabilised, or perhaps the proximal ends of *all* microtubules, attached to initiating complexes. NFs may be connected by occasional crosslinks to the tubules and may contribute to their stabilisation; an associated meshwork of filaments, together with cold-stable tubule ends, could explain why the overall arrangement and polarity of axonal microtubules recovers readily after cold depolymerisation.

10.5.4   Plasticity

Branches of neurites, even whole axons, can be retracted after growing out. The cytoskeleton can evidently be dismantled. This is quite probably achieved by detaching much of the crosslinking between the polymers. Hollenbeck and colleagues have observed small 'packets' of cytoskeletal proteins being transported bidirectionally at rapid speeds.[45] Occasional large-scale rearrangements are quite compatible with a model for slow transport that involves prefabricated structures.

The microtubules of mature axons become more stable because of changes in the composition of their associated proteins. At the same time, and presumably as a result of these changes, the axons lose their plasticity.

10.5.5 SCa transport mechanisms[10]

Since the membrane of an axon apparently remains stationary after being laid down near the growth cone, the cytoskeletal components must move relative to the cell surface, perhaps by motors which engage with proteins bound to the membrane or else by leapfrogging over one another (Figure 10.8a). It seems likely that neurofilaments are dragged along passively by neurotubules. This conclusion is derived from experiments with a neurotoxin called **IDPN**, which apparently destabilises NF crosslinks and leaves the filaments stranded in the cell body, close to the base of the axon. Microtubule transport is unaffected.

Figure 10.8

(a) Four modes of relative sliding between microtubules. Only two basic classes of motors (anterograde and retrograde) are needed to produce these four classes of sliding and the transport of other components (organelles, SCb complexes, etc.) along microtubules, provided there is a range of adaptors to attach the inactive ends of the motors to different surfaces. However, there may turn out to be a number of distinct motors within each class. Mode (iv) is equivalent to flagellar doublet microtubules and dynein; mode (i) is equivalent to central spindle microtubules and an unknown anterograde motor (see Section 11.6). Slow axoplasmic transport might correspond to modes (ii) or (iv), and could involve any of the known motors (Sections 8.5 & 8.6). (b) A possible explanation for the behaviour of microtubule-containing gels *in vitro*.[34] The microtubules are randomly orientated and the motor molecules anchored to protein aggregates ('particles'). The latter will move to a particular end of any microtubules with which they can interact, producing an 'aster' of microtubules. Movement will continue until the particles are concentrated in the centres of such asters; all the components will thus be concentrated in a smaller total volume.

10.5.6 Slow component b (SCb)[10]

SCb, moving 5–10 times faster, is a far bigger group than SCa. SCb includes actin, clathrin, spectrin, myosin, some so-called **carrier proteins** (200 kD and 500 kD), and even enzymes normally regarded as soluble. The enzymes travel as a wave, suggesting some sort of association with a carrier structure, but leave a significant trail, suggesting some exchange with a non-motile pool. Actin behaves similarly, while other components, such as clathrin, travel without leaving a trail. SCb may represent the movement of large complexes (including crosslinked actin filaments) using microtubules or neurofilaments as rails (Figure 10.7b).

10.5.7 Microtubule gelation and contraction *in vitro*[34]

A motor for slow transport may soon be identified, following experiments by Weisenberg and colleagues with microtubules purified from whole mammalian brain. These workers put the brain extract through three cycles of assembly and disassembly (see Section 7.3.2) with minor modifications from standard procedures. (The modified procedure gives a higher yield of microtubules, with a greater proportion of associated proteins.)

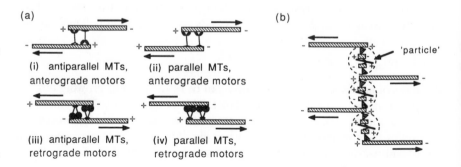

(a)

(i) antiparallel MTs, anterograde motors

(ii) parallel MTs, anterograde motors

(iii) antiparallel MTs, retrograde motors

(iv) parallel MTs, retrograde motors

(b)

'particle'

When they added ATP to the assembled microtubules, the result was a gradual gelation, followed by slow contraction of the gel, often to 25 per cent or less of the original volume. Electron microscopy shows that such preparations consist of a mixture of microtubules and apparently amorphous particles. It has been suggested that the particles might represent the moving aggregates postulated in the 'on-site assembly' model of axonal transport (see Section 10.5.1), but they could equally well be derived from stable segments of normal microtubules (Section 10.5.3). The nature of the particles is still obscure; their main component appears to be tubulin (possibly denatured), but they also include significant amounts of the NF polypeptides together with brain spectrin and high-molecular-weight MAPs (i.e. they are apparently equivalent to SCa).

Gelation and contraction have been followed by video-enhanced light microscopy. During the early gelation phase, microtubules form aster-like structures and particles move slowly towards their centres. Thereafter, the asters are gradually drawn together, at a rate of around 1 μm/min, and the microtubules gradually align into bundles. During this contraction relative sliding between microtubules must occur, though perhaps via some intermediate structure represented among the particles. When isolated particles are added to microtubules consisting of highly purified tubulin, the mixture is capable of ATP-dependent gelation and contraction. Obviously, some component of this mixture is a motor for microtubule-associated motility. It may be one of those already known, or something new.

Microtubules in nerves do not appear to produce contraction. Perhaps the difference between behaviour *in vivo* and *in vitro* is that *in vivo* various sorts of crosslinks between the parallel-orientated microtubules in an axon can become organised so that relative sliding temporarily leads to telescoping (one step in a leapfrogging process, perhaps), whereas *in vitro* unregulated sliding between randomly orientated microtubules will have the net effect of drawing them into a smaller volume (Figure 10.8b), just as mixtures of actin and myosin filaments precipitate *in vitro* when ATP is added.

10.5.8 Dendrites[15,16]

Neurons may put out many processes (**neurites**), only one of which (the longest) develops into an axon. The others, called **dendrites**, stay shorter, are more branched than axons, and have tapering ends. Once this differentiation has occurred, differences are detectable in the microtubules of the two kinds of process. As mentioned in Section 8.1, microtubules in axons contain tau protein but those in dendrites are decorated with MAP 2.

Another difference is that dendritic microtubules do not all have the same polarity: only about 60 per cent have their plus ends away from the cell body, the others are orientated with their plus ends *towards* the cell body. It has been suggested that this second set of tubules allows various categories of particle whose transport depends on 'retrograde' motors to be carried from the cell body into dendrites but not into axons. Such organelles include Golgi elements and ribosomes, both of which are observed in dendrites but not in axons. It is also known that, although most proteins including tubulin and tau are synthesised in the cell body and then transported into the extremities, MAP 2 is translated from message within dendrites. It is not yet known how many other dendrite-specific proteins are produced on local ribosomes.

10.6   Organisation of cellular membrane systems[35-41]

The bidirectional sliding of mitochondria along microtubules has already been mentioned in the context of fast transport; it seems probable that they are moved around many other cells types by similar mechanisms. The interiors of most cells contain extensive arrays of other types of membranous organelle. Current research is revealing that in most cases their positions within the cytoplasm appear to be determined by the microtubule network.

The **endoplasmic reticulum (ER)** is the name given to the membranous network which appears to branch out from the nuclear membrane and to extend in the form of continuous tubules or lamellae, almost as far as the plasma membrane. Integral membrane proteins and proteins destined for secretion through the plasma membrane are translated on the ribosomes attached to the so-called **rough ER** and extruded into its internal compartments. The remaining **smooth ER** probably includes a variety of distinct compartments, with different functions that may vary from one cell type to another.

There is a strong correlation between the positions of the ER membranous tubules and of microtubules. Also, depolymerisation of microtubules causes a slow retraction of the ER network towards the cell centre. It seems likely that the ER is normally kept in an extended state by motors that pull it in an anterograde direction along the microtubules. This idea is supported by the observations of Dabora and Sheetz on cell extracts containing microtubules and ER-like membranous tubules. By video-enhanced light microscopy, initially aggregated membrane was seen to spread out as a tubular network along stationary microtubules (Figure 10.9).

Some distinctive membrane systems behave in the opposite manner. The **Golgi system** is a cluster of membrane compartments, usually in the vicinity of the nucleus. If microtubules are depolymerised, the elements of the system disperse throughout the cytoplasm; if the microtubules are allowed to repolymerise, the cluster re-forms in less than an hour. It seems likely that Golgi membranes bind a retrograde microtubule-associated motor.

Lysosomes also appear to congregate around the cell centre, although they and the **endosomes** that are responsible for an earlier stage in the uptake and recycling of proteins both exhibit bidirectional movement. Again, in the absence of microtubules, the organelles become more widely dispersed. Although they are capable of moving in either direction along the microtubules, retrograde motility must dominate to keep the organelles clustered near the nucleus. Some recent experiments suggest that the direction of transport may be dependent on pH; this may well be important physiologically, since the functioning of lysosomes is dependent on their low internal pH.

10.6.1   Sliding of the plasma membrane relative to microtubules[54-6]

Membrane does not need to be in the form of vesicles or vesicular tubules to move along microtubules. In many protozoa, bridges have been observed by EM between the plasma membrane and underlying arrays of microtubules. Particles attached to the outside surface of *Chlamydomonas* flagella have been seen to move bidirectionally, apparently following the track of the underlying

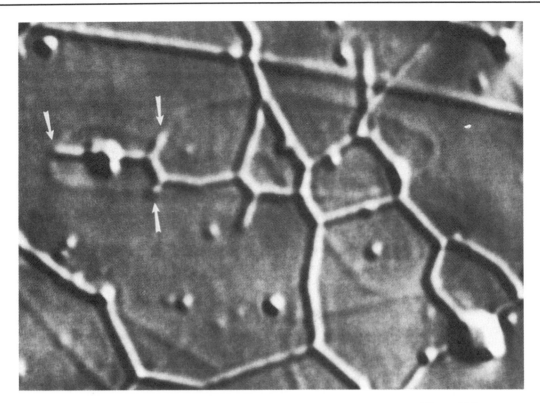

Figure 10.9

Video micrograph of a network of membrane tubules stretched out over a network of microtubules (seen as features of lower contrast in the background). An initially amorphous membrane extract has been spread out along purified microtubules by means of ATP-dependent components in the extract. × 5000. [Reproduced with permission from Sheetz, M. P. (1989), in *Cell Movement*, vol. 2, ed. Warner, F. D. and McIntosh, J. R., pp. 277–285. New York: Alan R. Liss.]

axoneme. A 350 kD integral membrane glycoprotein seems to be involved; possibly this is a membrane-bound form of dynein.

In the case of the oral apparatus mentioned in Section 10.1.3, food particles attached to the outer surface of the plasma membrane are dragged down into the gulley along with the membrane, which subsequently pinches off to form a digestive vacuole.

10.7   Pigment-granule movement

Chromatophores are pigment-containing cells that allow creatures such as frogs and fish to change their skin colour rapidly. In species with red pigment they are called **erythrophores**, whereas cells with black pigment are called **melanophores**. In response to external stimuli (which also vary according to species), the granules either migrate into a small volume in the centre of the cell, causing a lightening of the overall colour, or spread out, causing darkening. The movements follow an array of microtubules that radiate out from the centre of the cell. In many ways, the behaviour has the features of a synchronised version of fast axonal transport. Transport can be extremely rapid (3–6 μm/s).

Experiments on intact cells have tended to give complex and contradictory results but recent observations on demembranated cytoskeletons of fish chromatophores by McNiven and colleagues are more easily interpretable. Pigment transport can be reactivated if the membrane is removed by a suitably gentle detergent treatment. Both dispersion (anterograde particle transport) and aggregation (retrograde transport) took place along microtubules; retrograde transport could be induced with 1 mM MgATP and 0.1 μM free Ca^{2+}, anterograde movement with 1 mM MgATP and reduction of the free Ca^{2+} to 0.01 μM. It is not yet known whether there are two kinds of motor, or one with reversible gears (as suspected for *Reticulomyxa*, see Section 8.6.3). However, Rozdzial and Haimo reported that the mechanisms of transport in the two directions appear to be different, since aggregation and dispersion are differentially affected by ATP analogues, such as AMPPNP or ATPγS. Most likely the Ca^{2+}-activated switch disables one motor whilst activating another.

It is also uncertain whether Ca^{2+} acts directly on the motor molecules or indirectly, as in the case of smooth-muscle myosin (see Section 5.5). Rozdzial and Haimo and others found that moving granules in permeabilised cells could be made to reverse direction by adding or removing cAMP; its presence caused dispersion, its absence allowed aggregation. cAMP appears to activate a cAMP-dependent protein kinase that phosphorylates a 57 kD granule-associated polypeptide, subsequently dephosphorylated during aggregation. McNiven and Ward found no effect for cAMP. This nucleotide may activate an endogenous membrane system involved in controlling the level of free Ca^{2+}; such a system may have been destroyed by detergent treatment in the preparations of McNiven and Ward.

An interesting additional observation by McNiven's group is that over-lapping microtubule networks from different cells could interact and pull together in the retrograde reactivation buffer, after the pigment granules had aggregated in the centre of each network. This is a similar phenomenon to contraction of the brain microtubule preparation described in Section 10.5.7, but faster (0.5–1.0 μm/s, as opposed to 1.0 μm/min). Assuming this interaction was produced by the motors that move the pigment granules and that pigment granules were not involved (i.e. not acting as the 'particles' in Figure 10.8b), it suggests that the retrograde motors are soluble, like kinesin or brain dynein, and that their ATP-insensitive ends may have the capacity to bind to microtubules, as well as to pigment granules.

10.8   Summary

Microtubule behaviour in cells suggests varying degrees of dynamic instability, depending on the type of cell they are in, the organelle they are part of (if any), and the stage of the cell cycle. Dynamic behaviour is particularly marked during cell division and appears to be very important for the mechanism of mitosis (see Chapter 11); the radial network of microtubules found during interphase is more stable but dynamic instability still exists. It allows the microtubules in an interphase cell to adapt their arrangement to changes in the cell's environment and ensures that they are always evenly distributed.

Differentiated cells often have specialised microtubules that are very stable, especially when crosslinked into bundles; such microtubules may be important in maintaining a certain cell shape. Neuronal cells are a good example of cells apparently maintained in an extreme shape by their constituent microtubules; their role in directed transport may, however, be more important here than a purely structural one.

One of the most important functions of microtubules is to provide tracks for transporting packaged substances from one part of a cell to another. In axons and probably in most other animal cells, bidirectional fast vesicle transport is along microtubules. Functions related to transport are those of keeping the endoplasmic reticulum spread throughout the cell and the Golgi complex concentrated close to the centre of the cell. Microtubule polarity can be used to specify the direction of transport if appropriate motor molecules are activated. How transport direction is controlled may vary in different cell types and is not yet clear for any case.

Other cytoskeletal components are probably also transported along microtubules. This includes the proteins of slow axonal component b, including actin, moving along an axon to the growth cone or nerve ending. Microtubules also slide over one another, providing a means of effecting changes in the shapes of organelles constructed from microtubules, or of pushing objects apart. Sliding between microtubules is a likely mechanism for the slowest component of axonal transport, with NFs being carried along passively.

10.9   Questions

10.1 What features of the microtubule array in an animal cell make it suitable for distributing, to the cell periphery, newly synthesised proteins that are destined for secretion, and by what mechanism might they travel there?

10.2 What substances travel along axons by processes slower than vesicle transport but still faster than diffusion? Might this behaviour be relevant to non-neuronal cells, given what is known (Chapter 1) about cytoplasmic pore sizes?

10.3 Discuss the statement that dynamically unstable microtubules will tend to radiate out from the cell centre to give an even distribution throughout the cytoplasm, whereas an asymmetric arrangement needs microtubules stabilised at both ends.

10.10   Further reading

Reviews

1 Dustin, P. (1984). *Microtubules*, 2nd edn. New York: Springer-Verlag.

2 Porter, K. E. (1976). Introduction. In *Cell Motility*, ed. Goldman, C. R., Pollard, T. & Rosenbaum, J. New York: Cold Spring Harbor Press.
   (Some details are out of date, but this text is good for breadth of view.)

3 Roberts, K. R. & Hyams, J. S., eds (1979). *Microtubules*. London: Academic Press.
   (Comprehensive to 1978.)

4 Kirschner, M. & Mitchison, T. (1986). Beyond self-assembly: from microtubules to morphogenesis. *Cell* **45**, 329–342.

5 Vale, R. D. (1987). Intracellular transport using microtubule-based motors. *Annu. Rev. Cell Biol.* **3**, 347–378.

– Sheetz, M. P., Vale, R. D., Schroer, T. A. & Reese, T. S. (1986). Movements of vesicles on microtubules. *Ann. N.Y. Acad. Sci., USA* **493**, 409–416.

6 Schnapp, B. J. & Reese, T. S. (1986). New developments in understanding rapid axonal transport. *Trends Neurosci.* **9**, 155–162.

7 Kreis, T. E. (1990). Role of microtubules in the organisation of the Golgi apparatus. *Cell Motil. & Cytoskel.* **15**, 67–70.

– Terasaki, M. (1990). Recent progress on structural interactions of the endoplasmic reticulum. *Cell Motil. & Cytoskel.* **15**, 71–75.

8 Lloyd, C. W. & Seagull, R. W. (1985). A new spring for plant cell biology: microtubules as dynamic helices. *Trends Biochem. Sci.* **10**, 476–478.

9 Hollenbeck, P. J. (1989). The transport and assembly of the axonal cytoskeleton. *J. Cell Biol.* **108**, 223–227.

10 Lasek, R. L., Garner, J. A. & Brady, S. T. (1984). Axonal transport of the cytoplasmic matrix. *J. Cell Biol.* **99**, 212s–221s.

11 Grafstein, B. & Forman, D. S. (1980). Intracellular transport in neurons. *Physiol. Rev.* **60**, 1167–1283.

12 Ochs, S. (1982). On the mechanism of axoplasmic transport. In *Axoplasmic Transport*, ed. D. G. Weiss, pp. 342–350. New York: Springer-Verlag.

13 Bunge, M. B. (1986). The axonal cytoskeleton: its role in generating and maintaining cell form. *Trends Neurosci.* **9**, 477–482.

14 Hirokawa, N. (1986). Quick-frozen, deep-etch visualisation of the axonal cytoskeleton. *Trends Neurosci.* **9**, 67–71.

15 Black, M. M. & Baas, P. W. (1989). The basis of polarity in neurons. *Trends Neurosci.* **12**, 211–214.

16 Sargent, P. B. (1989). What distinguishes axons from dendrites? Neurons know more than we do. *Trends Neurosci.* **12**, 203–205.

17 Allen, R. D. (1987). The microtubule as an intracellular engine. *Sci. Am.* **255**(2), 42–49.

18 Amos, L. A. & Amos, W. B. (1987). Cytoplasmic transport in axons. *J. Cell Sci.* **87**, 1–2.

19 Brinkley, B. R. (1985). Microtubule organizing centers. *Annu. Rev. Cell Biol.* **1**, 145–172.

20 Tucker, J. B. (1984). Spatial organization of microtubule-organizing centers and microtubules. *J. Cell Biol.* **99**, 55s–62s.

21 Sleigh, M. (1973). *The Biology of Protozoa.* London: Edward Arnold.

Original papers, for brief scanning

22 Soltys, B. J. & Borisy, G. G. (1985). Polymerization of tubulin *in vivo*: direct evidence for assembly onto microtubule ends and from centrosomes. *J. Cell Biol.* **100**, 1682–1689.

23 Schultze, E. & Kirschner, M. (1988). New features of microtubule behaviour observed *in vivo*. *Nature* **334**, 356–359.

24 Sammak, P. J. & Borisy, G. G. (1988). Direct observation of microtubule dynamics in living cells. *Nature* **332**, 724–726.

25 Tao, W., Walter, R. J. & Berns, M. W. (1988). Laser-transected microtubules exhibit individuality of regrowth; however, most free new ends of the microtubules are stable. *J. Cell Biol.* **107**, 1025–1035.

26 Keith, C. H. (1987). Slow transport of tubulin in the neurites of differentiated PC12 cells. *Science* **235**, 337–339.

27 Okabe, S. & Hirokawa, N. (1988). Microtubule dynamics in nerve cells: analysis using microinjection of biotinylated tubulin into PC12 cells. *J. Cell Biol.* **107**, 651–663.

28 Smith, R. S. (1988). Studies on the mechanism of the reversal of rapid

organelle transport in myelinated axons of *Xenopus laevis*. *Cell Motil. & Cytoskel.* **10**, 296–308.

29 Pfister, K. K., Wagner, M. C., Stenoien, D. L., Brady, S. T. & Bloom, G. S. (1989). Monoclonal antibodies to kinesin heavy and light chains stain vesicle-like structures but not microtubules, in cultured cells. *J. Cell Biol.* **108**, 1453–1463.

30 Schroer, T. A., Schnapp, B. J., Reese, T. S. & Sheetz, M. P. (1988). The role of kinesin and other soluble factors in organelle movement along microtubules. *J. Cell Biol.* **107**, 1785–1792.

31 Schroer, T. A., Steuer, E. R. & Sheetz, M. P. (1989). Cytoplasmic dynein is a minus end-directed motor for membranous organelles. *Cell* **56**, 937–946.

32 Schnapp, B. J. & Reese, T. S. (1989). Dynein is the motor for retrograde axonal transport of organelles. *Proc. Natl. Acad. Sci., USA* **86**, 1548–1552.

33 Gilbert, S. P. & Sloboda, R. D. (1989). A squid dynein isoform promotes axoplasmic vesicle translocation. *J. Cell Biol.* **109**, 2379–2394.

34 Weisenberg, R. C., Flynn, J., Gao, B. & Awodi, S. (1988). Microtubule gelation-contraction *in vitro* and its relationship to component *a* of slow axonal transport. *Cell Motil. & Cytoskel.* **10**, 331–340.

35 Ellisman, M. H. & Lindsey, J. D. (1983). The axoplasmic reticulum within myelinated axons is not transported rapidly. *J. Neurocytol.* **12**, 393–411. *(Smooth ER in axons.)*

36 Sheetz, M. P. (1989). Kinesin structure and function. In *Cell Movement*, vol. 2, ed. Warner, F. D. & McIntosh, J. R., pp. 277–285. New York: Alan R. Liss.

37 Dabora, S. L. & Sheetz, M. P. (1988). The microtubule-dependent formation of a tubulovesicular network with characteristics of the ER from cultured cell extracts. *Cell* **54**, 27–35.

38 Lee, C. & Chen, L. B. (1988). Dynamic behavior of endoplasmic reticulum in living cells. *Cell* **54**, 37–46.

39 Vale, R. D. & Hotani, H. (1988). Formation of membrane networks *in vitro* by kinesin-driven microtubule movement. *J. Cell Biol.* **107**, 2233–2242.

40 Matteoni, R. & Kreis, T. E. (1987). Translation and clustering of endosomes and lysosomes depends on microtubules. *J. Cell Biol.* **105**, 1253–1265.

41 Heuser, J. (1989). Changes in lysosome shape and distribution correlated with changes in cytoplasmic pH. *J. Cell Biol.* **108**, 855–864.

42 McNiven, M. A. & Ward, J. B. (1988). Calcium regulation of pigment transport *in vitro*. *J. Cell Biol.* **106**, 111–125.

43 Rozdzial, M. M. & Haimo, L. T. (1986). Reactivated melanophore motility: differential regulation and nucleotide requirements of bidirectional pigment granule transport. *J. Cell Biol.* **103**, 2755–2764.

44 Koonce, M. P. & Schliwa, M. (1986). Reactivation of organelle movements along the cytoskeletal framework of a giant freshwater amoeba. *J. Cell Biol.* **103**, 605–612.

45 Hollenbeck, P. J. & Bray, D. (1987). Rapidly transported organelles containing membrane and cytoskeletal components: their relation to axonal growth. *J. Cell Biol.* **105**, 2827–2835.

46 Schulze, E., Asai, D. J., Bulinski, J. C. & Kirschner, M. W. (1987). Post-translational modification and microtubule stability. *J. Cell Biol.* **105**, 2167–2177.

47 Barra, H. S., Arce, C. A. & Argarana, C. E. (1988). Post-translational tyrosination and detyrosination of tubulin. *Molec. Neurobiol.* **2**, 133–153.

48 Gundersen, G. G., Khawja, D. & Bulinski, J. C. (1987). Postpolymerization detyrosination of α-tubulin: a mechanism for subcellular differentiation of microtubules. *J. Cell Biol.* **105**, 251–264.

49 Maruta, H., Greer, K. & Rosenbaum, J. L. (1986). The acetylation of α-tubulin and its relationship to the assembly and disassembly of microtubules. *J. Cell Biol.* **103**, 571–579.

50 Brinkley, B. R., Cox, S. M., Pepper, D. A., Wible, L., Brenner, S. L. & Pardue, R. L. (1981). Tubulin assembly sites and the organization of cytoplasmic microtubules in cultured mammalian cells. *J. Cell Biol.* **90**, 554–562.

51 Bré, M.-H., Kreis, T. E. & Karsenti, E. (1987). Control of microtubule nucleation and stability in Madin-Darby canine kidney cells: the occurrence of non-centrosomal, stable detyrosinated microtubules. *J. Cell Biol.* **105**, 1283–1296.

52 Mogensen, M. M. & Tucker, J. B. (1987). Evidence for microtubule nucleation at plasma membrane-associated sites in *Drosophila. J. Cell Sci.* **88**, 95–107.

53 Woodrum, D. T. & Linck, R. W. (1980). Structural basis of motility in the microtubular axostyle: implications for cytoplasmic microtubule structure and function. *J. Cell Biol.* **87**, 404–414.

54 Bloodgood, R. A. (1988). Gliding motility and the dynamics of flagellar membrane glycoproteins in *Chlamydomonas reinhardtii. J. Protozool.* **35**, 552–558.

55 Tucker, J. B. (1968). Fine structure and function of the cytopharyngeal basket in the ciliate *Nassula. J. Cell Sci.* **3**, 493–514.

56 Tucker, J. B. (1972). Microtubule-arms and propulsion of food particles inside a large feeding organelle in the ciliate *Phascolodon vorticella. J. Cell Sci.* **10**, 883–903.

57 Scholey, J. M. (1990). Multiple microtubule motors. *Nature* **343**, 118–120.

58 Kelly, R. B. (1990). Associations between microtubules and intracellular organelles. *Curr. Opinion Cell Biol.* **2**, 105–108.

CHANGES IN THE 11
CYTOSKELETON
DURING CELL DIVISION

11.1 Introduction

The onset of cell division brings about greater changes in the structure of the cytoskeleton than occur at any other point in the cell cycle. Microtubules and bundles of actin-containing filaments are disassembled into subunits or small oligomers, ready for use in building new structures. Many intermediate filaments, on the other hand, do not disassemble but collapse into a meshwork closely associated with the nuclear membrane. They appear to have no active role in nuclear division or cytokinesis but are passively shared between the daughter cells.

11.1.1 Stages of mitosis and cytokinesis[1,3–5]

The various stages in eukaryotic cell division are illustrated in Figure 11.1.

Nuclear division in all cells involves some method of physically separating two sets of chromosomes. In most somatic cells, the two sets contain identical copies of DNA sequences and the process of separating them is called **mitosis**. Germ-line cells separate pairs of non-identical homologous chromosomes in a process called **meiosis**. As regards the role of the cytoskeleton, there is probably little difference between the two types of event.

In general, mitosis or meiosis is followed by separation of the cytoplasm into two compartments (**cytokinesis**). But mitosis and cytokinesis are quite distinct and independent processes. Treatments which affect microtubule polymerisation disrupt mitosis in eukaryotes. In contrast, treatments which disrupt actin filaments (such as exposure to the cytochalasin alkaloids) or interfere with their function (such as microinjection of antibodies to myosin) allow normal nuclear events in mitosis but block division of the cytoplasm, producing a multinucleate mass of cytoplasm.[14–16]

Moreover, there are examples of developmental stages where nuclei undergo many successive divisions whilst remaining in a shared cytoplasmic compartment. This occurs in a wide range of organisms, including insect embryos (studied in detail in *Drosophila*, for example); in fungi; in some algae, such as *Chara* and *Acetabularia*; and in *Physarum*, the non-cellular myxomycete type of slime mould. The cytoplasm is divided up subsequently in some cases, such as the insect embryos, to produce cells with single nuclei.

Figure 11.1

Stages in the division of an animal cell.[1-5]

Prophase: Paired chromatids condense. The two poles start to move apart.

Prometaphase: The nuclear membrane breaks down. The spindle grows, pushing the poles further apart.

Metaphase: Chromosomes line up as opposing forces become balanced.

Anaphase: Chromosomes move towards the poles (anaphase A). Poles may continue to move apart (anaphase B).

Telophase: Chromosomes decondense and the nuclear envelope re-forms around them. The cleavage furrow deepens.

Cytokinesis: The cleavage furrow closes to form a contractile ring, which contracts around remaining central-spindle microtubules to produce the midbody.

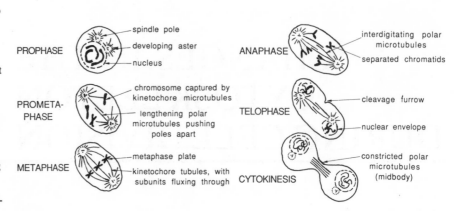

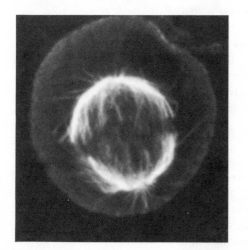

(a)

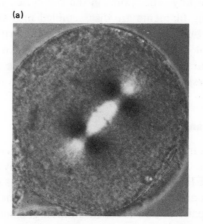

(b)

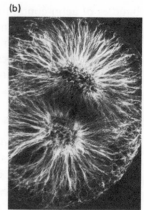

Figure 11.2

An optical section through a mitotic HeLa cell whose microtubules have been labelled with anti-tubulin antibodies and fluorescent secondary antibodies. The image was obtained by confocal fluorescence microscopy. [Reproduced from Amos, W. B. (1988). *Cell Motil. & Cytoskel.* **10**, 54–61.]

Figure 11.3

Sea-urchin egg spindles as seen by two kinds of light microscopy. In (a), the birefringence of the spindle components is shown by polarisation microscopy. The specimen is oriented so that the central spindle is maximally bright. Those astral microtubules that lie at right angles to this direction show up maximally dark. In (b), a dividing egg has been permeabilised and the microtubules visualised by confocal microscopy of secondary immunofluorescence.

11.1.2   Methods of chromosome separation[18]

In prokaryotes and primitive dinoflagellates such as *Gyrodinium*, chromosome separation appears to be brought about by forces produced on membranes. In the prokaryotes it is the cell membrane that is involved, in *Gyrodinium* it is the nuclear membrane. In most other organisms chromosome separation involves a specially constructed arrangement of microtubules and other material, the **spindle** (see Figures 11.2 & 11.3), which owes its name to its resemblance in shape to the spindle once used in spinning yarn. The poles of the spindle are anchored, either directly or indirectly, to the cell membrane and the two sets of chromosomes move towards their respective poles.

11.2   Centrosomes[1,13]

The **centrosome** of an animal cell is the main centre of microtubule polymerisation and, as such, important in organising events in the cytoplasm. During interphase the structure normally contains two **centrioles** (see Figure 9.13) surrounded by a mass of granular material.

11.2.1   Centrosome division[4]

Towards the end of interphase the centrosome doubles in the manner described in Section 9.6 for the basal bodies in *Chlamydomonas*. The *four* centriolar cylinders proceed to move slowly apart in pairs, by a mechanism that can be blocked by colchicine. Presumably under normal conditions microtubules start to grow out from two centres within the duplicated centrosome. If these two sets of microtubules interact in such a way that the two daughter centrosomes are pushed apart, the beginnings of a mitotic spindle will be formed. Centrosome separation is also reported to be retarded by cytochalasin D, so microfilaments may also play a part.

The active components of centrioles were first investigated by Mazia in 1960. He studied the effects of **mercaptoethanol** on the division of sea-urchin eggs and concluded that a normal pole contains just two entities that are capable of generating a pole. Mercaptoethanol blocks the duplication of these entities irreversibly (Figure 11.4). Although the cylinder of triplet microtubules looks a good candidate for Mazia's entity, the visible structure is not essential to centrosomal function: many organisms (including some vertebrate cells and all higher plant cells) lack this type of structure at the poles. There is no great gulf between organisms with triplet centrioles and those without: in *Physarum* the two conditions alternate during the life cycle.

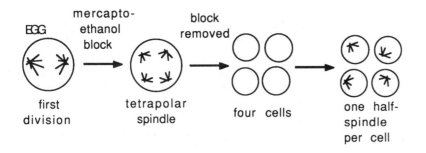

Figure 11.4

The effects of mercaptoethanol on a dividing sea-urchin egg.[4] The presence of this agent in the medium probably interferes with many biochemical processes in the egg; one seems to be involved in centrosome duplication. Cell division, but not centrosome division, continues after removal of the block. At the four-cell stage, each cell forms only a monoaster instead of a normal spindle.

Attention has therefore been directed towards the indistinct granular material around the centrioles. Antibody-labelling studies have shown that various polypeptides are specifically located at the centrosome. The isolation of a centrosomal protein with aster-forming properties has been reported (see Section 8.3). However, such a major component, present in a large number of copies, is unlikely to be what determines the *number* of centrosomes. The location of the centrosomes within a cell is also determined precisely, by mechanisms which are not yet clear.

11.2.2   Centrosome division and nuclear chromosome division

If the nucleus is removed from a fertilised sea-urchin egg, the centrosome continues to divide, up to three times, with roughly the normal cycle time. Each daughter centrosome produces an astral array of microtubules and there are abortive attempts at cytokinesis. Similar effects are observed when DNA synthesis is blocked with the drug **aphidicolin**. Nuclear chromosome division is halted in *Drosophila* embryos treated with the drug, but centrosome division continues.

11.2.3   Does the centrosomal region contain nucleic acid?[32]

Basal bodies from *Tetrahymena* have been found to cause the formation of multiple asters when injected into *Xenopus* eggs. The effect is abolished by pre-incubation with RNase though not with DNase. Supporting evidence has come from treating live cells with specific nucleic-acid crosslinking agents called **psoralens**. The psoralens were introduced into the cells in the dark and then activated by local microbeam irradiation of the cell. Types of psoralen which normally crosslink only DNA were without effect, but those which are also active on RNA blocked centrosome separation and aster formation after irradiation.

Recently, nucleic-acid hybridisation studies appear to have located specific sequences in the basal bodies of *Chlamydomonas* (see Section 9.6.3). The significance of this finding will not be really clear until similar studies have been made on centrosomes of other cell types (including those lacking recognisable centrioles). However, it seems distinctly possible that the number of centrosomes in a cell is controlled in some way by centrosomal nucleic acid.

11.3   Prophase

This is the stage in its cycle at which a cell begins to rearrange its contents to prepare for division. The **chromatin**, already duplicated during the S-phase of the cycle, slowly condenses into recognisable chromosomes inside the nucleus. Each consists of two **chromatids**, joined together at a specialised region containing some very repetitive DNA sequences (non-transcribable **heterochromatin**) and known as the **centromere**. Chromosome movements during prophase have been very little studied. They appear to become attached to the nuclear membrane and move relative to it. These movements, which are unaffected by treatments which disassemble microtubules, may be analogous to those which suffice to separate chromosomes in primitive dinoflagellates (see 11.1.2).

In the case of higher eukaryotes, the interphase microtubular array is broken down at the start of prophase but, after division of the centrosome into two, polymerisation near both these **mitotic centres**, or **poles**, becomes even more active: in one set of observations, the persistence of a tubule shortened to an average of 30 s and new ones were initiated at a rate of 1000 per pole per minute. The spindle which forms between the poles, being a large bundle of orientated tubules, is birefringent (Figure 11.3), and the birefringence fluctuates rapidly with time as a result of the extremely dynamic state of the structure.[31] The increase in microtubule activity may be due to modification of the properties of some of the MAPs by phosphorylation.[38]

An **aster** of radially-arranged microtubules appears around each pole. Some microtubules from opposite poles which meet and form lateral interactions are referred to as **polar microtubules**. The lateral interactions apparently help to stabilise these particular microtubules and they grow preferentially. The same interactions are probably responsible for pushing the poles apart from one another as the spindle develops alongside the nucleus. Eventually the asters become less prominent, with fewer microtubules apparently ending freely; microtubules radiating from the poles in directions away from the central spindle become sparser and shorter.

11.3.1 Closed mitosis

In yeast cells, the spindle forms inside rather than outside the nuclear membrane (Figure 11.6a). In place of a centrosome, there is a specialized region of the nuclear membrane known as a **spindle pole body**. Before division, the region splits into two and a bundle of microtubules forms between them. The spindle pole bodies appear to be pushed apart and the nucleus elongates. The membrane never breaks down but is eventually pinched into two after the chromosomes have been segregated into opposite ends.

In some protozoa (see Figure 11.9) the spindle forms outside the nuclear membrane but still separates the two enclosed sets of chromosomes within an intact, though distended, nuclear membrane. The spindle may lie alongside the nucleus or, as in the case of the dinoflagellates, run in a tunnel through the centre. In some cases, it is known that kinetochore microtubules end at kinetochores in the nuclear membrane and may operate in an analogous way to kinetochore microtubules in open mitosis (see Section 11.6). In other cases, it is possible that transmembrane motor assemblies simply run along microtubules that lie parallel to the nuclear membrane, in order to move the chromosomes apart.

11.4 Prometaphase in animal and plant cells

Except in some of the unicellular eukaryotes discussed above, the nuclear envelope breaks down suddenly into small membrane fragments, probably as a consequence of the disassembly of the underlying nuclear lamina. The three IF-like lamin proteins become heavily phosphorylated at this point of the cell cycle and this may induce their disassembly. One of the lamins remains associated with the nuclear membrane; the other two become soluble.

11.4.1   Microtubule capture by kinetochores[11]

In the absence of the nuclear envelope, the spindle becomes surrounded by the condensed chromosomes. A complex multilayered pair of structures called **kinetochores** assemble symmetrically onto the two sides of each chromosomal centromere and become connected to the poles by **kinetochore microtubules** (Figure 11.5a). The small bundles of kinetochore microtubules connecting each chromatid to its pole are also referred to as **spindle fibres**. The observed polarity of these microtubules suggests that the kinetochores may 'capture' the plus ends of microtubules which have grown out from the poles. This process has indeed been observed, in a remarkable feat of light microscopy, in the large flat cells of cultured newt lung.[39]

It may be that no particular set of astral microtubules is directly targeted at the various kinetochores but rather that, out of the hundreds of tubules extending randomly from the centrosome, some will touch each kinetochore. Thus, the selection of both polar and kinetochore microtubules from the mass of astral microtubules may be a random process. It is also probable that some are initiated from the kinetochores, since this has been observed *in vitro*. Such microtubules could persist if they were in the correct orientation and were stabilised by suitable interactions with neighbouring microtubules.

The mechanism suggested for microtubule capture by a kinetochore involves a temporary association between the kinetochore and any part of the surface of a microtubule which makes contact with it; subsequent shortening of the tubule may bring its end to the kinetochore or the chromosome may reach the plus end by sliding there. Chromosomes have also been observed moving rapidly towards the pole when first captured.[39] There is evidence for the presence of dynein associated with kinetochores[8] but not, as yet, kinesin-like molecules. However, as mentioned in Sections 8.5–8.9, dynein-like molecules may possibly be capable of bidirectional forces.

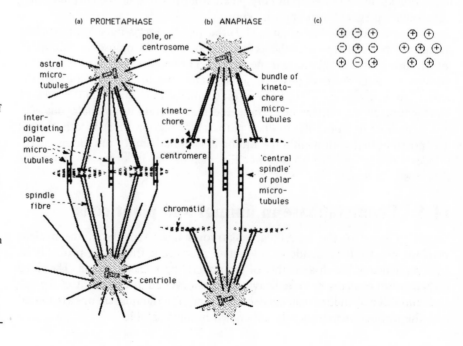

Figure 11.5

(a) & (b) Spindles at the stages of forming the correct bipolar arrangement and of using the arrangement to separate pairs of chromatids.[1-5] The various classes of microtubule and their interactions with each other, with the centrosomes and the kinetochores are shown schematically. (c) Packing of microtubules: square packing optimises the associations between equal numbers of oppositely-orientated polymers; near-hexagonal close-packing of equivalently orientated polymers maximises their mutual interactions.

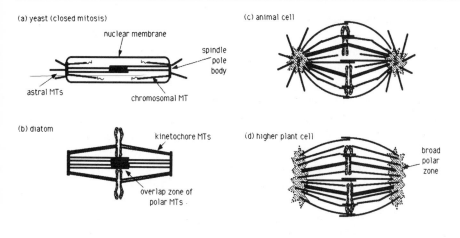

(a) yeast (closed mitosis)

(b) diatom

(c) animal cell

(d) higher plant cell

Figure 11.6

Variations in spindle structure in different eukaryotes. (a) A yeast spindle is mainly intranuclear and fairly simple. Microtubules (MTs) appear to run in parallel from one spindle pole body to another. The bodies are pushed apart by elongation of these MTs, which may slide actively relative to one another. Chromosomes (not condensed as in higher eukaryotes) are attached to single MTs and pulled to opposite ends of the nucleus during mitosis. (b) A diatom spindle is dominated by overlapping polar microtubules, which have been shown to grow and slide relative to one another. Another set of microtubules attaches to the chromosomes but there are no 'superfluous' astral microtubules. (c) The spindle of an animal cell has many more microtubules and appears less ordered than (a) or (b). Microtubules grow out radially from the centrosomes, so there are many astral microtubules that neither attach to chromosomes nor interact with microtubules from the opposite pole. The requirements in terms of pre-existing arrangements within the cell for the successful assembly of such a spindle are probably not very stringent. (d) The spindles of higher plants resemble those of animal cells but the MTs grow out from an extended centrosomal area rather than being focused towards a pole at each end.

It is assumed that after first being captured the tubule moves over the surface of the kinetochore but, when the kinetochore reaches the end of the tubule, the latter is transferred to fit into a circular channel through the outer kinetochore plaque (see Figure 11.8c & d). Experiments involving disruption of microtubules in cells show that, like the bundles of polar microtubules, kinetochore microtubules are more stable during prometaphase than simple astral microtubules.[17] The stability is presumably a result of the interaction between the plus end and the kinetochore. The attachment itself is stabilised in some unknown way by tension (just as resistance increases the force generated by a muscle); this has been demonstrated by experiments involving micro-manipulation of chromosomes. Thus, microtubules captured by one kinetochore may be stabilised by interactions between the opposing kinetochore on the sister-chromatid and microtubules from the opposite pole. They still exhibit dynamic instability to some extent, however, and may occasionally disassemble right back to the pole. Mitchison has estimated that this happens at less than a tenth of the rate of loss of non-kinetochore microtubules. The kinetochore may then capture another microtubule in place of one that has been lost.

11.4.2   Variation in the spindles of different species[1,18]

The overall shapes of mitotic spindles vary from the neat parallel bundles of diatoms (Figure 11.6b) to apparently disordered, nearly spherical assemblies of animal cells (Figure 11.6c). Nevertheless, the various sets of microtubules are thought to be homologous in all cases. In diatoms, the two parallel interdigitating sets of polar microtubules form a well-ordered **central spindle** that dominates the structure. In animal cells, the central spindle is a more minor component and there are more kinetochore and astral microtubules.

11.5   Congression of chromosomes onto the metaphase plate[2,33]

Many models have been proposed for the mechanism by which the chromosomes assemble at the centre (or equator) of the spindle. All agree that a balance of opposing forces is involved. The motion of the chromosomes appears to be disorderly but the net result is that they become oriented so that one kinetochore faces each pole and the chromosomes are gradually pushed (or pulled) to the centre of the spindle. The time-point when balance is achieved and the chromosomes are merely oscillating about a plane equidistant from the two poles (the metaphase plate) is known as **metaphase** (Figure 11.5a).

11.5.1   The flux of tubulin from kinetochore to pole[20-5]

Mitchison and colleagues injected a pulse of biotinylated tubulin (see Section 10.2.2) into mitotic cells and investigated the sites of tubulin assembly by electron microscopy. Gold-labelled antibodies to the biotin tag showed tubulin addition onto microtubule ends attached to kinetochores (Figures 11.7a & 11.8c). If some time was allowed to elapse between the microinjection and fixation of the cell, the label appeared to move along the microtubules. The results suggest a flux of tubulin subunits along the kinetochore tubules in the direction of the pole, at a rate that would allow the label to pass through a microtubule in about a minute. The process might be regarded as an example of 'assisted' treadmilling in microtubules (see Sections 7.4.7 & 7.6.6).

More recent experiments by Mitchison, who labelled short segments of microtubule fluorescently and watched the fluorescent spots move towards the poles (Appendix A.1.2), seem to rule out the possibility that labelling at the plus end is simply the result of dynamic instability (Section 11.4.1).[25] Subunits apparently continue to add on to the plus end after capture and the kinetochore presumably moves in the plus end direction.

11.5.2   Polewards transport of kinetochore microtubules[1,2,6-8,19,22]

Growth onto the active end of a kinetochore microtubule, whilst the kinetochore itself remains at the metaphase plate, requires that the microtubule be moved towards the pole and subunits be lost from the minus end. There is at present only indirect evidence to suggest how this movement is achieved.

It is an important observation that the polewards force seems to depend on the summed lengths of the kinetochore microtubules; if the kinetochore on one side of a pair of chromatids is experimentally damaged (e.g. by laser microsurgery), the remaining microtubules attached to this side elongate while

Figure 11.7

Events associated with microtubules in individual half-spindles:[2,17-23] all plus ends, even those associated with kinetochores, are subject to dynamic instability. (a) During the events involved in forming the metaphase spindle, the stabilising effect of the kinetochore, continually moving towards the plus ends (see Figure 11.9c), ensures net growth at these ends of the kinetochore microtubules (KMTs). A shearing force operating between KMTs and neighbouring astral tubules (AMTs) may push the KMTs towards the pole causing tubulin loss from the minus ends, either due to pressure-induced depolymerisation or breakage. (b) Conditions change at anaphase so that there is a net loss of tubulin from all plus ends. The kinetochore motor either stops working or reverses its direction. Interactions between KMTs and AMTs may continue as in (a).

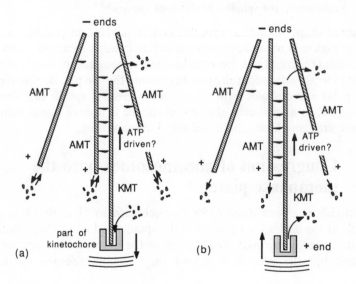

the full number on the opposite side shorten, as if in order to restore the balance of forces. This is most consistent with models in which the polewards force is generated, as in Figure 11.7a, by motor molecules acting all along the kinetochore microtubules.

The motors might work between the kinetochore microtubules and either adjacent astral microtubules or other components of the spindle matrix. The motor molecules could be kinesin, which has been localised near the spindle poles in sea-urchin eggs and is apparently associated with the spindle matrix in an ATP-independent manner. Immunofluorescent staining has shown cytoplasmic dynein in association with spindle microtubules in the entire zone between chromosomes and their poles; kinesin has been located at and near the centrosomes. In fact any motor, whether anterograde or retrograde, acting directly between adjacent microtubules (Figure 11.8a) could be arranged to produce a pushing force at the poles. Thus, kinesin, dynein and dynamin could all be involved in this process.

Alternatively, elastic components acting between the kinetochore and pole might generate a force proportional to the number and lengths of kinetochore microtubules if one such component were associated with each microtubule. But it is difficult to imagine how they would be maintained in position by dynamically unstable microtubules.

Kirschner has suggested that compression at the minus end of the microtubule might cause subunits to be lost here. Alternatively, the minus ends may be pushed beyond lateral stabilising interactions and become free to lose subunits. The longer the kinetochore microtubules, the greater the pushing force and the faster the subunit loss from the minus ends. The bundle of kinetochore microtubules presumably reaches its equilibrium length (and the polewards force its correct magnitude) when this rate of loss from the minus ends is the same as the rate of addition to the plus ends.

The balance of forces will be disturbed to some extent by the dynamic behaviour of individual microtubules. After the loss of a whole microtubule, the remaining ones may tend to lengthen until a replacement is captured, when they will shorten again. This may explain the observed 'jostling for position' by chromosomes at the equator. The above explanation for the balancing of opposing forces on the chromosomes also satisfies the requirement that forces be balanced within individual half spindles. Bayer and colleagues have observed that chromosomes attached to *monopoles* in abnormal cells show the same congressional behaviour as normal bipolar spindles.

11.5.3   Elimination forces[33]

Alternative models for congression have been based on a balance between a position-dependent polewards force and so-called **elimination forces**. The latter are weak repellent forces between chromosomes and spindle components. If part of a chromosome arm is severed by laser microsurgery, it slowly slides out of the spindle while the remaining attached segment of chromosome moves closer to its pole. These repellent forces must contribute to the net balance, but they are unlikely to be crucial as Brinkley and colleagues have found that kinetochores which have lost their chromosomes still promote the assembly of normal spindles.[26]

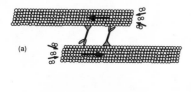

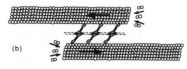

Figure 11.8

(a) & (b) Possible anaphase B motors responsible for producing sliding apart of interdigitating microtubules in the overlap zone.[10] (a) Independent anterograde motors attached randomly to either microtubule: binding by the non-sliding end cannot be permanent— presumably it is favoured if the other end is interacting with a microtubule of opposite polarity. (b) Anterograde motors semi-permanently attached together (possibly by filaments) in a bipolar array.

11.6 Anaphase

In response to some specific signal, the pairs of chromosomes separate suddenly and start to move towards opposite poles (Figure 11.5b). Anaphase movement is relatively slow (see Table 10.1, page 196), being roughly equivalent in speed to continental drift. It actually consists of two distinct processes. Anaphase A is a process by which the newly-separated chromatids move closer to their respective poles. This involves shortening of the kinetochore microtubules, so microtubule depolymerisation must play some part in any suggested mechanism (Figure 11.7b). Anaphase B is a process of moving the two poles further apart than in the original spindle structure and requires relative sliding of polar microtubules from opposite poles (Figure 11.8).

Both processes have the effect of moving the two sets of chromatids apart. Most cells separate their chromosomes with a combination of the two processes but there are cases where one or other process appears to dominate. In some cases, anaphase B movement is slight but in others (such as in diatoms), the pole-to-pole distance may increase by up to five times and obviously requires a considerable lengthening of the polar microtubules during anaphase.

11.6.1 Anaphase B: pushing poles apart[12,28]

Of the two types of movement which separate the chromosomes, anaphase B is perhaps the easier to understand. Sliding between the two sets of polar microtubules has been demonstrated by careful EM reconstruction studies by McIntosh's group and others. The extent of overlap between the opposing sets of microtubules decreases in synchrony with the increase in the pole-to-pole distance during anaphase. In some cells the polar microtubules clearly continue to elongate during anaphase. The site and extent of growth has been measured by injecting labelled tubulin. Fresh tubulin is always added onto the end furthest from the pole, the plus end. If growth here is allowed for, the measurements suggest that pole separation is always accompanied by relative sliding of polar microtubules.

There is good evidence that ATP is required to effect the sliding process. This has been shown particularly well by Cande and colleagues for the case of diatoms (Figure 11.6b) permeabilised during mitosis and reactivated in exogenous buffers. Under these conditions, tubulin incorporation and sliding can be uncoupled. Labelled tubulin adds onto the ends of the microtubules but spindle elongation continues only when ATP is added.

The motor for the sliding mechanism is thought to be located in the overlap zone, since microbeam irradiation of this region blocks elongation *in vivo*, whilst irradiation nearer the poles does not. By EM, the interdigitating polar microtubules appear to be embedded in **osmiophilic material**. The microtubules in the overlap zone seem to prefer neighbours of opposite polarity (evidence for a specific interaction) and, at least in diatom central spindles, are ordered such that they form a square-packed lattice in cross-section (Figure 11.5c). Elsewhere, microtubules (of equivalent polarity) tend to be close-packed to produce a roughly hexagonal arrangement. The osmiophilic material remains in the overlap zone as its length decreases. Reactivated diatom spindles continued to elongate even when the sections of microtubule in the overlap zone were composed entirely of labelled tubulin, suggesting that the motor molecules are able to diffuse and bind to new regions of microtubules which have antiparallel neighbours.

The exact nature of the osmiophilic material is at present uncertain. No antibodies to kinesin or to dynein have yet been found to label this region of a spindle. As illustrated in Figure 11.8a and b, the direction of force production corresponds to that of kinesin, rather than of flagellar or mammalian brain dynein *in vitro*, so some member of a kinesin superfamily[7] may be responsible. A mutant of the fungus *Aspergillus nidulans*, which has been found to be deficient in a kinesin-like protein (Table 8.1), is unable to push its spindle pole bodies apart. Dynamin, isolated from vertebrate cells (Section 8.6.3), is a possible anaphase B motor, though this protein is reported to produce hexagonally-packed microtubule bundles rather than the square packing observed in sectioned diatom spindles.

11.6.2 Anaphase A: polewards transport of chromosomes

Two main possible mechanisms for anaphase A movement have been suggested: microtubule shortening might in itself provide the polewards force on each chromatid, or the chromatids might be dragged towards the pole by ATP-dependent motors. The latter might be retrograde motors included in the kinetochore complex (possibly the same motors that operate during metaphase, put into reverse gear), or molecules that pull the kinetochore microtubules, which in turn pull the chromosomes. A further possibility is that there are elastic fibres of some sort, which produce tension between the chromosomes and the poles.

Inoué and his colleagues have argued for many years that motor molecules are unnecessary and that a microtubule-depolymerisation mechanism would be sufficient.[5,35] This idea was based on observations of living mitotic cells. Since microtubules produce birefringence, changes in the distribution and amount of polymer can be observed directly. After mild microtubule-destabilising treatments the distances between poles and chromosomes shrink in parallel with the loss of birefringence, due to depolymerisation of many of the microtubules.[30] On the other hand treatments that tend to stabilise microtubules, such as warming the cells or incubating them in heavy water (D_2O) or taxol, were found to enhance spindle birefringence and seemed to inhibit chromosome separation.

Those who prefer a mechanism involving a specific motor or elastic components have argued that microtubule depolymerisation may just be the *rate-limiting* process, and that changes assisting disassembly would allow the motor to work faster, while changes inhibiting disassembly would prevent motor-driven movements. Harsher microtubule-depolymerising conditions that cause most of the birefringence to disappear rapidly leave the chromosomes motionless near the metaphase plate; this observation weighs against the idea of a merely elastic component producing the movements.[36] It does not distinguish between the other possibilities. However, a major piece of evidence in favour of some sort of poleward-pulling motor is the observation by Rieder and colleagues of maximum speeds of around 1 μm/s for the movement of individual chromosomes towards their pole. Such rates seem to be significantly faster than those for rapid disassembly of subunits from microtubule ends (Table 10.1, page

196). Moreover, antibodies to cytoplasmic dynein have been found to bind to mitotic spindles and kinetochores.[8]

11.6.3 The loss of tubulin subunits at the kinetochore during anaphase[23]

The *in vivo* labelling experiments of Mitchison and colleagues, described in Section 11.5, show that subunits are lost during anaphase from the ends of microtubules embedded in the kinetochores (as shown schematically in Figures 11.7b & 11.9b). In the experiments, spindle microtubules were allowed to incorporate labelled tubulin during prometaphase and then investigated by immunoelectron microscopy at various intervals after the incorporation. Although labelled subunits accumulated on ends attached to the kinetochores during metaphase, the labelled lengths shortened again during anaphase. Subunit loss at the pole ends, thought to take place during metaphase, may also continue in anaphase but has not been measured.

11.6.4 ATP and chromosome sliding *in vitro*[24]

Observations of chromosome separation in lysed cells, and calcium-dependent shrinkage of isolated spindles in the absence of any added nucleotide triphosphates, have both tended to support Inoué's hypothesis. Experiments with isolated chromosomes and purified tubulin provide much stronger evidence that microtubule depolymerisation *can* produce a force. Koshland, Mitchison and Kirschner crosslinked short microtubule segments to form stable 'seeds' and tagged them with biotin so that they could subsequently be identified with a fluorescent anti-biotin antibody. Unlabelled tubulin was polymerised onto the plus ends to produce long microtubules. When purified chromosomes were added, many became attached to free plus ends and their behaviour was monitored under changing conditions. When the free tubulin level in the mixture was high and no ATP was present, there was no change in the distance between kinetochore and labelled seed. If, on the other hand, the free tubulin concentration was reduced by diluting the mixture with buffer, the kinetochore-to-seed distances shortened without any requirement for ATP. The crosslinked seeds remained in place since their minus ends were unable to lose subunits.

These results indicate that retrograde movement of a kinetochore does not necessarily require a retrograde motor bound to the complex. A loss of subunits, induced *in vitro* by lowering the free tubulin concentration, is apparently sufficient to cause sliding towards the minus end; neither GTP nor ATP is absolutely *required* to provide energy for sliding. The results also indicate that a kinetochore has some mechanism for holding onto the plus end of a microtubule even while that end is shrinking. A kinetochore with a sleeve-like structure could attach to the sides and slide along the microtubules while the exposed ends lose material (Figure 11.9b).

11.6.5 Thermal-diffusion model[24]

An interesting model postulated by Koshland *et al.* has each microtubule 'diffusing' under thermal forces, within a close-fitting channel in the kinetochore. If only a few subunits in the microtubule were bound to sites within the kinetochore by weak attractive forces, it might be able to 'jump' by a unit distance in either direction using only random thermal energy, and bring other subunits to the binding sites.

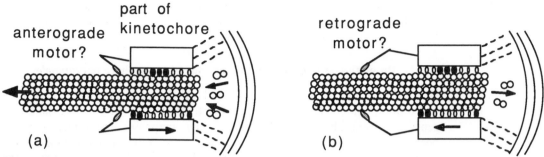

Figure 11.9

Possible events at the kinetochore shown schematically. (a) During prometaphase, an anterograde motor keeps the kinetochore at the growing end.[17-23] The initial capture of microtubules by kinetochores[9,20] is best explained if the motor molecules are on the outer surface. (b) During anaphase, the motor is switched off (or may perhaps work in reverse).[22] Weak bonds between sites within the kinetochore and the tubulin lattice hold the microtubule inside its sleeve. Even random shifting of the kinetochore would allow tubulin loss from the microtubule end.

The tubulin-binding sites in the kinetochore are shown here with a longitudinal periodicity, different from that of the microtubule lattice and producing something like a Vernier scale. The black particles are shown making optimal contact with the microtubule lattice, but a slight shift would put other particles in better contact instead. Thus no position is stable, in spite of specific binding between the particles and the microtubule. A more likely three-dimensional arrangement is one or more 13-fold *rings* of tubulin-binding sites inside the kinetochore, interacting with the shallow helices of the microtubule lattice—imagine the horizontal line in Figure 7.4a sliding up and down to contact monomers in different protofilaments.

Jumping would be more likely if there were competing binding sites within the kinetochore channel that could not be filled simultaneously, as in Figure 11.9. The microtubule would thus tend to remain attached always but its precise position would be unstable. Sliding beyond the plus end would be unlikely because of the attractive forces, but steps towards the minus end could occur freely. If exposed subunits readily disassembled from the plus ends, the kinetochore would gradually move towards the minus ends. The laws of thermodynamics do not allow the energy for the directed movement (see Section 11.12 for an estimate of the amount needed) to be obtained simply from heat but require some other source, as discussed in the next two sections.

11.7 Requirement for ATP during anaphase A *in vivo*[28,29]

Evidence produced by other workers studying lysed cells suggests that ATP *is* needed *in vivo*. Experiments carried out by the groups of both Pickett-Heaps and Cande show that chromosome separation can continue in mitotic cells lysed during anaphase in suitable buffers. If the buffer conditions are such as to stabilise the spindle microtubules, the movement requires ATP. If microtubule-disassembly is favoured, for example by chilling the preparations or adding calcium ions, separation continues in the absence of ATP. It has been suggested, therefore, that the supply of nucleotide might be needed, not for chromosome movement *per se*, but to promote tubulin disassembly, as would be required to satisfy the thermal-diffusion model (to bias net sliding in the direction of the pole).

However, the maximum speeds observed for chromosome movement make it likely that ATP *is* used *in vivo* for both purposes. The polewards force postulated to operate during prometaphase may continue during anaphase, as suggested in Figure 11.7b, which would mean that shortening of the kinetochore bundles under normal conditions was the combined result of depolymerisation at both the plus and minus ends. In addition, cytoplasmic dynein may pull each kinetochore along its microtubule tracks.

11.7.1   Promotion of disassembly at plus ends of kinetochore microtubules[2]

How disassembly is suddenly promoted at the plus ends of kinetochore microtubules *in vivo* is not yet clear. McIntosh has pointed out that there is a general disassembly of microtubules during anaphase; astral microtubules become shorter and fewer. There are many ways in which a cell might raise the critical concentration of tubulin required for assembly and reduce the total amount of polymer. It is possible that assembly-promoting MAPs are further modified at this point in the cell cycle (for example by increased levels of phosphorylation, for which a supply of ATP would be needed) to be less effective, which would lead to a lower level of assembly; other possible types of control include the release in the cytoplasm (perhaps very locally) of destabilising agents such as calcium ions, or the activation of tubulin- or MAP-sequestering proteins, if these exist.

The kinetochore itself may also possess a specific mechanism for biasing the effects of dynamic instability to favour growth during prometaphase and shrinkage during anaphase.[34] Its binding to plus ends could have a stabilising effect during prometaphase. A change at anaphase might switch on some sort of ATP-dependent depolymerising activity in each kinetochore. Subunits extending beyond a certain point in the kinetochore might be actively excised. Alternatively, the structure might act as a selective gate letting GDP–tubulin subunits escape but preventing the entry of fresh GTP-dimers and thus reducing the concentration of GTP–tubulin subunits in the volume around the microtubule end to below the critical concentration.

Any of these energy-consuming processes would be equivalent to increasing the critical concentration of free subunits required for assembly. In the *in vitro* experiment of Koshland *et al.*, energy is provided by the experimenter in lowering the tubulin concentration, producing a net loss of subunits once the protein concentration is below the critical value. Similarly, *in vivo*, conditions in the cytoplasm appear to be changed at the start of anaphase, to favour disassembly.

11.8   Summary of forces involved in chromosome movement

Spindle formation and chromosome separation appear to be the results of balancing the forces produced by several distinct mechanisms. As summarised in Figure 11.7, a kinetochore moves bidirectionally, relative to its attached bundle of microtubules; net movement is away from its ultimate destination during prometaphase and metaphase, and may be driven by an anterograde motor attached to the kinetochore. During anaphase the movement is towards the pole; sliding along the microtubule is probably not simply the result of tubulin loss from the microtubule ends embedded in the kinetochore but actively produced by a retrograde motor attached to the kinetochore.

Kinetochore microtubules themselves also appear to move persistently towards the pole with which their minus ends are associated during prometaphase and possibly during metaphase and anaphase. This movement is most likely to be due to microtubule motors acting between kinetochore microtubules and neighbouring structures. Dynein fixed to the kinetochore tubules may slide along neighbouring microtubules, or kinesin fixed to spindle-matrix material may slide along the kinetochore tubules.

Additional movement results from the entire half spindle being pushed away from the other pole throughout mitosis by the interdigitating polar microtubules (Figure 11.5b). The latter force appears to be responsible for initially setting up the bipolar form of the spindle, whilst the pairs of chromatids are still joined together. The interdigitating (polar) microtubules may become bowed through pushing against each other in metaphase if pole separation is prevented at this stage by the kinetochore microtubules; this would explain the rotund shape often developed by spindles in animal cells.

11.9   Telophase

At telophase nuclear membranes re-form around each set of separated chromosomes. Repolymerisation of the lamina proteins, which are dephosphorylated at this stage, is thought to be an important process in restoring the correct structure. The kinetochores in the daughter nuclei seem in some (perhaps all) nuclei to remain close together, attached to a small region of the nuclear membrane. This may be essential for organisation of the chromosomes during the next mitosis.

11.10   Cytokinesis[13-16]

Nuclear division is normally followed by division of the cytoplasm into two cells, each containing one of the daughter nuclei. In animal cells, cytokinesis is produced by the formation of a furrow in the cell surface, which gradually cleaves the cytoplasm into two compartments. The furrow is thought to be produced by localised contraction within the actin-rich cortical layer of cytoplasm (see Section 6.4). The cleavage furrow usually begins to form during anaphase. The cleft parallels the metaphase plate, even when the spindle is positioned asymmetrically in the cytoplasm. It appears first where the spindle equator is closest to the cell cortex, and gradually deepens and extends over a greater proportion of the cell's circumference. In many cell types it forms a closed ring, known as a **contractile ring**, which closes up like a purse-string. By the end of telophase the connection between the two halves of the cell has become a narrow tube of membrane. The interdigitating spindle microtubules run through this connection, forming a dense structure known as the **midbody**.

11.10.1   The contractile ring

The formation of a contractile ring in the cortical layer of cytoplasm requires the presence of viable myosin, as discussed in Chapter 6. By the time it is identifiable as a distinct structure it has the appearance, by electron microscopy, of a ring-shaped dense bundle of actin filaments. It can be labelled with anti-myosin antibodies, though thick filaments have not been identified by EM, nor is there an obvious sarcomeric organisation.

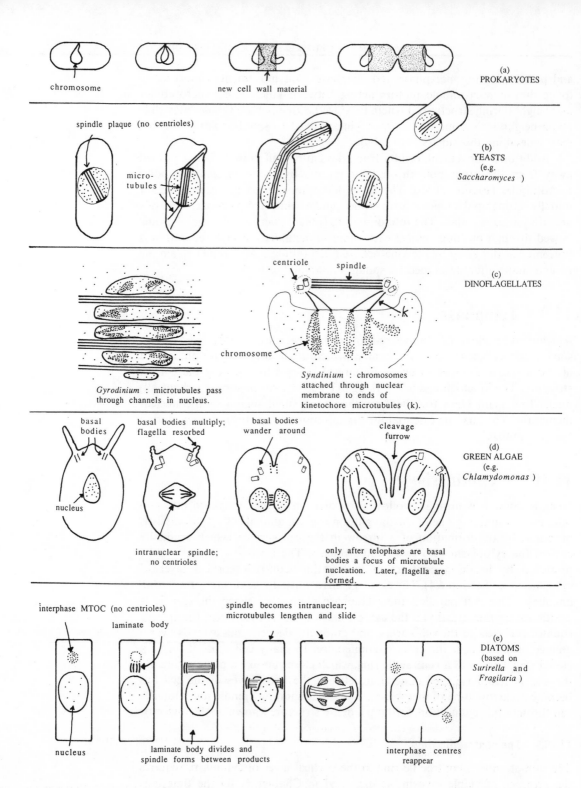

Figure 11.10

A summary of the various forms of mitosis, ranging from the simple method used by prokaryotes to the elaborate mechanisms of higher eukaryotes. The apparently non-essential involvement of centrioles in mitosis is also illustrated.

(a) A bacterial genome is connected to the cell wall. After duplication of the DNA, the copies are separated by insertion of new material into the cell wall and eventual division of the cytoplasm into two cells.

226

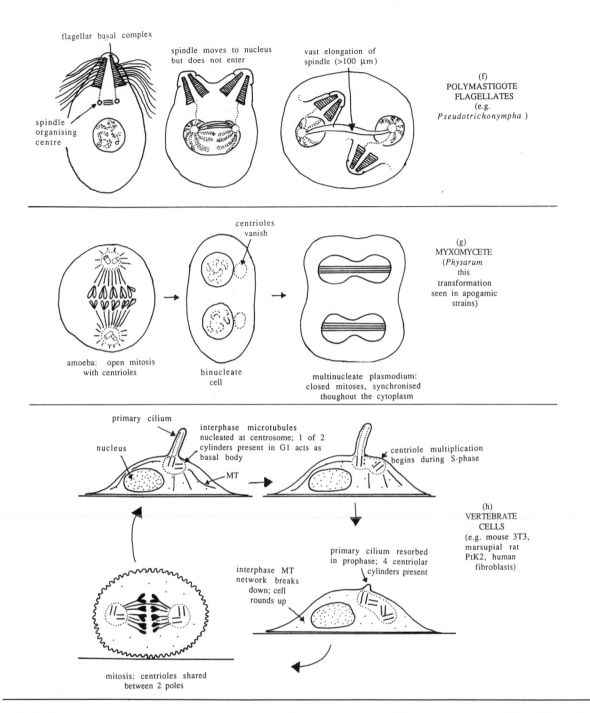

flagellar basal complex

spindle moves to nucleus but does not enter

vast elongation of spindle (>100 μm)

spindle organising centre

(f)
POLYMASTIGOTE FLAGELLATES
(e.g. *Pseudotrichonympha*)

centrioles vanish

(g)
MYXOMYCETE
(*Physarum*
this
transformation
seen in apogamic
strains)

amoeba: open mitosis with centrioles

binucleate cell

multinucleate plasmodium: closed mitoses, synchronised thoughout the cytoplasm

primary cilium

interphase microtubules nucleated at centrosome; 1 of 2 cylinders present in G1 acts as basal body

nucleus

MT

centriole multiplication begins during S-phase

(h)
VERTEBRATE CELLS
(e.g. mouse 3T3, marsupial rat PtK2, human fibroblasts)

primary cilium resorbed in prophase; 4 centriolar cylinders present

interphase MT network breaks down; cell rounds up

mitosis: centrioles shared between 2 poles

(b)–(f) Most lower eukaryotes separate their chromosomes within an intact nucleus, which is pinched into two before division of the cell as a whole (see Section 11.3). Microtubule arrays apparently replace the rigid cell walls of the prokaryotes as structural supports. In yeast cells (b), separation of the two sets of chromosomes by intranuclear microtubules is organised from spindle pole bodies, or plaques, embedded in the nuclear membrane. In dinoflagellates (c) and polymastigote flagellates (f), spindle microtubules lying *outside* the nuclear membrane interact with chromosomes via uncharacterised structures in the nuclear membrane to push the two sets apart. In some diatoms, a spindle formed originally in the cytoplasm is taken inside the nucleus to perform its function.

(g) The myxomycete *Physarum* exhibits both open and closed mitosis at different stages in its life cycle, with centrioles present in the first case but not in the second.

(h) So-called open mitosis, involving total breakdown and re-formation of the nuclear membrane, is discussed in detail in the text. The behaviour of centrioles in animal cells is shown here (see also Sections 9.6 & 11.2). In green algae such as *Chlamydomonas* (d), the centriolar cylinders follow a similar cycle without any clear involvement in nuclear division.

227

11.10.2   What determines the position of the contractile ring?

What induces the ring to form? The available evidence suggests it is the response to some sort of message from the equatorial region of the spindle. Physical obstructions placed between the metaphase plate and the cell cortex by micromanipulation can inhibit furrow formation. The message is conveyed by spindle microtubules, since even a single aster close to the cell surface can produce cortical contraction and cleavage furrows.

The form of the message is not yet known. Do kinesin-driven vesicles move along the astral microtubules towards the ring of membrane around the metaphase plate and fuse with the cell membrane to increase the surface area here? Does the inserted membrane include specific protein components that cause calcium ions to be released from sequestered stores or let in from the external medium and start the processes of filament bundling and contraction? There are, as yet, few answers in this area.

Plant cells do not divide by means of a contractile ring. Instead, large numbers of vesicles gradually accumulate at the equatorial plane of the central spindle after the chromosomes have moved towards the poles. The vesicles travel along a complex structure known as the **phragmoplast**, which includes microtubules, and finally coalesce to form a partition between the two daughter cells. Kinesin-driven transport along the microtubules also seems a likely candidate in this process. Apart from the contribution of the contractile cortex, the processes of cell division in animal and plant cells may be more related than is apparent at first glance.

11.11   Summary

In most eukaryotic cells, duplicated chromosomes are separated by 'spindle' structures assembled from microtubules. Division of the cell's microtubule organising centre into two is a crucial preliminary step. Microtubules then assemble from these two foci, or poles. Interactions between the microtubules and specialised features on the chromosomes known as kinetochores, and also between the two sets of microtubules, creates an ordered arrangement. Further types of interaction between these components cause the two sets of chromosomes to move apart. The spindle arrangement also directs division of the cell into two daughter cells, probably by transporting active components along the microtubules. In animal cells, division is achieved by constricting the cell with a contractile ring of actin filaments; in plant cells, a new barrier is built to separate the mother cell into two compartments.

11.12   Questions

11.1 What is the evidence *against* the involvement of actin filaments and *for* the involvement of microtubules in mitosis? What about cytokinesis?

11.2 To what extent can change and movements during cell division be explained by the known dynamic properties of microtubules?

11.3 Where and when might dynein and kinesin-like molecules be involved in the mechanisms of spindle formation and chromosome separation?

11.4 If the energy required to move a chromosome *in vivo* is that given below*, how many molecules of nucleotide need to be split to transport a single chromosome from metaphase plate to pole, a distance of perhaps $5\,\mu m$? (Assume that the free energy released by splitting ATP or GTP is $30-60\,\text{kJ mol}^{-1}$; Avogadro's number is 6×10^{23}.) If the process takes 10 min, compare the rate of nucleotide hydrolysis with that in an axoneme (about 4 per second per dynein arm).

---

* Nicklas (1983, *J. Cell Biol.* **97**, 542–548) has calculated that a force of 7×10^{-10} newtons per kinetochore, or 2×10^{-11} newtons per microtubule end is required *in vivo* to account for chromosome movement towards the pole. In the *in vitro* experiments of Koshland *et al.* (Section 11.6), lower viscous resistance would mean that even smaller forces were needed. Thus, assuming that microtubules shrink by 0.6 nm for each dimer lost (8 nm/13), a force of 2×10^{-11} newtons per microtubule would require the release of only 7.2 kJ (force × distance × Avogadro's number) of free energy per mole of tubulin dimers.

11.13   Further reading

Reviews

1 Hyams, J. S. & Brinkley, B. R., eds (1989). *Mitosis: molecules and mechanisms.* San Diego: Academic Press.
   (Comprehensive and up-to-date.)
2 McIntosh, J. R. (1989). Assembly and disassembly of mitotic spindle microtubules. In *Cell Movement*, vol. 2, ed. Warner, F. D. & McIntosh, J. R. New York: Alan R. Liss.
   (Excellent, balanced and up-to-date discussion of motile mechanisms.)
3 Wolfe, S. L. (1981). *Biology of the Cell*, 2nd edn., pp. 404–425. Belmont, CA: Wadsworth.
   (A good short account of mitosis in general though not up to date on tubulin dynamics.)
4 Mazia, D. (1961). Mitosis and the physiology of cell division. In *The Cell*, vol. 3, ed. Brachet, J. & Mirsky, G, Chapter 2. New York: Academic Press.
   (A classic summary of early work.)
5 Inoué, S. (1981). Cell division and the mitotic spindle. *J. Cell Biol.* **91**, 131s–147s.
6 Mitchison, T. J. (1988). Microtubule dynamics and kinetochore function in mitosis. *Annu. Rev. Cell Biol.* **4**, 527–550.
7 Sawin, K. & Mitchison, T. (1990). Motoring in the spindle. *Nature* **345**, 22–23.
8 Vallee, R. (1990). Dynein and the kinetochore. *Nature* **345**, 206–207.
9 McIntosh, J. R. & Koonce, M. P. (1989). Mitosis. *Science* **246**, 622–628.
10 Mitchison, T. & Hyman, A. (1988). Kinetochores on the move. *Nature* **336**, 200–201.
11 Rieder, C. L. (1982). The formation, structure and composition of the mammalian kinetochore and kinetochore fiber. *Int. Rev. Cytol.* **79**, 1–58.
12 Cande, W. Z. & Hogan, C. J. (1989). The mechanism of anaphase spindle elongation. *Bioessays* **11**, 5–8.
   (A useful review of anaphase B.)

13 Kellogg, D. R. (1989). Centrosomes: organizing cytoplasmic events. *Nature* **340**, 99–100.

14 Salmon, E. D. (1989). Cytokinesis in animal cells. *Curr. Opinion Cell Biol.* **1**, 541–547.
     (A lucid short review, including recent work.)

15 Rappaport, R. (1986). Establishment of the mechanism of cytokinesis in animal cells. *Int. Rev. Cytol.* **105**, 245–281.

16 Mabuchi, I. (1986). Biochemical aspects of cytokinesis. *Int. Rev. Cytol.* **101**, 175–213.

17 De Brabander, M., Geuens, G., Nuydens, R., Willebords, R., Aerts, F. & De Mey, J. (1986). Microtubule dynamics during the cell cycle: the effects of taxol and nocodazole on the microtubule system of Ptk2 cells at different stages of the mitotic cycle. *Int. Rev. Cytol.* **101**, 215–274.

18 Kubai, D. (1975). Unusual forms of mitosis. *Int. Rev. Cytol.* **43**, 167–227.

Original papers providing more detail

19 Gorbsky, G. L., Sammak, P. J. & Borisy, G. G. (1987). Chromosomes move poleward in anaphase along stationary microtubules that coordinately disassemble from their kinetochore ends. *J. Cell Biol.* **104**, 9–18.

20 Gorbsky, G. L., Sammak, P. J. & Borisy, G. G. (1988). Microtubule dynamics and chromosome motion visualized in living anaphase cells. *J. Cell Biol.* **106**, 1185–1192.

21 Gorbsky, G. J. & Borisy, G. G. (1989). Microtubules of the kinetochore fiber turn over in metaphase but not in anaphase. *J. Cell Biol.* **109**, 653–662.

22 Mitchison, T. J. and Kirschner, M. W. (1985). Properties of the kinetochore *in vitro.* II: Microtubule capture and ATP-dependent translocation. *J. Cell Biol.* **101**, 766–777.

23 Mitchison, T. L., Evans, L., Schultze, E. & Kirschner, M. (1986). Sites of microtubule assembly and disassembly in the mitotic spindle. *Cell* **45**, 515–527.

24 Koshland, D. E., Mitchison, T. J. & Kirschner, M. W. (1988). Polewards chromosome movement driven by microtubule depolymerization *in vitro.* *Nature* **331**, 499–504.

25 Mitchison, T. J. (1989). Polewards microtubule flux in the mitotic spindle: evidence from photoactivation of fluorescence. *J. Cell Biol.* **109**, 637–652.

26 Brinkley, B. R., Zinkowski, R. P., Mollon, W. L., Davis, F. M., Pisegna, M. A., Pershouse, M. & Rao, P. N. (1989). Movement and segregation of kinetochores experimentally detached from mammalian chromosomes. *Nature* **336**, 251–254.

27 Nicklas, R. B. (1983). Measurements of the force produced by the mitotic spindle in anaphase. *J. Cell Biol.* **97**, 542–548.

28 Masuda, H., McDonald, K. L. & Cande, W. Z. (1988). The mechanism of anaphase spindle elongation: uncoupling of tubulin incorporation and microtubule sliding during *in vitro* spindle reactivation. *J. Cell Biol.* **107**, 623–633.

29 Spurck, T. P. & Pickett-Heaps, J. D. (1987). On the mechanism of anaphase A: evidence that ATP is needed for microtubule disassembly and not generation of polewards force. *J. Cell Biol.* **105**, 1691–1705.

30 Salmon, E. D. & Segall, R. R. (1980). Calcium-labile mitotic spindles isolated from sea urchin eggs *(Lytechinus variegatus)*. *J. Cell Biol.* **86**, 355–365.

31 Salmon, E. D., McKeel, M. & Hays, T. (1984). Rapid rate of tubulin dissociation from microtubules in the mitotic spindle *in vivo* measured by blocking polymerization with colchicine. *J. Cell Biol.* **99**, 1066–1075.

32 Peterson, S. P. & Berns, M. W. (1978). Evidence for centriolar region RNA functioning in spindle formation in dividing PTK_2 cells. *J. Cell Sci.* **34**, 289–301.

33 Salmon, E. D. (1989). Metaphase chromosome congression and anaphase poleward movement. In *Cell Movement*, vol. 2, ed. Warner, F. D. & McIntosh, J. R. pp. 431–440. New York: Alan R. Liss.

34 Mitchison, T. J. (1989). Chromosome alignment at mitotic metaphase: balanced forces or smart kinetochores? In *Cell Movement*, vol. 2, ed. Warner, F. D. & McIntosh, J. R. pp. 421–430. New York: Alan R. Liss.

35 Inoué, S. (1976). Chromosome movement by reversible assembly of microtubules. In *Cell Motility*, ed. Goldman, C. R., Pollard, T. & Rosenbaum, J., pp. 1329–1342. New York: Cold Spring Harbor Laboratory.

36 Forer, A. (1965). Local reduction of spindle fiber birefringence in living *Nephrotoma surturalis* (Leow) spermatocytes induced by ultraviolet microbeam irradiation. *J. Cell Biol.* **25**, 95–117.

37 Amos, W. B. (1988). Results obtained with a sensitive confocal scanning system designed for epifluorescence. *Cell Motil. & Cytoskel.* **10**, 54–61.

Late additions

38 Verde, F., Labbe, J.-C., Doree, M. & Karsenti, E. (1990). Regulation of microtubule dynamics by cdc2 protein kinase in cell-free extracts of *Xenopus* eggs. *Nature* **343**, 233–238.

39 Rieder, C. L. & Alexander, S. P. (1990). Kinetochores are transported poleward along a single astral microtubule during chromosome attachment to the spindle in newt lung cells. *J. Cell Biol.* **110**, 81–96.

APPENDIX: SOME GENERAL METHODS OF STUDYING CYTOSKELETAL COMPONENTS

A.1 Modern modes of light microscopy[1,2,16]

The light microscope has been used for a couple of centuries in studying cells but it is only in the past two decades that techniques have been developed that allow individual components of the cytoskeleton to be studied. The improvements are not in **resolution**, which is limited by the wavelengths of visible light (400–700 nm), but in methods of detection.

A.1.1 Immunofluorescence[3,7]

The use of **specific antibodies** with **fluorescent labels** bound to them has revolutionised the study of the cytoskeleton. To produce images such as those in Figure 1.2, cells are first fixed in such a way as to leave their membranes permeable (plunging them into methanol at $-20\,°C$ is often quite adequate), then incubated with suitable antibodies. The fluorescent label may be attached directly to these antibodies; usually it is more convenient to label a *second* antibody, specific to IgG molecules of the first type. Thus, if the anti-cytoskeletal protein antibody was raised in a rabbit, for example, the secondary label is attached to anti-rabbit IgG obtained in another species.

The label that makes the chosen structures stand out from their background does not have to be fluorescent; another fairly popular ploy is to attach the second antibodies to small (5–50 nm diameter) **gold particles**, which show up *en masse* by light microscopy. Gold has the great advantage that it can be seen in the electron microscope also (Section A.2.4). Another method, capable of less precise resolution, consists of labelling the second antibody with an enzyme, usually **peroxidase**. The location of the enzyme is then revealed by supplying it with a substrate which it converts into a coloured end-product.

If one wishes to label more than one cytoskeletal component in the same cell (as in Figure 1.2), the primary antibodies must be raised in species that do not cross-react. One can soon exhaust available ranges of experimental animals. Fortunately, there are a number of **chemical reagents** that also label certain structures specifically. A notable example is the F-actin-specific drug **phalloidin** (Section 3.3), which can be labelled fluorescently. There is a growing number of reagents that bind to membranes, including some which are fairly specific for endoplasmic reticulum (e.g. **DiOC$_6$**) and Golgi elements (e.g. **NBD-C$_6$-ceramide**). And both DNA and RNA can easily be labelled; **Hoechst** or **DAPI** staining are most commonly used for chromatin.

A.1.2 Microinjection of fluorescent proteins[9-15]

For studying changes in the distribution of a particular protein with time, one needs to label the cytoplasm without harming the cell or impairing the function of the protein molecules being labelled. This seems to be possible in some cases; some modified cytoskeletal proteins have been found to polymerise normally and exhibit normal enzymatic activity. Injection of minute quantities of protein into the interiors of individual cells without damaging them is nowadays a fairly routine procedure.

Observation of **fluorescence recovery after photobleaching (FRAP)** is a way of following the movement of fluorescent protein after injection into cells. It has been used in particular to study the dynamic instability of microtubules in mitotic spindles and in interphase cells. Bleached areas become invaded by molecules which are still fluorescent as the microtubules shrink and regrow (see Chapters 7, 10 & 11). More recently, Mitchison[15] has introduced the use of a **photoactivated fluorescent label** for microinjected tubulin; initially non-fluorescent modified protein is injected into a cell and molecules in a selected region are photoactivated with ultraviolet (365 nm) light. Fluorescence is observed at a longer wavelength. The advantage of this technique is that molecules in the selected area show up bright against a dark background, rather than vice versa, so their movements can be followed with much more precision.

A.1.3 Resolving filaments and tubules by light microscopy[16]

The diameter of an individual actin filament or a microtubule is well below the resolution of the light microscope; their 8–25 nm diameters are at least an order of magnitude less than visible light wavelengths (300–600 nm). Well-separated filaments can, however, be visualised by a number of light microscopy methods, as objects with greatly expanded apparent diameters. The contrast between them and the background is correspondingly reduced (Figure A.1). If the background is nearly black, then the objects can be seen without much difficulty. This can be achieved using an **epifluorescence microscope** and fluorescently-labelled specimens. Light of a suitable wavelength to excite the fluorophor is shone down onto the specimens and only light that returns upwards at the

Figure A.1

An object whose diameter (d) is less than the wavelength of visible light may still scatter light passing through a microscope but will appear spread out in the resulting image. Recording the image electronically allows one to subtract a general background level (b) and magnify the remaining intensity variations, to make the object apparent to the eye.

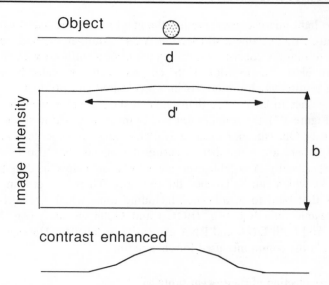

emission wavelength goes into the eyepiece. Unlabelled filaments can be seen by **dark-field light microscopy**; in this case, a dark central stop prevents any direct light from reaching the eyepiece and only rays scattered by the filaments contribute to the image. Individual actin filaments have been visualised by both these methods.

A.1.4   Video-enhanced microscopy[4-6]

A light-microscope image of any type can be fed into a video camera and processed electronically to increase greatly the range of contrast. This means that minute differences that cannot be seen by the naked eye against a *light* background become detectable. The average background level can be subtracted electronically to give a result equivalent to a dark-field image (see Section A.1.3). The remaining variation can be enhanced. These improvements can be applied even to optically-produced dark-field images, to great advantage.

Video-enhancement is most often applied to images obtained by **differential interference contrast (DIC)**, such as are obtained with the Nomarski optics available for most modern microscopes (see Figures 8.15 & 10.9). These optical systems are useful because only a narrow depth range within the specimen is in focus at any setting; everything else merely contributes to an out-of-focus blur, which can mostly be removed by the background subtraction process. For thicker specimens, therefore, it is preferable to a dark-field image, which shows filaments at all levels superimposed.

The extent of contrast enhancement is limited by optical imperfections in the light source, in the lenses, and so on. The imperfections, when enhanced, produce a 'mottle' pattern that may obliterate the image of the specimens. The mottle is fairly constant for a particular optical set-up, so it is possible to subtract it to leave a cleaned-up image of the specimens. This is usually done by means of an on-line computer; a mottle image obtained with the specimen out of focus is stored in the memory and subtracted from each video frame as it is collected. This procedure can produce very nice images of microtubules but actin filaments have not yet been imaged in this way, perhaps because mottle-subtraction cannot be carried out with sufficient accuracy.

A.1.5 Confocal laser-scanning microscopy[8]

The types of **confocal microscope** used in biology are essentially extended versions of the epifluorescence microscope: instead of continuously bathing the whole specimen in light, however, an image is built up by focusing a movable spot of **laser** light (of a suitable colour for excitation) at a series of points on a two-dimensional **raster**. Fluorescently-emitted light that travels back along the same path is collected in a **photomultiplier** and the intensity value for each point stored in a computer. The spot of light is focused at a chosen level in the specimen and an aperture in front of the detector restricts the collected light to that from the same small volume of the specimen eliminating background glare and providing an 'optical section'. A series of images at different focal levels can provide a three-dimensional raster image of the specimen. The photomultiplier setting can be adjusted to give optimal background subtraction (as defined in Section A.1.4) and contrast enhanced electronically.

Furthermore, it is possible to collect emitted light at more than one wavelength simultaneously, filtered into different photomultipliers. This means that two separate images from 'double-labelled' specimens can be superimposed exactly (a problem in the case of conventional fluorescence microscopy, since different sets of filters produce differential shifts in the focused image). Fluorescence images can also be compared exactly with transmitted light images if another light collector is positioned on the other side of the specimen (i.e. below the condenser lens).

A.2 Electron microscopy[17,21-2]

Resolution to atomic level (tenths of a nanometre, or **Ångstroms**) is technically possible in an **electron microscope (EM)**. The problem with biological material is that it is destroyed by electrons and the number of electrons required to give a clear image of a protein molecule is far greater than the number needed for destruction. Proteins are, therefore, usually coated with some form of heavy metal, and it is the *metal* that you study in the EM. Resolutions are then limited by the fidelity with which the metal follows the outline of the protein.

A.2.1 Negative staining[17]

For isolated particles, one of the simplest procedures is to immerse them in a solution of a **uranium salt**, which then dries like toffee in the form of a thin layer, supported on the surface of a 10–20 nm film of **carbon**. Since the protein particles appear as white holes in the electron-dense stain layer, the technique is known as **negative staining**. Examples are shown throughout the book, including Figures 1.4 and 1.5a. Features as small as 1.5 nm have been resolved from such images, though 3–5 nm is the more usual sort of resolution available. Because an EM necessarily has a very large depth of focus, it is not possible to focus on different depths within the specimen; images are therefore essentially projected views of the whole three-dimensional density distribution. This sometimes makes visual interpretation quite difficult, since we are not accustomed to looking *through* objects in everyday life.

A.2.2 Low-angle shadowing[22]

Evaporating metal (usually **platinum**) from a low angle (5–20° from the horizontal) leads to more easily interpretable images of surface features. If the metal arrives from a single direction it leaves a metal-free shadow behind each particle, producing an impression of objects lit from one side. Figure 1.5b shows an example. Note that in normal positive images (prints from the negatives) it is the *shadow* that appears bright. For this reason, researchers often study photographic *negatives* whilst interpreting the features. The specimen may instead be rotated while the metal is evaporating, a process known as **rotary shadowing**. This gives a better all-round view of particulate specimens and can be used where there is no ambiguity over convex and concave surfaces. (These are a problem when studying complex topologies such as fractures through cells and membranes.) 4–5 nm resolution is considered 'good' for contrast with shadowing.

Unlike specimens treated with negative stain, proteins must be dry *before* shadowing with metal. Whilst a solution of protein molecules is drying down onto a surface, the three-dimensional configurations of the molecules are in danger of being totally destroyed by surface tension and obliterated by salts crystallising out of the medium. Diluting proteins into a mixture of water and **glycerol** has been found to improve the results; removing the water (by evaporation under vacuum) in the presence of glycerol seems to leave the molecules in fairly good shape. A superior though less easy method, perfected by J. Heuser and colleagues, involves **freezing** the specimens rapidly (within microseconds, to avoid the formation of structure-damaging ice crystals) in a low-salt medium and then subliming the amorphous ice away under vacuum.

A.2.3 Visualisation of structures within cells

Cells can be split open, for example by **freeze-fracturing**, and **shadowed replicas** made of the exposed surfaces whilst they are still frozen; some of the water is usually sublimed away before shadowing. A layer of carbon is evaporated onto the replica to support it, and the tissue is completely dissolved away. This technique has produced some very striking electron micrographs of cell interiors. In the case of very thin layers of cytoplasm, another alternative is to dissolve away just the cell membrane with **detergent** (in the presence of **fixatives** to preserve the structure of the underlying cytoskeleton) and image the specimen itself, contrasted by shadowing or negative staining in a transmission EM. Examples are shown in Figures 1.4, 1.7 and 1.8, as well as in later chapters.

However, most of our detailed knowledge about the way cytoskeletal components are arranged in cells comes from **thin sectioning** of fixed material embedded in **plastic resins**. In this case, the contrast is usually produced by **positive staining**. Metal ions that bind to certain groups on proteins whilst fixed cells are left soaking in heavy-metal salt solutions remain bound even after washing. Thus, areas where protein is more dense appear darker. For some cytoskeletal components a greater contrast (and consequently a higher effective resolution) can be achieved using **tannic acid**. When fixed material is soaked in a solution of tannic acid, it forms a layer around the exposed surfaces of proteins such as tubulin and actin. If the cells are subsequently stained, **uranyl ions** bind avidly to the tannic acid, producing a negative-staining effect around the protein subunits. 5 nm diameter microtubule protofilaments show up very clearly using this technique (see Figures 8.1b & 9.1 for examples). Under most conditions, the resolution obtainable from thin sections is rather worse than this.

High-voltage electrons are able to penetrate much thicker specimens than the usual 60–100 kV beams: because they interact less with the material the contrast is correspondingly reduced, so lower voltages are better for thin specimens. But thick specimens provide a better understanding of the 3-D relationships between different structures. **Stereo pairs** of electron micrographs are usually needed to appreciate the arrangement of overlapping filaments in a thick section through the cytoskeleton.

A.2.4 Immunoelectron microscopy

Gold particles coated with specific antibodies can also be seen by EM and thus provide a very useful method of identifying particular components. Surprisingly, many proteins retain their antigenicity, even after fixation with **glutaraldehyde**, as is required for the preservation of structure at the EM level (Figures 1.7 & 1.8).

A.2.5 Scanning transmission electron microscopy (STEM)

If the specimen is scanned on a raster with a narrow beam of electrons focused to a spot, an image equivalent to the normal transmission image can be built up (using diffracted electrons that retain their initial wavelength/energy). It is also possible to collect, for each point on the raster, electrons that have lost energy through interacting with the matter the beam has had to penetrate, which are mostly excluded from the normal transmission image. This allows an estimate to be made of the *amount of matter* at each point in the image and is very useful for determining the probable masses of large individual molecules, such as dynein.[20]

A.3 Molecular substructure

None of the biophysical methods of studying protein molecules is peculiar to research into cytoskeletal proteins. Accounts of the use of **ultracentrifugation** and **chromatography** for investigating the sizes and shapes of molecules can be found in standard textbooks of biochemistry. Substructure at an atomic level can be studied by means of specialised techniques such as **X-ray diffraction** and **nuclear magnetic resonance (NMR)**. Currently, a popular approach requiring much less specialist knowledge is the **prediction of secondary structure** from a protein's *sequence* (usually deduced from a nucleic-acid sequence). This has been particularly fruitful in the case of cytoskeletal proteins with extensive α-helical domains (see Chapter 2, for example).

The relative proportions of α-helix and other types of secondary structure in a solution of protein can be estimated by measurements of its **circular dichroism (CD)**, a form of absorption spectroscopy.[18] CD is defined as the difference in the levels of left- and right-circularly polarised light absorbed by the sample. The arrangement of chemical bonds in α-helical, extended (β-sheet), U-turn and random-coil configurations varies sufficiently to produce significantly distinctive spectra, with peaks and troughs at different wavelengths.

A.4   The use of antibodies[23-6]

As should already be clear, antibodies are an essential tool in the study of cytoskeletal proteins. In many cases there is no way of identifying a band on an SDS gel other than by testing its affinity for specific antibodies. Similarly, the distribution of structural proteins within cells may be difficult to determine using other reagents.

It is worth keeping in mind that antibodies can, however, give misleading results. Through sheer bad luck, the experimental animals may possess an irrelevant antibody which happens to react with protein in the specimen. This can be checked by testing **pre-immune serum** (collected before the animal had any protein injected into it). With monoclonal antibodies, there is danger of false positives, simply because the tiny region, or **epitope**, recognised by the antibody occurs on other, perhaps totally unrelated, proteins. A **polyclonal serum**, purified by methods which exploit the affinity of the relevant antibody for the antigen, is often the best reagent to use.

Besides their use in detecting the presence and distributions of proteins, antibodies can be injected into living cells, where they may interfere with the function of the antigen. This is a useful way of obtaining information about the functions of a cytoskeletal protein *in vivo*. A classic example is the inhibition of cytokinesis that was obtained by injecting antibodies to myosin into developing starfish embryos.

Finally, specific antibodies play a crucial role in the molecular biology of cytoskeletal proteins. They are used to identify DNA sequences from gene libraries by detecting their protein products. This key step allows the genes to be cloned, sequenced, expressed in *in vitro* systems, and even reintroduced in modified form back into cells (see Section A.6).

A.5   Optical diffraction and image analysis[27-8]

Although the structures in EM images are often hard to see by eye (e.g. because of radiation damage or overlying material), they are very frequently arranged in a regular way. This means that the EM negative will behave as a diffraction grating when a beam of light is shone through it: the resulting **optical diffraction pattern** reveals the underlying regularity. By the use of screens with regularly positioned holes to select light beams diffracted by the repeating components of the image, an **optically-filtered image**, free of irregular noise, can be created.

The entire process can be simulated in a computer, after digitising the image using a **densitometer**, and provides a powerful method for squeezing every last drop of information from an EM picture.

A.6   Genetic approaches to function[29-33]

Apart from treating cells with specific drugs or injecting antibodies that will bind to the target protein, the current ways of judging the functional importance of a particular protein involve interfering with its expression in a cell. Until quite recently it has been necessary to search for naturally-occurring or randomly-induced **mutants** that either fail to make the protein or produce an altered version of it. Much information has been gathered in this way from studies of simple organisms, such as yeasts (especially *Saccharomyces*), fungi (e.g. *Aspergillus*), amoebae (especially the cellular slime mould, *Dictyostelium*), and somewhat more complex creatures (including fruit flies, *Drosophila*, and nematode worms, notably *Caenorhabditis elegans*).

More recently, methods have been developed for specifically deleting targetted genes in simple organisms or replacing them with altered sequences; also for disrupting expression of a chosen protein in organisms where removing the original gene is more difficult. In the latter case, a DNA sequence is introduced into the cell which causes the production of mRNA that is complementary to the normal mRNA (**antisense RNA**). Pairing between complementary strands inhibits their translation into protein. The technique is known as **antisense pseudogenetics** and has been applied with particular success to the disruption of myosin expression in *Dictyostelium* (see Section 6.4.1).

A related technique is the expression of cytoskeletal proteins in bacteria (usually *E. coli*) by introducing cloned genes incorporated in plasmids. This provides the possibility of investigating the *in vitro* properties of proteins in great detail. It may be possible to generate manageable quantities of proteins that are only very minor components of normal eukaryotic cells, for example. In other cases, alterations can be made to normal sequences to investigate the properties of individual domains in the protein.

A.7   Further reading

1 Bradbury, S. (1984). *An Introduction to the Optical Microscope*. Oxford: Oxford University Press.

2 Spencer, M. (1982). *Fundamentals of Light Microscopy*. Cambridge: Cambridge University Press.

3 Ploem, J. S. & Tanke, H. J. (1987). *Introduction to Fluorescence Microscopy* (Royal Microscopical Society Handbook No. 10). Oxford: Oxford Scientific Publications.

4 Salmon, E. D., Walker, R. A. & Pryer, N. K. (1989). Video-enhanced differential interference contrast light microscopy. *Biotechniques* 7, 624–633.

5 Inoué, S. (1988). *Video Microscopy*. New York: Plenum Press.

6 Allen, R. D. (1985). New observations on cell architecture and dynamics by video-enhanced contrast optical microscopy. *Annu. Rev. Biophys. Chem.* **14**, 265–290.

7 Taylor, D. L. & Salmon, E. D. (1988). Basic fluorescence microscopy. In *Fluorescence Microscopy of Living Cells in Culture. Part A: Fluorescent analogs, labeling cells, and basic microscopy*. Methods in Cell Biology, vol. 29, pp. 208–238. San Diego & London: Academic Press.

8 White, J. G., Amos, W. B. & Fordham, M. (1987). An evaluation of confocal versus conventional imaging of biological structures by fluorescent light microscopy. *J. Cell Biol.* **105**, 41–48.

9 Wang, K., Feramisco, J. & Ash, J. (1982). Fluorescent localization of contractile proteins in tissue culture cells. *Methods Enzymol.* **85**, 514–562,

10 Wang, Y.-L., Heiple, J. M. & Taylor, D. L. (1982). Fluorescent analogue cytochemistry of contractile proteins. *Methods Cell Biol.* **25**, 1–11.

11 Amato, P. A. & Taylor, D. L. (1986). Probing the mechanism of incorporation of fluorescently labeled actin into stress fibers. *J. Cell Biol.* **102**, 1074–1084.

12 Salmon, E. D., Leslie, R. J., Saxton, W. M., Karow, M. L. & McIntosh, J. R. (1984). Spindle microtubule dynamics in sea urchin embryos: analysis using a fluorescein labeled tubulin and fluorescence redistribution after laser photobleaching. *J. Cell Biol.* **99**, 2165–2174.

13 Sammak, P. J. & Borisy, G. G. (1988). Detection of single fluorescent microtubules and methods for determining their dynamics in living cells. *Cell Motil. & Cytoskel.* **10**, 1–9.

14 Vigers, G. P., Coue, M. & McIntosh, J. R. (1988). Fluorescent microtubules break up under illumination. *J. Cell Biol.* **102**, 1032–1038.

15 Mitchison, T. J. (1989). Polewards microtubule flux in the mitotic spindle: evidence from photoactivation of fluorescence. *J. Cell Biol.* **109**, 637–652.

16 Rebhun, L. I., Taylor, D. L. & Condeelis, J. S., eds (1988). Optical approaches to the dynamics of cellular motility. *Cell Motil. & Cytoskel.* **10**, 1–348.

17 Sommerville, J. & Scheer, U., eds (1987). *Electron Microscopy in Molecular Biology.* Oxford & Washington DC: IRL Press.

18 Johnson, W. C. (1988). Secondary structure of proteins through circular dichroism spectroscopy. *Annu. Rev. Biophys. Biophys. Chem.* **17**, 145–166.

19 Small, J. V., Rinnerthaler, G. & Hinssen, H. (1982). Organization of actin meshworks in cultured cells: the leading edge. *Cold Spring Harbor Symp. Quant. Biol.* **46**, 599–610.
     (Electron microscopy of cells grown on EM grids.)

20 Johnson, K. E. & Wall, J. S. (1983). Structure and molecular weight of the dynein ATPase. *J. Cell Biol.* **96**, 669–678.
     (Example of STEM of unstained biological particles.)

21 Chiu, W. (1986). Electron microscopy of frozen hydrated specimens. *Annu. Rev. Biophys. Biophys. Chem.* **15**, 237–257.

22 Heuser, J. (1981). Quick-freeze, deep-etch preparation of samples for 3-D electron microscopy. *Trends Biochem. Sci.* **6**, 64–68.

23 Milstein, C. (1980). Monoclonal antibodies. *Sci. Am.* **243**(4), 66–74.

24 Yelton, D. E. & Scharff, M. D. (1981). Monoclonal antibodies: a powerful new tool in biology and medicine. *Annu. Rev. Biochem.* **50**, 657–680.

25 Hudson, L. & Hay, F. C. (1980). *Practical Immunology*, 2nd edn. Oxford: Blackwell Scientific Publications.

26 Roitt, I. M. (1977). *Essential Immunology*, 3rd edn. Oxford: Blackwell Scientific Publications.

27 Stewart, M. S. (1986). Computer analysis of ordered microbiological objects. Chapter 12 in *Ultrastructure Techniques for Microorganisms*, ed. Aldrich, H. C. & Todd, W. J. New York & London: Plenum.

28 Crowther, R. A. & Klug, A. (1975). Structural analysis of macromolecular assemblies by image reconstruction from electron micrographs. *Annu. Rev. Biochem.* **44**, 161–182.

29 Weintraub, H., Izant, J. G. & Harland, R. M. (1985). Anti-sense RNA as a molecular tool for genetic analysis. *Trends Genet.* **1**, 22–25.

30 Weintraub, H. M. (1990). Antisense RNA and DNA. *Sci. Am.* **262**(1), 34–41.

31 Spudich, J. A., ed. (1989). Molecular genetic approaches to protein structure and function: applications to cell and developmental biology. *Cell Motil. & Cytoskel.* **14**(1), 1–175.
     (Special issue devoted to a range of reviews focused on cell motility and cytoskeleton.)

32 Segall, J. E. & Gerisch, G. (1989). Genetic approaches to cytoskeleton function and the control of cell motility. *Curr. Opinion Cell Biol.* **1**, 44–50.

33 Bray, D. & Vasiliev, J. (1989). Cell motility: networks from mutants. *Nature* **338**, 203–204.

BRIEF ANSWERS TO QUESTIONS

1.1 Stiff gel; centrifugation, introduction of magnetic particles, microinjection of diffusible particles, FRAP.

1.2 Light microscopy: phase, DIC, specific fluorescent labelling (e.g. immuno-fluorescence). Electron microscopy: fixed, sectioned cells; detergent-extracted cell mounts; freeze-fractured, deep-etched material.

1.3 (a) IFs and (to a lesser extent) F-actin: tensile strength. (b) Microtubules and actin bundles: protrusive effect. (c) Microtubules and actin filaments can be highly dynamic, pioneers in cell shape changes, thus organising cell polarity. (d) Actin and tubulin both interact with motor proteins that produce intracellular transport.

1.4 More variable minor components are mainly responsible.

1.5 (a) Relative sliding of filaments/tubules, to give muscle contraction and eukaryotic flagellar propulsion. (b) Coordinated conformational changes by many molecules and/or coiling of filaments change the overall shape of some protozoan organelles.

2.1 Both form 10 nm filaments but cytokeratin filaments are smooth, with no obvious projections, though their tendency to bundle may be due to protruding domains; neurofilaments have long projections, which appear to hold filaments apart rather than bundling them together, and probably interact with microtubules and F-actin. Cytokeratin filaments are made up of type I plus type II dimers (heterotetramers); different epithelial cells have different varieties. Neurofilaments, found only in neurons, consist of triplets of different polypeptides; structural subunits may be homo- or hetero-tetramers.

2.2 Tetramers are assembled end-to-end to form 2–3 nm protofilaments (pfs); pairs of pfs, probably half-staggered, form 4.5 nm protofibrils and four of these are twisted together into a 10 nm filament. Alternatively, seven or eight pfs may simply surround one central pf. The dimers in a tetramer are probably antiparallel, so homopolymers are probably bipolar. Heteropolymers may be polar, however, if the two types of dimer making up a heterotetramer are consistently arranged.

2.3 Myosin II rods have twice the cross-sectional area of an IF. A pair of large N-terminal domains on each myosin dimer must protrude from the surface of a myosin filament; N-terminal domains of IF dimers are probably involved in pf-assembly but, in the case of NFs, C-terminal domains project from the filaments. Myosin filaments are easily soluble compared with IFs.

3.1 Different critical concentrations at the two ends; reactions are different at plus and minus ends because ATP hydrolysis makes disassembly not simply the reverse of assembly.

3.2 Rate of addition (c) $= k_{on} \times c - k_{off}$
- (a) Barbed (+) end rate $(1.0 \, \mu M) = 11.6 \times 1.0 - 1.4 = 10.2 \, s^{-1}$
  Pointed (−) end rate $(1.0 \, \mu M) = 1.3 \times 1.0 - 0.8 = 0.5 \, s^{-1}$
  Both ends grow, the plus end much faster.
- (b) +rate $(0.62 \, \mu M) = 11.6 \times 0.62 - 1.4 = 5.8 \, s^{-1}$
  −rate $(0.62 \, \mu M) = 1.3 \times 0.62 - 0.8 = 0.006$
  The minus end is almost stationary.
- (c) +rate $(0.5 \, \mu M) = 4.4 \, s^{-1}$
  −rate $(0.5 \, \mu M) = -0.15 \, s^{-1}$ (shrinking)
- (d) +rate $(0.17 \, \mu M) = 0.57 \, s^{-1}$
  −rate $(0.17 \, \mu M) = -0.58 \, s^{-1}$
  Growth at plus end is balanced by loss at minus end (treadmilling).
- (e) +rate $(1.0 \, \mu M) = -0.24 \, s^{-1}$
  −rate $(1.0 \, \mu M) = -0.67 \, s^{-1}$
  Both ends shrink.
- (f) Severed barbed end rate $(1.0 \, \mu M) = 11.6 \times 1.0 - 3.8 = 7.8 \, s^{-1}$
  Severed pointed end rate $(1.0 \, \mu M) = 1.3 \times 1.0 - 0.27 = 1.03 \, s^{-1}$
  Both ends rapidly acquire ATP-actin caps again (and then grow).
  sev. + rate $(0.17 \, \mu M) = 11.6 \times 0.17 - 3.8 = -1.8 \, s^{-1}$
  sev. − rate $(0.17 \, \mu M) = 1.3 \times 0.17 - 0.27 = -0.05 \, s^{-1}$
  Both ends shrink, until barbed end is rescued by addition of ATP-actin.

3.3 If rate $= 100/2.8 = 35.7$ subunit/s $= 11.6$, c $= 3.2 \, \mu M$: no difficulty in explaining motility at this rate in terms of actin polymerisation.
If rate $= 1000/2.8 = 357$ subunit/s, c $= 30.9 \, \mu M$: special circumstances required to produce monomer levels sufficient to explain this rate of motility.

4.1 (a) Pointed-end capping proteins prevent further depolymerisation once the monomer concentration is above the barbed end cc $(>0.12 \, \mu M)$. Barbed-end capping proteins prevent further depolymerisation provided the monomer concentration is below the pointed end cc $(<0.6 \, \mu M)$.
(b) Monomer sequestration to reduce the effective cc; addition of barbed-end capping proteins to treadmilling filaments (monomer conc. $0.17 \, \mu M$ — see Qu. 3.2) will cause net depolymeristion. Addition of pointed-end capping proteins to treadmilling filaments will induce net polymerisation.

4.2 Depends on the type of molecule: bipolar dimers with two equivalent actin-binding sites tend to crosslink antiparallel filaments; monomer cross-links with two (or more) distinct actin-binding sites can be used for polar bundles. Filaments will be more strictly parallel if crosslinks are short and/or stiff; longer or more flexible crosslinks will produce a more random mesh (crosslinked gel). Other factors: pH dependence, Ca^{2+}-sensitive binding, severing activity, actin-dependent ATPase activity and ATP-dependent binding.

4.3 (a) Control of actin-activated myosin ATPase: (i) by Ca^{2+}-binding to RLC of invertebrate muscle myosin heads, (ii) by Ca^{2+}-binding to troponin-C or to caldesmon to regulate muscle thin filaments, (iii) by controlling (via Ca^{2+}-calmodulin) phosphorylation of non-muscle myosin light chains (and heavy chains). (b) Severing of F-actin (gelsolin, severin, etc.). (c) Binding of non-muscle α-actinin, etc. to F-actin.

4.4 (a) Actin–profilin: IP_3 causes dissociation of complex; (b) actin–gelsolin: similar; (c) actin–myosin I: C-terminus may bind to lipids as well as to actin in a non-ATP dependent manner; (d) actin–ponticulin (integral membrane protein); (e) actin–α-actinin/talin, etc. in strong membrane attachments.

4.5 Double-headed, long C-terminal tail; single-headed, small C-terminal domain; *Nina C* single-headed molecule has extra N-terminal enzymatic domain.

5.1 Both are bipolar bundles of myosin II filaments and F-actin, together with tropomyosin. α-Actinin holds the bundles together, only at the sarcomere ends (Z-discs) in muscle; along the whole length of stress fibres but in greater concentration at the adhesion plaque. The latter is equivalent to special Z-discs at the ends of muscle fibres, where filaments are firmly anchored to the cell membrane.

5.2 (a) Thin-filament regulation, as in vertebrate striated muscle; the accessibility of the myosin-binding site on actin is determined by the position of tropomyosin, which is controlled by the state of troponin, which in turn depends on Ca^{2+}-binding to troponin C. (b) Direct regulation of thick filaments by Ca^{2+}, as in invertebrate striated muscle; Ca^{2+} binds to RLC and switches on myosin II. (c) Myosin ATPase is regulated by the state of phosphorylation, as in smooth muscle and non-muscle cells.

5.3 The cytoskeletal network on the cytoplasmic side of the cell membrane determines its shape. Spectrin, actin and associated proteins form a roughly hexagonal network. Spectrin's acidic composition gives an extended conformation at lower ionic strengths and presumably produces an expansive force on the network.

5.4 Binding sites at the two ends of a rigid crosslinking molecule need to be fairly accurately parallel or anti-parallel to form oriented bundles. It helps if crosslinks are fairly closely spaced axially and bind in a cooperative fashion. Polar bundles that have been studied are formed by monomer crosslinks that bind to complementary sites on opposite sides of F-actin. For a random mesh, smaller quantities of flexible (or even deliberately skewed), non-cooperative crosslinking molecules will be effective.

6.1 All are protrusions from the cell surface formed by the growth of a polar bundle of F-actin, apparently pushing against the cell membrane. The acrosomal filament derives from an identifiable store of monomeric actin, kept in that state by the presence of profilin and by conditions within the sperm head. The acrosome can be activated *in vitro*. The speed of growth is so high that diffusion-limited F-actin assembly can be discounted as the only driving mechanism. Fairly accurate measurements of volume changes are possible, showing a large influx of water. Such observations are more difficult for more complex cells where different processes take place simultaneously in different places.

6.2 Contraction of actin bundles: towards the rear of a moving cell (retraction fibres); in stress fibres throughout a resting cell dislodged from its substrate (e.g. by trypsin treatment). (b) Gelation of actin networks (ectoplasm): in the leading lamella and other protrusions from the cell. (c) Contraction: in the zone behind the leading lamella and over the cell cortex; solation; as the contracted material moves into the middle of the cell. (d) Assembly of F-actin from monomers: notably at the cell surface. especially in the leading lamella; monomers insert between the barbed-end of F-actin and the attachment point on the membrane. (e) Osmotically-driven protrusion: probably also occurs here.

6.3 Particles become stuck to extracellular domains of glycoproteins. Rather than simply flowing with the tide of membrane components being endo-cytosed (membrane flow theory) recent observations on nerve growth cones and keratocytes (Sections 6.4.3 & 6.4.4) support the alternative theory that intracellular domains of the glycoproteins are transiently connected in some way to membrane-associated F-actin and carried backwards as the filaments polymerise at the leading edge. The particles may become concentrated on the surface as the actin meshwork contracts, and then phagocytosed by invagination of the membrane.

6.4 Adhesion plaques form when glycoproteins, such as integrin, interact with components of the substratum (Section 5.6.2) and aggregations of actin-binding proteins, such as α-actinin and vinculin, bind to intracellular domains of integrin. The cell moves over the attachment in an analogous way to transporting a free particle over its surface.

7.1 Rate (c) = $k_{on} \times c - k_{off}$

(a) 10 μM dimer:
+end, growth phase: $8.9 \times 10 - 44 = 45$ s^{-1}
+end, shrinking phase: -733 s^{-1}

$$\text{net rate} = \frac{45 \times 250 - 733 \times 40}{290} = -62.3 \text{ s}^{-1}$$

−end, growth phase: $4.3 \times 10 - 23 = 20$ s^{-1}
−end, shrinking phase: -915 s^{-1}

$$\text{net rate} = \frac{20 \times 400 - 915 \times 7}{407} = +3.9 \text{ s}^{-1}$$

Slow addition to − end, but much faster loss from + end.

(b) 15 μM dimer:

$$+\text{end, net rate} = \frac{89.5 \times 750 - 733 \times 25}{775} = +63.0 \text{ s}^{-1}$$

$$-\text{end, net rate} = \frac{41.5 \times 1250 - 915 \times 3.5}{1253.5} = +38.8 \text{ s}^{-1}$$

Net growth at both ends. There could be steady treadmilling at an intermediate dimer concentration, with subunit flux from net addition at − ends and net loss at + ends.

7.2 Dimer concentration; entropy (temperature, state of water, pressure); buffer conditions (including pH, ionic strength, etc.); GTP-hydrolysis; MAPs.

7.3 (a) End-to-end binding is strongest and most conserved: longitudinal protofilaments (pfs). (b) Side-to-side binding is fairly well conserved, but small changes allow for variations in pf number. Complete switch if Zn^{2+} ions are bound. (c) Side-to-front binding: hook formation, as in flagellar doublet junction.

8.1 MAPs promote the assembled state of tubulin and thereby stabilise microtubules to varying extents. In some cases the microtubules become quite stable; others remain highly dynamic. The effectiveness of some kinds of MAP can be changed by phosphorylation. Since each MAP molecule covers several tubulin dimers, the density of assembled microtubules can be controlled by modifying a minor component of their structure.

Apart from microtubule-binding segments, many MAPs have domains that project outwards. These may have functions specific to particular cells or regions of a cell, which could explain why tau is found in axons, MAP 2 in dendrites.

8.2 The main feature is a series of high-molecular-weight polypeptides or 'heavy chains', some derived from the outer row of dynein arms, others from inner arms. Depending on the species, an arm resembles a bouquet of two or three flowers. Each band comes from one dynein globular 'head' containing ATPase activity on the end of a short stem. A fine protrusion from the top of the head also seems to be part of the heavy chain. A variety of 'medium chains' (around 80 kD) are thought to be associated with the stems of different heads. The location of a number of 'light chains' (around 20 kD) is unknown.

8.3 Brain dynein and kinesin are both soluble molecules with an ATPase activity that is enhanced in the presence of tubulin. Both cause microtubules to slide over glass and are involved in moving particles along microtubules; brain kinesin travels towards the + end and is probably the main effector of fast transport away from the cell centre; brain dynein moves towards the − end of microtubules and is thought to transport particles towards the interior of the cell.

9.1 Dynein arms produce sliding between adjacent doublet tubules but if there are fixed, elastically extensible crossbridges (nexin links) between the tubules, the net displacements will be limited and any additional sliding will produce bending. Links between outer doublets and the central structure also contribute. Radial spokes extending inwards from the doublets, instead of being extensible, are apparently capable of bending to a limited extent whilst remaining attached to the central structure, but they detach when under greater strain.

9.2 The axial repeat of accessory structures is most probably measured out by long protein molecules lying end-to-end along the length of axonemal microtubules. The projecting structures may then simply bind to equivalent sites on the spacer molecules. Tektin filaments, consisting of IF-like protein molecules, probably represent such a system. One tektin polypeptide at least seems to be expressed in limiting amounts, so the axoneme may not extend stably when this protein has been used up. To assemble a basal body, the initial '9 + 2' arrangement must be laid out by unique components that self-assemble into a flat 'cartwheel' structure. Something attached to this determines the final length of the triplet tubules that assemble onto it.

10.1 The microtubule array grows out radially from a central organising site close to the nucleus. Proteins to be secreted are synthesised in the ER, which is closely aligned with the microtubule array, probably owing to interactions involving kinesin. ER tubules or vesicles derived from them can be moved to the nearest point on the cell surface by travelling towards the + end of a nearby microtubule.

10.2 Cytoskeletal proteins and associated components travel at rates of $0.01-0.1$ μm/s away from the neuronal cell body. Tubulin and NF protein travel at the slower rate; actin, clathrin, spectrin, myosin and numerous enzymes at the faster rate. Since anti-microtubule agents interfere with all such transport, it seems likely that relative sliding between microtubules and sliding of other components along microtubules are involved in the process. Movement of cytoplasmic components besides membrane-bounded organelles and vesicles along microtubule tracks in other types of animal cell may ensure the redistribution of severed F-actin fragments, for example, through the dense cytoskeletal meshwork.

10.3 Microtubules initiated at a central site will tend to continue growing until the environment of the + end has a tubulin dimer concentration approaching the critical value. This will occur sooner if many + ends are growing close together. Thus, more isolated microtubules will grow longer and be more stable. In equilibrium, therefore, microtubule ends will be distributed as far from each other as possible. Specific factors in a local region of a cell may stabilise more microtubules there.

11.1 Anti-F-actin agents, such as cytochalasin, and anti-myosin antibodies stop cytokinesis but not mitosis in animal cells. Mitosis can be arrested by a variety of anti-microtubule treatments, including colcemid, taxol, chilling, and pressure, most of which have little effect on cytokinesis.

11.2 The rapid breakdown of interphase microtubule networks and the establishment of a completely new structure (the mitotic spindle). The efficiency of kinetochore capture. The high turnover of tubulin dimers, even during apparently stationary metaphase; the often extensive spindle elongation during chromosome separation.

11.3 Kinesin-like molecules may produce relative sliding of microtubules emanating from opposing poles and thereby assist pole separation, both during spindle formation and chromosome separation (anaphase B). Other kinesin-like molecules (or reversed dynein) may act between kinetochore microtubules and chromosomes to transport the latter to the + ends during prometaphase (congression). Dynein probably moves chromosomes towards the – end of kinetochore microtubules in anaphase A (assisted by microtubule depolymerisation).

11.4 (a) Work done = 7×10^{-10} newtons $\times 5 \, \mu m = 35 \times 10^{-16}$ joule.

Total number of nucleotides hydrolysed =

$$\frac{35 \times 10^{-16} \times 6 \times 10^{23}}{(30 \text{ to } 60) \times 10^{3}} = 35,000 \text{ to } 70,000$$

(i.e. 1000 to 3000 per kinetochore microtubule).

(b) Hydrolysis rate during 10 min = 58 to 116 per second per kinetochore. (Assuming synchronous events on each microtubule, estimated distance moved per event = $5 \, mm/(3000 \text{ to } 1000) = 1.7$ to $5 \, nm$.) If kinetochore motors worked at same fast rate as axonemal dynein (probably untrue), only 14 to 29 arms per kinetochore would be needed (less than one per microtubule!).

A $5 \, \mu m$ long axoneme has $(5000/24) \times 9 \times 2$ dynein arms. Rate of hydrolysis = no. of arms $\times 4 = 15\,000 \, s^{-1}$: a considerably more powerful machine than a mitotic spindle.

Index and glossary

Abercrombie, 103

Acanthamoeba (species of amoeba that can be cultured in bulk), proteins of, 13, 58–59, 63

accessory fibres in sperm axonemes, 10, 176

accessory proteins, 13, 35, 56ff, 142ff

Acetabularia, 212

acidic residues (negatively-charged amino-acid residues; attract H⁺ and produce net acidity, e.g. aspartic (D), glutamic (E)), 27, 34, 125, 143

acrosome of sperm head, 10; of *Limulus*, 81; of *Thyone*, 98

actin, in cells, 80ff, 98ff; actin network, 5, 102, 109

actin, links to membrane, 74, 82ff, 94, 109

actin activation of myosin, 89

actin-binding protein (ABP), 58, 72

actin depolymerising factor (ADF), 64

actin filaments (F-actin), 4–7, 13, 42ff; bundles of, 6, 84ff, 93, 98ff

actin monomers, 13, 42ff, 99

actinogelin, 58

actobindin, 62

actophorin, 59, 64

Adams and Pollard, 68

ADF (actin depolymerising factor) adhesion plaque, 102; proteins, 95

ADP (adenosine diphosphate), 51, 89

ADP-ribosylated actin, 58

Aebi *et al.*, 28

Afzelius, 176

A-lattice, 119

aldolase, 58, 73

algal cells, *see Nitella, Chara*

alpha-actinin (α-actinin), 33, 58, 70, 94; in Z-disc, 71, 93

alpha-helix (α-helix), α-helical proteins, 23ff, 57, 69, 72, 147

amino-acid sequences, in one-letter code, 147, 148

amoebae, 9, 16

Amoeba chaos, 103

Amoeba proteus, 103

AMPPNP (non-hydrolysable analogue of ATP), 197

anaphase, 212, 220ff; anaphase A, 221; anaphase B, 220; thermal-diffusion model, 222; depolymerisation hypothesis of, 221

animal cells, 10–11

ankyrin, 37, 60, 74, 82, 149

annealing of F-actin, 52

annexins, 58, 74

anterograde (= orthograde) transport (away from cell centre), 164

antibodies, 232; attached to gold particles, 6, 7; fluorescently labelled, 3, 232ff

antisense pseudogenetics, antisense RNA, 106, 239

aphidicolin, 214

apical surface of polar cell, 11

Aplysia (sea slug) neurones, growth cone of, 108

arrowhead appearance of decorated F-actin, 47

Asakura, 182

Aspergillus, 238; mutant with defective kinesin, 221

assembly of filaments, MTs, 16; of myosin filaments, 30; of axonemes, 184; of basal bodies, 185

aster of MTs, 212, 215

aster-forming protein, 144

astrocytes (non-neural cells of brain), 32

ATP (adenosine triphosphate), 10; and chromosome movement, 222ff

ATPase activity/ATP hydrolysis, 25, 50, 176

ATP in actin, 46

ATPγS (hydrolysable analogue of ATP), 206

attachment plaque, attachment zone, 9, 37, 94

A-tubule, 119, 172; model, 120

axolinin, 144, 152

axon, 9, 196

axonal transport, 16, 196ff

axoneme of a cilium or flagellum, '9 + 2' arrangement, 9, 16, 18, 171; assembly, 184

axoplasm (cytoplasm in an axon), 197ff

axopods of heliozoa, 16

axostyle of flagellates, 16; of *Saccinobaculus*, 195

bacteria, comparison with eukaryotes, 15, 226; bacterial flagella, 181

band 3, 82

band 4.1, 60, 82

band 4.9, 59

barbed (plus) end of F-actin, 46; barbed end-capping proteins, 59, 60, 61

bare zone of myosin filaments, 29–30, 66

basal body, 10, 17, 171, 226; assembly, 185ff

basal cell, 2

basal surface, basolateral surfaces of polar cells, 11

basement membrane (extracellular structure supporting tissue), 94

basic residues (positively-charged amino acids; attract negative counter ions and produce net alkalinity, e.g. histidine (H), lysine (K), arginine (R)), 27, 35

Bennett, 83

benomyl, 130

beta-actinin (β-actinin), 60

beta-sheet (β-sheet) (protein secondary structure in which stretches of extended polypeptide lie in anti-parallel), 72

BHK cells (cell line from baby hamster kidney cancer), 36

bim C gene product (kinesin-like protein), 144

biotin: vitamin used experimentally to link fluorescent labels because of its extreme affinity for avidin

bipolar filament-bundling proteins, 58

bipolar myosin filaments, 29

bipolar tetramers of IFs, 26–27

B-lattice, 119

Bretscher, 104

brevin, 59

bridges 5/6 of axoneme, 176

Brokaw, 181

brush border, 84; 110 kD myosin I, 58

B-tubule, 119, 172; model, 175

Bullitt et al., 74

bundling of keratin filaments, 31

bundling proteins for F-actin, 58; for MTs, 144

buttonin, 144

Ca²⁺ (calcium ions), 17–18, 36, 206

caⁿᵈ⁺ gene product (kinesin-like protein), 144

calcium-binding proteins, 18, 56, 71, 72, 91

caldesmon, 58, 72, 93

calelectrins, 58, 74

Calladine, 182

calmodulin, 17, 56, 57, 60

calpactin, 58, 74

cAMP (cyclic nucleotide with important role as intracellular messenger), 18, 206

CAMs (cell-adhesion molecules), 95

Cande, 223

cap of ATP-actin on F-actin, 51; GTP-tubulin on MTs, 135, 185

capping of cells, 107

capping of filaments, 14, 103; capping proteins, 57–59

cap Z, 59

carrier proteins in axons, 202

catastrophe (dynamic change from assembly to disassembly of MTs), 135

'catch mechanism' of muscle, 73

CD (circular dichroism), 237; of tubulin, 122

CDR, see calmodulin

cell-adhesion molecules (CAMs), 95

cell division, 36, 211ff

cells, 8–9

cellulose in plant cell walls, 11

central pair microtubules, 172

central sheath, 172

centriole, 190, 213ff

centrohelidians, 192

cGMP (cyclic guanosine monophosphate), 18

Chara (giant alga), 16, 87, 212

charged amino-acid residues (see also basic and acidic), 24, 72

chartins, 144

Chlamydomonas (motile alga with pair of flagella), 178

chloroplast, 2, 12

chromatids, 214

chromatophores, 205

chromosome movement, 16, 212ff

cilia, ciliated cells, 2, 9

cilium, primary, 227

circular dichroism (CD), 237

clathrin, 104, 105

cofilin, 59, 64

coiled coils, 23

colcemid, 130

colchicine, 37, 130

collagen, see connective tissue

compartments in cell, 1, 11, 17

concanavalin A, 103

confocal microscopy, 235

conformational change in protein, 15, 17, 48–49

connectin (= titin), 60

connective tissue (extracellular material), 2

contractile ring, 235

contraction, 14–18, 85, 88ff, 110

corrals in membrane, 109

cortical layer of cytoplasm (underlying cell membrane, ectoplasm), 9, 37, 80, 82

crawling of cells, 16, 100ff

critical concentration for assembly, 49, 93, 134

crossbridge cycle in muscle, 88

crossbridges, 14, 88; between membrane and MTs, 204

crossover in F-actin, 42

crystals containing actin, 43; of tubulin, 131

C-terminal domain of IF protein, 25; of NFs, 29, 34; of tubulin, 123–125

C-terminus: carboxyl (COOH-) end of polypeptide, translated last on ribosome

C-tubules of triplet MTs, 185

C₁ and C₂ tubules of axonemal central pair, 172

cytochalasin, 37, 52, 102, 211

cytokeratins, 25, 31

cytokinesis, 107, 211, 225ff

cytoplasm, 1–4

cytoplasmic motors, 156, 166

cytoskeleton, 3, 23; role in locomotion, 106ff

cytosol, 3

Dabora and Sheetz, 204ff

DAPI (fluorescent dye for DNA), 233

darkfield microscopy, 234

DEAE tubulin (purified by DEAE ion-exchange chromatography: MAPs elute first, then tubulin, in increasing salt gradient), 127

dendrites, 9, 203

dense bodies, 33

depactin, 46, 59, 64

depolymerising factors (actin-associated proteins), 57

DeRosier, 85

desmin, 25, 32–33

desmoplakin, 36

desmosomes, 36

destrin, 59

dextran particles, 3–4

DG (diacylglycerol), 110

diatoms, 217, 226

DIC (differential interference contrast), optics, 165; and MTs, 165

Dictyostelium (cellular slime mould), proteins of, 58, 59, 72, 102; mutants, 75, 106

differential interference contrast (DIC), 165, 205, 234

Difflugia (amoeba), 103

dinoflagellates, 215, 226

disulphide bonds in keratin, 32

DiOC₆, fluorescent dye, 233

DNA in centrosomes, 214

DNase I, 59

D₂O (heavy water), effect on MTs, 222

domains, of IF proteins, 25; of actin, 43ff; of myosin head, 67; of tubulin, 123ff

doublet MTs, 172

Drosophila (fruit fly), 212

dynamic behaviour of actin, 52

dynamic instability of MTs, 132, 194, 217
dynamic structural change, 2
dynamin, 144, 162
dynein, cytoplasmic, 148, 154, 156ff, 216; axonemal, 153ff, 157ff
dynein arms, 172, 177, 178, 180
dystrophin, 71

Eagles, 123
ectoplasm, 9, 81
EF hand, 56, 57
Eisenberg, 90
ELC ('essential' light chain of myosin), 65, 67
electron microscopy (EM), 4–8, 235
elimination forces, 220
elongation rate, 49, 194, 196
EM (electron microscopy), 235
endocytosis, 104
endoplasm, 9, 81
endoplasmic reticulum (ER): system of extensive intracellular membrane, 1, 204
endosome, 2, 204
enzymatic activity of myosin, 89ff
epidermis texture, 31
epifluorescence microscopy, 233
epinemin (intermediate filament-associated protein), 7, 35
epithelial cells, epithelium (cells covering a surface in an organism), 2, 11
epitope (region recognised by antibody), 238
ER (endoplasmic reticulum, rough and smooth varieties), 204
erythrocyte (red blood cell), 82, 83
erythrophores, 205
E-site of tubulin dimer, 123ff
'essential' light chain of myosin (ELC), 65, 67
eukaryotic cell, eukaryote (cell with nucleus surrounded by double membrane; nuclear division and cell division are separate events), 1
exocytosis (fusion of vesicle with plasma membrane, secretion of contents into medium), 104
expression of genes in cells, 19

F-actin (filamentous actin), 48
fascin, 58, 73, 85, 86
fast axonal transport, 16, 196ff
fibroblast, 2, 9–11, 16, 87, 226; motility 100ff
fibronectin, 94, 95
filaggrin (IF-associated protein), 32, 35

filaments, 2–5 nm, 8; see also F-actin, IFs
filament-severing agents, 61
filamin, 33, 58, 93
fimbrin, 59, 61, 73
Finch, 181
flagellar mutants, 178; beating, 10, 16; bacterial, 181
flagellin (subunit protein of bacterial flagella), 181
fluorescein (dye excited by blue light, emits green light)
fluorescence, 232ff
fluorescent particles, 3; beads coated with myosin, 66, 87
flux of actin subunits, 52; of tubulin dimers, 218
fodrin, 69ff, 145
footprint of cell, 102
Forscher and Smith, 107
fragmin, 59
FRAP (fluorescence recovery after photobleaching), 4, 233
freeze-fracture method, 183, 236

G-actin (globular, monomeric actin), 47, 48
gelation, 112; factors, 58; of actin, 68
Gelles et al., 164
gelsolin, 59, 61
genetic analysis of flagella, 178
Gerisch et al., 75
GFAP (glial filament acidic protein), 32–33
Gilbert and Sloboda, 159, 198
glial filaments, 25, 33
glutaraldehyde fixative, 237
glycerol, effect on membrane, 110
glycophorin, 82
goblet cell, 2
Goldman et al., 36
gold particles, 232; for labelling antibodies, 6–7
Golgi apparatus/system, 2, 11, 204
Goodenough, 183; and Heuser, 153
green algae, 226
Greuet, 12
growth cone, 9, 16, 103, 107–108
Gyrodinium, 213, 226

HA1 actin-associated protein, 59, 60
hard keratins in hair, hoof, horn, 31
Harvey and Marsland, 3
head domains of rod proteins, 25
helices, 23, 42–43; helical family, 118
heptapeptide/heptad repeat, 24
heterochromatin, 214
heterodimers, of spectrin, 69; of tubulin, 117

Heuser, 129, 177
hexagonal packing, 86, 216, 221
HMM (heavy meromyosin), 50
HMWP of Physarum, 58
Hoechst 33342 (fluorescent dye for DNA), excited by UV, 233
Hollenbeck et al., 37, 201
Horio and Hotani, 138
horseshoe crab, see Limulus
HVEM (high-voltage electron microscopy), 8
hyalin (clear) layer of cytoplasm, ectoplasm, 9
hydrophobic amino-acid residues (especially isoleucine (I), leucine (L), methionine (M), valine (V)), 24
hydrophobic bonds, 24, 125–126

IDPN (a neurotoxin), 202
IFAP (intermediate filament-associated protein), 32, 35
IF network collapse, 37, 202
IF proteins, 12, 25ff, 32
IFs (intermediate filaments), 7, 25ff
image analysis of micrographs, 43
immunofluorescence, 35, 232
initiation complex for assembly, 50, 128
inositol lipids, 110
inositol triphosphate (IP_3), lipid derivative with signalling role, 18, 111
Inoué, 99, 221
integrin, 94
intermediate filaments (IFs), 4–7, 13, 23ff
intestinal epithelium, see brush border
ion transport, 11
IP_3 (inositol triphosphate), 111
isotropic networks of F-actin, 68; network-forming proteins, 58

Jensen and Smaill, 151
Johnson, 163
juxtanuclear cap of IFs, 36

KAR3 gene product (kinesin-like protein), 144
keratins, 28, 32, 38
keratocytes, of fish, 109
Kilmartin, 191
kinase, of myosin light chain, 60
kinesins, 23, 25, 144, 160ff
kinetics of actin assembly, 48ff; of myosin ATPase, 89ff; behaviour of dynein and kinesin, 163
kinetochore, 82, 216ff; KMTs, MTs, attached to kinetochore, 216
Kirschner, 194, 219
Koshland et al., 222, 234

lag time in assembly of F-actin, 49
lamella, leading edge of cell, 9, 100
lamina, of nucleus, prometaphase breakdown of, 215
laminin (extracellular matrix protein), 94
lamins, proteins on nuclear membrane, 23, 25, 32, 34, 36, 215
Lasek *et al.*, 200
laser, 235, severing of stress fibres, 93
latex, use in measuring force of cell attachment, 94
lattice of MTs, 121
Lawson, 6, 7
leucine zipper, 24
Lewin, 178
light chains of myosin, 91
light microscopy, 233ff
Limax (amoeba), 16, 103
Limulus acrosome 81, acrosomal filaments, 50; proteins of, 58, 74
Linck, 172, 173, 174
linker regions within IF rod domain, 26
lipid-flow theory, 103, 105ff
lobopodia of amoebae, 103
Luby-Phelps *et al.*, 4
Luck, 179
lysosome, 2, 204

magnetic particles, 3
MAP 2, 147; in dendrites but not axons, 203; associated with actin, 59
MAPs (microtubule-associated proteins), 142ff; effect on tubulin assembly, 138; structure, 149ff
Mayorella (amoeba), 102
Mazia, 212
McIntosh, 160, 224
McNiven and Ward, 205
MDCK cells (cell line derived from dog kidney epithelium, unusual in forming polarised epithelial sheets in culture), 191
meiosis, 211ff
melanophores, 205
membrane-bound proteins, associated with actin, 58
membrane junctions, 11, 94
membrane of cell, during motility, 103; insertion of, 109
membrane protrusion, 16; flow theory, 103ff
mercaptoethanol, 213
metal salts, used as electron-dense stain, 4–8, 235
metaphase, 212, 217ff
microbeam, laser, 93, 193

microfilaments (actin-containing filaments), 8
microfilopodia, 100
microspike, 100
microtrabeculae, microtrabecular lattice, 8
microtubule-associated proteins (MAPs), 142
microtubules (MTs), 4–7, 13, 120ff; lattice, 121; polarity, 132; dynamic instability, 132; model, 120; GTP caps, 135, 185; hooks, 128; rotation, 165; networks in cells, 190ff; functions in cells, 194ff; gelation and slow contraction, 202; square and hexagonal lattices, 220
microvilli, 84, 85
midbody in cytokinesis, 225
midpiece of sperm, 10
minus (inactive) end of MT, 134
Mitchison, 132, 217
mitochondria, 2, 10, 34
mitosis, 36, 211ff
mitotic centres, 215
mitotic spindle, 16, 212ff
MLCK (myosin light-chain kinase), 57, 60
M-line (of muscle, structure joining centres of bipolar myosin filaments in register within a sarcomere), 15, 17
models of flagellar motility, 181
monomer-sequestering proteins, 57, 63
motility 10–11, 94, 100ff
motor molecules, myosins, 58; associated with MTs, 156, 166
MSP (major sperm protein, of nematodes), 16, 113
MTOC (microtubule-organising centre), 152, 201, 226
MTs, *see* microtubules
mucus transport, 2, 11
muscle, 13, 15–17, 88ff
mutants, 75, 178, 238
myofibril (bundle of myofilaments within a muscle cell or myocyte), 33
myoneme (contractile organelle of vorticellids), 16, 35
myosin light-chain kinase (MLCK), 60
myosins, 58; myosin I, 67, 106, 107; myosin II, 64ff, 84, 94, 106, 107; filaments, 15, 17, 29–31; head, 46; tail, 23, 25, folded 65, of scallop, 66
myxomycete, *see Physarum*

Nassula, 192, 193
NBD-C$_6$-ceramide (fluorescent dye for Golgi), 233
nebulin, 60
negative staining, 4, 235
nematode sperm/gametes, 8, 16, 113
nerve cells (neural cells, neurons), 2, 9
N-ethylmaleimide (NEM), 157
neurites, neural process, 9–10, 16, 103, 203
neurofilaments (NFs), 25, 199
newt-lung cell cultures, 216
nexin, 172, 180
NF-projections, 34, 38
NFs (neurofilaments): NF-L, light chain; NF-M, medium chain; NF-H, heavy chain of the NF protein 'triplet', 29, 33
Nitella (giant alga), 16, 68, 87
10 nm filaments (IFs), 7, 25, 27–28
NMR (nuclear magnetic resonance), 237
nocodazole, 130
N-site of tubulin dimer, 123ff
N-terminal domain, of IF protein, 25; of tubulin, 123–125
N-terminus: amino (NH$_3$) end of polypeptide
nuclear membrane, nuclear lamina, 1, 34
nuclear pore, 1, 36
nucleated red blood cells, 191
nucleation of assembly, 49, 62, 127ff

ocelloid of *Erythropsidium*, 12
on-rate (k_{on}), off-rate (k_{off}) for polymerisation, 49
Oosawa and Asakura, 48
optical diffraction pattern, 238
optical filtration, 238
osmiophilic material (between spindle microtubules), 220
osmotic/hydraulic theory of cell extension, 99, 110
Oster, 99, 110

paracrystals of F-actin, 45, 47
paramyosin (accessory protein of invertebrate thick filaments), 23
paranemin, 32, 35
particle transport, 16, 34, 107, 108, 164, 197
parvalbumin, 56
patching of cell surface, 107
PC tubulin (tubulin dimers purified by phosphocellulose (ion-exchange) chromatography; tubulin passes straight through, MAPs bind until released by higher salt concentration, 127

Peranema, 192
pericentriolar material, 190, 214
peripherin, 32–33
peroxidase, 233
pfs, *see* protofilaments
PGGG motif in MAP sequences, 147
phagocytosis, 107
phalloidin, 53, 80, 98, 233
phosphatase, 18
phosphate (P$_i$), release after
    hydrolysis of nucleotide, 52
phosphatidyl inositol
    4,5-bisphosphate (PIP$_2$), 63, 110
phosphorylation, 18, 34, 68, 92, 93,
    146, 148, 215, 224
photoactivated labels, 218, 233
photobleaching, 3–4, 233
phragmoplast, 228
Physarum (acellular slime mould),
    cycles of cytoplasmic flow, 113;
    proteins of, 58, 59; mitosis, 212, 227
Pickett-Heaps, 223
pigment granules, 205
pinocytosis, 104
PIP$_2$ (phosphatidyl inositol
    4,5-bisphosphate), 63, 110
pitch of helix, 23, 43
plant cells, 11–12, 191; spindle, 216
plasma membrane (membrane
    enclosing cell), 103
plastids (precursor organelles
    becoming mitochondria,
    chloroplasts), 2, 12
platelets, 100, 191
platinum (*see* shadowing), 236
plectin, 32, 35
plus (active) end of MT, 133, 193ff
plus and minus strains of
    Chlamydomonas, 178
pointed (minus) end of F-actin, 46;
    pointed-end capping proteins, 57
polar cells, 11
polar dimers of IF protein, 25
polar MTs in spindle, 215
polarity of F-actin, 81; of MTs, 132
pole body of yeast, 215
poles, mitotic, 215
Pollard, 51, 68
polyclonal antibodies, sera, 238
polyethylene glycol (PEG), effect on
    membrane, 110
polymeric proteins, assembly and
    disassembly, 13, 15, 47
ponticulin, 58, 63, 74, 95
pore size in cytoplasm, 3, 34
positive staining, 7, 236
pre-immune serum, 238
primary cilium, 226
probasal body, 186

profilactin (1:1 complex of profilin
    and actin), 98
profilin, 43, 45, 46, 59, 63ff, 100
prometaphase, 212, 214ff
prophase, 212, 214ff
protein kinases, 18
protists (protozoa, organisms with
    organisation as single cells), 9,
    11–12, 17, 226; microtubular
    organelles in, 192, 193
protofibril of IFs, 28
protofilaments (pfs), of MTs, 5, 117,
    177; of IFs, 27–28; of bacterial
    flagella, 182
protrusive/extensive forces, 11, 14, 98ff
pseudopodia, 11
Pseudotrichonympha, 227
psoralens, 214
PtK2 cell line, derived from rat
    kangaroo, useful because cells stay
    relatively flat in mitosis, 227

radial links, 174, 180
rapid-freeze technique, 177, 236
raster scan, 235
rat kidney cells, 72
red blood cell, 82, 83, 191
regulatory light chain (RLC) of
    myosin, 91
replica, made of carbon and
    platinum, for EM of tissue, 6–7
resolution (as distinct from
    detection), 232
Reticulomyxa, 197; HMWP, 159
retrograde transport (towards cell
    centre), 164, 196ff
rhodamine (fluorescent dye
    chemically similar to fluorescein
    but excited by green light to emit
    red), 194
ribbon-shaped filaments of smooth
    muscle and non-muscle myosin,
    30–31
ribosomes (associated with
    cytoskeleton), 4, 38
RLC (regulatory light chain of
    myosin), 57
rod domain of IF protein, 26
rootlets associated with basal bodies,
    17
Rozdzial and Haimo, 206
ruffling, 9, 100ff

S-1 subfragment of myosin, 47
Saccharomyces (yeast), 226
sarcomeres, 15, 17, 93
sarcoplasmic reticulum, 91
SCa and SCb (slow components of
    axonal transport), 202

scanning, confocal optical, 235;
    transmission EM, 237
Schliwa, 159, 195
Schnapp, 163
Scholey *et al.*, 163
Schultze and Kirschner, 194
sea cucumber (*Thyone*), 99
secretion, mucus, 2; *see* exocytosis
seed, *see* initiation complex
severin, 59
severing and annealing of filaments,
    13, 52
shadowing with platinum, 5, 236
Sheetz, 66, 87, 108
Shen *et al.*, 83
single-headed myosin (myosin I), 67
sliding filaments, 14–18; active
    sliding, 179; speeds of myosins, 91
slow axonal/axoplasmic transport, 16
Small *et al.*, 3, 5
smooth muscle, 33; smooth-muscle
    myosin, 93
soft keratins, cytokeratins, 31
solation, 112
solid-phase chemistry, 4
spasmoneme (giant myoneme), 16–18
spectrin, 37, 58, 69ff, 83, 84
spectrofluorometric assay of actin
    assembly, 48
sperm, spermatozoa, 12; accessory
    fibres, 176
spindle, 213, 216ff, pole body, 215,
    217; plaque, 226
spokes, 174, 180
Spudich, 66, 85, 87
square packing of MTs, 216, 221
STEM (scanning transmission electron
    microscopy), 153, 237
Stephens, 181
stereocilia, 85
STOPs (stable tubule-only proteins),
    144, 147, 201
streaming of cytoplasm, 9, 16, 87
stress fibres, 93ff
striated muscle, 15–17, 91
structural hypothesis of slow axonal
    transport, 200
Summers and Gibbons, 179
superlattice (regular arrangement
    (e.g. planar or curved lattice)
    superimposed on another
    arrangement with a smaller unit
    cell), of MAPs on a tubulin
    lattice, 150, 185
Surirella (diatom), 226
symbiotic prokaryotes (bacteria or
    blue-green algae taken into the
    cytoplasm of eukaryotic cells); *see
    also* plastids), 2

synapse, 9, 201
synapsin I, 59, 60, 144, 149
Syndinium (dinoflagellate), 226
synemin, 32, 35
syneresis, 103, 112

tail domains of polypeptides, 25
talin, 60, 74, 95
tannic acid, staining method, 236
tau protein, 185; in axons but not
    dendrites, 203
taxol, 130
Taylor, 107
tektins, 23, 26, 32, 35, 177, 185
telophase, 212, 225
tension, 14, 94
terminal web, 84
Tetrahymena (ciliated protist), 9, 178
tetramers of IF polypeptides, 26
thick-filament regulation, 91
thin and thick filaments of muscle,
    15, 17
thin-filament regulation, 91
Thyone (sea cucumber) acrosome,
    16, 98, 99
tight junction, 94
Tilney, 85, 99
titin (= connectin), 60
tonofilaments/tonofibrils, 35, 36
torsional flexibility of F-actin, 42
trachea (windpipe), 2
transcription of DNA, 19
translation of mRNA, 19
translocation of cell, 100

transparency of cells, 3
transport along filaments/MTs,
    14, 87, 196
transport directions in cells, 198ff;
    dependence on pH, 204
transport rates, associated with
    MTs, 196
treadmilling, 50, 134, 138
triplet MTs, 185
tropomyosin, 23, 46, 72, 92
troponin, 91; troponin C, 57, 91
trypanosomes (protists causing
    sleeping sickness), 149;
    trypanosome MAP, 144
trypsin treatment of cells, 36;
    digestion of axonemes, 179
tubulin, biotinylated, 218; dimers,
    13, 117ff; 6 S tubulin (dimers
    separated from tubulin–MAP
    oligomers by gel filtration), 127
    (*see also* PC tubulin, DEAE
    tubulin); crystals, 131;
    dimer-sequestering proteins, 152
tubulin concentration in cells, 13
Tucker, 193
turbidity of MTs, 127, 133
TW260/240 (spectrin), 58
two-dimensional crystals, of actin,
    43; of tubulin, 121

uranium salts, for staining, 235
UV (ultraviolet light), laser
    microbeam, 135, 193

vacuole of plant cell, 12, 191
Vale, 165
vanadate, 148, 153, 198, 206
Vernier scale, 223
vesikin, 159, 198
villin, 59, 61, 84
villipodia in nematode sperm, 113
vimentin, 25, 33
vinblastine, 130
vinculin, 60, 95
viscometry of F-actin, 48
viscosity, of cytoplasm, 3–4; of
    F-actin, 48
vitamin D-binding protein, 59
vorticellids, 8, 17

Walker *et al.*, 134
Warner and Satir, 180
Weisenberg, 202
Weiss, 165
Woodrum and Linck, 195

Xenopus eggs, multiple asters in, 214
X-ray diffraction, X-ray
    crystallography, 42–44, 237

yeast cell, 11, 217, 226

zinc-induced sheets of tubulin, 121
Z-line, Z-disc boundary between two
    sarcomeres where oppositely-
    pointing F-actin interdigitates,
    15, 17, 33
zona adherens, 11